Network Analysis for Management Decisions

International Series in
Management Science/Operations Research

Series Editor:

James P. Ignizio
The Pennsylvania State University, U.S.A.

Advisory Editors:

Thomas Saaty
University of Pittsburgh, U.S.A.
Katsundo Hitomi
Kyoto University, Japan
H.-J. Zimmermann
RWTH Aachen, West Germany
B.H.P. Rivett
University of Sussex, England

Sang M. Lee
University of Nebraska

Gerald L. Moeller
Management Consultant

Lester A. Digman
University of Nebraska

Network Analysis for Management Decisions

A Stochastic Approach

Kluwer • Nijhoff Publishing
Boston/The Hague/London

DISTRIBUTORS FOR NORTH AMERICA:
Kluwer Boston, Inc.
190 Old Derby Street
Hingham, Massachusetts 02043, U.S.A.

DISTRIBUTORS OUTSIDE NORTH AMERICA:
Kluwer Academic Publishers Group
Distribution Centre
P.O. Box 322
3300 AH Dordrecht, The Netherlands

Library of Congress Cataloging in Publication Data

Lee, Sang M., 1939–
   Network analysis for management decisions.

   (International series in management science/operations research)
   Bibliography: p.
   Includes index.
      1. Network-analysis (Planning) I. Moeller, Gerald L.   II. Digman, Lester A.
III. Title. IV. Series.
T57.85.L4   658.4′032   81-9473

ISBN 0-89838-077-4   AACR2

Copyright © 1982 by Kluwer • Nijhoff Publishing

No part of this book may be reproduced in any form by print, photoprint, microfilm, or any other means without written permission from the publisher.

Printed in the United States of America

CONTENTS

I INTRODUCTION TO NETWORK ANALYSIS FOR MANAGEMENT

This book is about network analysis techniques for managers. More specifically, it is about stochastic networking techniques that can be used by practicing managers in making decisions, whether the setting be in the private or public sector. While the book is written for the manager and adopts his or her perspective, enough detail is included to permit the staff analyst or specialist to implement a stochastic technique and to provide the manager with a completed risk analysis as input to the decision that only the manager can make.

Within this framework part I provides a background for analysis. Chapter 1 contains a discussion of management's analytical functions. Chapter 2 deals with the management of new projects or ventures. Chapter 3 provides an overview of the major types of network models for management use. In part II an advanced stochastic networking technique—the Venture Evaluation and Review Technique (VERT)—is investigated in detail. Included is a discussion of the technique's general concepts, network construction and logic, input requirements, output reports, and computer mechanics.

Part III applies VERT-3—the most up-to-date version of the technique—to a representative sample of application areas to illustrate both the use of the technique and its value in a variety of real-world management and strategic decision situations. Part IV contains conclusions and discusses future directions for both VERT and stochastic network analysis in general.

1 ANALYTICAL FUNCTIONS OF MANAGEMENT

Numerous management science techniques that purport to aid managers exist. The vast majority are designed to assist in the process of analyzing a particular strategic or operational situation or a particular problem arising out of special projects or new ventures. Management science techniques are intended to provide a rational, scientific approach or solution for managers. But the real-world manager deals with risk; the manager must be able to account for, deal with, and accept the fact that nothing is absolutely certain. We live in a stochastic world.

Most analytical techniques are deterministic—that is, they do not specifically incorporate probabilistic events or occurrences in their solutions. Certain techniques, such as decision trees, help the manager visualize the array of choices and outcomes available and possible. However, such techniques greatly oversimplify the interrelationships involved. They also tend to give a point estimate, or specific value, for each alternative. Network techniques, on the other hand, allow for systematic planning and control but rarely incorporate chance events, partial successes, and a host of other real-world managerial problems. Simulation provides the manager with an array of likely outcomes, but largely on an ad hoc basis. In recent years, however, stochastic networking techniques have been developed that combine key advantages of previous techniques and thus provide the managers with analyses more closely in tune with their needs. In fact, the stochastic methods are the only quantitative techniques that can accommodate "fuzzy" and uncertain future planning activities in a realistic manner.

In the past, managers were able to rely on judgment and intuition in areas where analytical techniques did not exist or in areas that the techniques were not able to treat. Problems, situations, and decisions are becoming progressively more complex and interrelated, however, and the manager needs progressively more help in dealing with these problems, situations, and decisions. It is not that judgment and intuition are no longer *necessary;* they are no longer *sufficient* for today's manager. More and more analytical input to decisions is required because of their increasing complexity and is possible because of recent advances in stochastic network analysis techniques.

FUNCTIONS OF THE MANAGER

Just as intuition and judgment are no longer sufficient (though still necessary) for effective management, neither are analytical techniques alone sufficient. Perhaps a brief glimpse at the manager's job will point out the complementary and dependent relationship between judgment and analytical techniques.

Traditionally, the manager's job has consisted of several universal functions—typically, planning, organizing, directing (or leading), and controlling. While generations of managers have been conditioned to think of their jobs in this way, few appear to relate to such generalized functions when describing their day-to-day tasks. With this dilemma in mind, Mintzberg (1975) has attempted to redefine the manager's job by observing actual tasks and activities that practicing managers perform and by classifying the activities into a more meaningful and descriptive framework. Mintzberg produced a series of managerial roles grouped into three categories:

1. *Interpersonal roles:* The manager acts as figurehead or leader and engages in liaison activities.
2. *Informational roles:* The manager acts as monitor, disseminator, and spokesperson.
3. *Decisional roles:* The manager acts as entrepreneur, disturbance handler, resource allocator, and negotiator.

While Mintzberg's framework undoubtedly adds to our understanding of what managers do, it is debatable whether individual managers can better relate on a day-to-day basis to his system of roles or to their traditional functions. The point is, however, that the manager's job defies description according to a logical, step-by-step, scientific framework. Much as we have heard about the "rational managers," precise decision-making approaches, detailed formal planning, or cascading Management by Objectives (MBO) systems, most management jobs—particularly at higher levels of organizations—involve a good deal of trial and error. Political scientists irreverently describe the process of winding one's way through uncharted territory—as managers must do in strategic planning, in dealing with an unforeseen crisis, or in tackling a new problem—as "muddling" (Lindblom, 1959). This term refers to the conscious but nonscientific approach human beings employ in attempting to structure the unstructured. Recent studies have confirmed that effective managers "tend to arrive at their strategic goals through highly incremental 'muddling' processes rather than through the kinds of structured analytical processes so often prescribed in the literature and 'required' according to management dogma" (Quinn 1977, p. 21). The processes managers actually use are purposeful, politically astute, and effective, given the unstructured type of situation, problem, or decision at hand. Once the problem or decision has been defined, or structured, the analytical processes can be employed in a specific resolution or implementation.

Successful management involves three key elements: managing reality, managing time, and managing risks (Uyterhoeven et al. 1977). The manager must determine and deal with the *real* situation, not a simplified, deterministic model of the real world. The manager is under time constraints; some decisions must be "made by Tuesday" and cannot wait for the results of a study or a dynamic programming solution. In addition, the manager must think in terms of, and be able to deal with, risks—which ones and how many are worth taking and which are not.

In general, management's responsibility is to assure that an organization functions effectively and efficiently in a real-world setting. That is, the manager must see that the organization possesses four key elements, each of which is required for continued long-term success:

1. *Mission and strategies:* The organization should have a clear idea of why it exists (purpose or mission), what it is attempting to accomplish in the long run and in the short run (goals and objectives), how it plans to reach its goals (strategies), and how it chooses to conduct its affairs (policies and procedures).
2. *Plans:* The organization must have strategic, operational, and implementation plans in order to turn mission, objectives, and strategies into reality. Time, cost, and performance schedules are essential, as are specific work assignments.

3. *Structure and systems:* The organization must be structured in such a way as to facilitate effective planning, work accomplishment, and control. It must rely on timely information systems and organizational processes.
4. *Resource management:* The organization must have in place effective and efficient systems of supervision and management to assure and encourage a high level of return from the organization's human, financial, and physical resources.

Against this backdrop management science techniques must be designed and applied in order to be truly effective. The more congruent such analytical techniques are with the real world of the manager, the more widely and effectively they will be implemented.

Role of Management Science

Beer (1968) has defined science as systematic knowledge about the world —not the way we would like the world to be, but the way it actually exists. Management science, then, is systematic knowledge dealing with organized activities like the ones we have discussed. Management science is much broader than the mere application of quantitative techniques to largely operational problems of organizations. It goes to the very heart of the management process—the measurement and development of hypotheses and theories about the full range of organizational activity. Management science involves trying to understand the underlying systems and processes of the organization, including missions and strategies, plans, structures, and resource management. From this understanding will develop techniques to assist managers in their various roles and functions. Those techniques more in tune with the reality of the managers' needs will, of course, be more widely employed.

The Decision-Making Function

Decision making is a critical responsibility of managers. Encompassed by the decision-making functions are strategic decisions, operating decisions, planning decisions, control or corrective action decisions, and a host of others. While a relatively small number of people in any given organization have the final responsibility for officially making key decisions, a much larger number of people take part in the process by collecting data, analyzing the situation, working with information systems, evaluating alternatives, and developing recommendations. In fact, it is often difficult to ascertain just where a given decision is actually *made,* because it is shaped by the various people, groups, and organizational entities involved in decision-support processes.

Quantitative techniques have found their greatest application in the general area of decision making because of the usefulness of management science/operations research models in objectively analyzing decision situations. That is, quantitative techniques provide objective information to decisionmakers to assist in the choice function.

It is helpful to look at decisions from two perspectives, that of planning and that of control. Most of what we consider to be problem-solving activities are related to the control function—that is, when we engage in these activities, we are attempting to bring the organization's performance back "under control." The implication is that performance has deviated from what is desired, and the manager is trying to bring things back to where they should have been all along. The other side of the control coin involves the "breakthrough" concept of Juran (1964). This concept is more proactive: Once things are "under control," the goal is to achieve a higher level of performance through conscious efforts. Reaching this goal involves certain changes that may reduce control in the short run but that are designed to improve performance (quality, output, effectiveness, efficiency, cost, and so forth) when things are brought under control at the new, higher level of performance. This approach involves risk and short-term disruption, but it recognizes that performance improvement, not control, is the manager's ultimate goal. This type of planning decision requires that the manager be able realistically to assess and evaluate the risks of the attempted "breakthrough" strategy. This type of managerial action is also more positive in that the manager is proactively involved in "opportunity-finding" activities rather than in the more reactive "problem-solving" approach. To be optimally effective, the analytical techniques employed must be amenable to this type of managerial environment.

DECISIONS AMENABLE TO
QUANTITATIVE ANALYSIS

It is natural for managers to use tools, quantitative or of other kinds, that are at their disposal. Where no such tools exist, or where tools do not fit the realities of the situation, managers must rely on less "scientific" approaches to do what they must as best they can. However, tools tend to be used (perhaps misused) just because they exist. In the words of one author, "If the only thing you have is a hammer, you tend to treat everything as though it were a nail'' (Maslow 1965, p. 111). Specialists in the technique areas tend to view problems and situations as opportunities to apply techniques rather than viewing techniques as means to solve problems. We tend to measure what is measurable and sometimes overlook what is not so measurable but is equally important. The tools available tend to affect the type of problems we attack and how we attack them; perhaps McLuhan (1964) was correct when he concluded that "the medium is the message."

Given the above caveats, let us not forget that this book deals with techniques to assist managers and analysts—techniques that come as close to the manager's actual decision-making and problem-solving processes as the current state of the art permits. Furthermore, let us not forget that all techniques, even advanced stochastic networking techniques, require that a given problem, decision, or situation be modeled. Not all situations can or should be modeled, regardless of the power of the technique. In order for modeling to be advisable (or even possible), several conditions must exist:

1. *Awareness:* The manager/analyst must be aware of the existence of a problem or opportunity situation and aware that quantitative techniques exist that could aid in the analysis of the specific situation.
2. *Time availability:* Sufficient lead-time must exist to permit implementation of the appropriate management science technique and must include times required for technique selection, problem formulation, model preparation or technique implementation, analysis of results, and implementation of results.
3. *Technique availability:* An appropriate technique must exist that is applicable to the problem or opportunity situation at hand.
4. *Resource availability:* The required human skills, computer time, money, and other resources must be available when needed.
5. *Data/information availability:* Most management science models/techniques require rather specific, quantitative data. The required information must exist in a form that will result in accurate, meaningful results. If such information does not exist (or if its accuracy is in doubt), it may not be advisable to perform the analysis since the results may be misleading. As the adage admonishes, "It is better to be roughly right than precisely wrong."

THE CONCEPT OF RISK ANALYSIS

The essence of managerial action involves dealing with risks. Risk is inherent in any activity because we live in a stochastic world; nothing is absolutely certain. The successful decisionmaker understands the existence of risk, is able to ascertain the degree of risk, evaluates the desirable and undesirable potential outcomes and their ratios, limits assumed risks to tolerable levels given the situation and resources, and is able to make decisions in this context. In short, the successful decisionmaker possesses what we typically call *judgment*—that is, knowing when and when not to pursue certain courses of action.

Simply stated, risk analysis involves determining the unfavorable outcomes that can occur as the result of a decision or action and evaluating the likelihood that one or more of these undesirable events will occur. Thus, risk implies the occurrence of an unfavorable outcome, and risk analysis involves the assessment of its likelihood. The likelihood that our venture will fail is information that every manager/analyst needs in order to make decisions about a venture. If the information does not exist, it must be judged, guessed at, appraised by experts, or obtained by other means; it cannot be ignored.

However, many people are unable to deal effectively with less-than-certain situations involving risks. The optimist will tend to minimize the likelihood of unfavorable outcomes and assume that the positive results that are possible will, in fact, occur; the pessimist sees the situation from the opposite perspective. The effective manager, however, is a realist who knows that we are not certain that positive or negative results will occur and who is able to think in terms of their relative likelihoods. Such a manager attaches values ranging from highly favorable to highly unfavorable to the spectrum of outcomes. By examining the values of various possible outcomes,

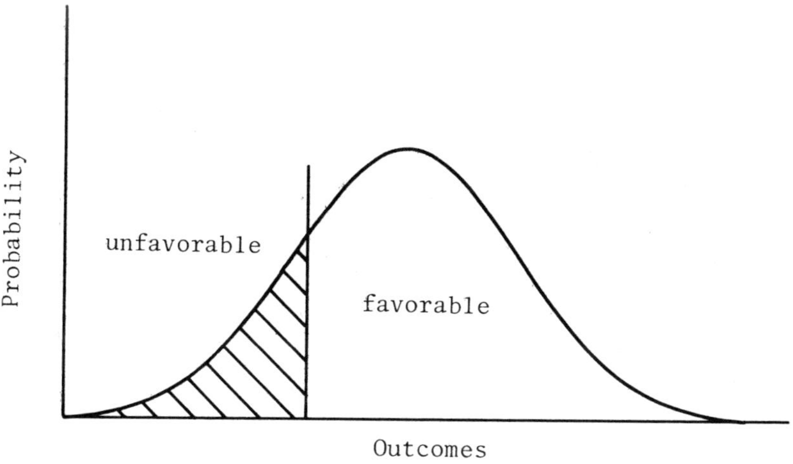

FIGURE 1.1.   Risk Analysis Diagram

coupled with the likelihood of their occurrence, the manager can assess the expected return from a venture. The prudent manager will weigh the risk-to-reward ratio to assure that the favorable outcomes and their likelihood (return) clearly exceed the unfavorable and their likelihood (risk). Furthermore, the manager will assure that the organization can afford the worst possible case—that is, if the most unfavorable set of outcomes does occur (which is, after all, possible), the organization must be able to afford the amount of resources at risk.

Various techniques exist to aid the manager/analyst in assessing risks. Probability theory, network theory, decision theory, and reliability engineering techniques have proven successful to a degree. In recent years advances in the state of the art have permitted integration of the most desirable features of each of these approaches and have resulted in advanced stochastic networking techniques. The most advanced allow the manager/analyst realistically to model a wide range of possible occurrences and their likelihoods and to present final cost/profit, schedule, and performance outcomes (and their interactions) against risk in a format similar to that shown in figure 1.1. Thus, the manager/analyst is able systematically to deal with complex, interactive, probabilistic situations in an objective, rational way that greatly assists his or her judgmental ability.

ANALYSIS OF STRATEGIC DECISIONS

Traditionally, quantitative techniques have been applied to relatively specific, operational-type problems of organizations. Certain management science techniques have permitted a broadening of this focus to include finance, marketing, human resources, and other areas, but the scope of application has been relatively specific compared to the breadth of concerns facing managers. Even most stochastic techniques, while enabling managers and analysts to deal with and include more realism in their models, have been limited because of their inability to deal on a fully integrated basis with the range of parameters of concern to managers.

One area that has received little more than ad hoc, piece-by-piece assistance from management science techniques is strategic planning. Thus, it has remained largely judgmental and qualitative in nature—supported, of course, by ad hoc quantitative data and analyses. But this area of management, strategic planning and decision making, has the greatest impact on the success of the organization. In the final analysis strategic decisions, not efficiency of operations, determine organizational and competitive success. What good is it to produce Edsels efficiently? The direction and environment resulting from strategic decisions are clearly the overriding factors in determining success. Again, "It is better to be roughly right than precisely wrong."

Thus, techniques that could truly assist managers in evaluating new ventures would be of significantly more value to an organization than any other application. Ventures such as new products, marketing strategies, mergers and acquisitions, and alternative development projects are fertile ground for systematic analysis, provided that the analysis techniques are able to deal with the variables the manager must consider. The technique used must be able to incorporate the major threats and opportunities in the external environment—that is, it must account for economic, competitive/market, political/legal, technological, and social factors. It must also be able to incorporate the organization's internal strengths and weaknesses, including financial, physical, and human resources. The resulting analysis must provide interrelated information in the form of likely costs, time, and results (performance) and must incorporate contingencies for risks and discontinuities. The manager must be provided with a realistic picture of what can happen and how likely the outcomes will be, as well as be enabled to answer "what if" questions. Stochastic network techniques are now able to provide the manager with this critical and long-awaited capability.

CONCLUSIONS

We have observed that management science techniques have been limited in their usefulness to management by their inability to deal with the manager's critical concerns and needs on an integrated basis. What are needed are techniques to deal with chance-related events and to incorporate probabilistic data. Furthermore, the techniques must include key parameters and their interactions and portray the full range of outcomes and their relative likelihoods. This type of information, valuable to the manager in strategic and project as well as operational areas, would be a giant step in closing the long-existing gap between the manager and the management scientist.

<h1>2 PROJECT/VENTURE MANAGEMENT</h1>

Network techniques for management use were popularized by the development of the Critical Path Method (CPM) and the Program Evaluation and Review Technique (PERT) in the mid- to late fifties. These techniques were designed to facilitate the planning and control of large-scale construction and engineering development projects and were soon proven to be very effective aids in the management of complex, one-of-a-kind undertakings (see *Work Scheduling Techniques,* 1968). In the ensuing years, the techniques found ready application to a number of projects—for example, the *Polaris* and other weapon system development efforts, NASA's Apollo program and lunar excursion, and a host of more down-to-earth industrial and commercial undertakings.

Very frequently the PERT/CPM networking techniques were employed in a separate, ad hoc organization entity, a project office, set up to complete a one-time task. This office, headed by a project manager, relied on the networking technique as its key planning and control system. However, from a management point of view, the PERT/CPM approach had several inherent weaknesses that resulted in modifications incorporating probabilistic branching (PERT/CPM networks are deterministic) and variable (stochastic) events and activities. These stochastic network techniques, such as the Graphical Evaluation and Review Technique (GERT), led the way for the development of techniques that enable a manager to model the range of alternatives and uncertainties present at the initiation of a new project, operation, strategy, or other venture. Thus, the application of the title "Venture Evaluation and Review Technique" (VERT) to the latest advance in the evolution of networking techniques.

The material contained in the following four subsections is based on an article by Digman and Green (1981).

THE CONCEPT OF PROJECT MANAGEMENT

Most successful projects progress through several phases (concept, design and development, production, and operation) during their life cycles. Each phase requires somewhat different information, information processing, and management decisions with regard to the parameters critical to management control: time, cost, performance, and risk. Various network techniques—PERT, CPM, PERT/Cost, Line of Balance (LOB), GERT, and VERT—have been employed to manage certain parameters during certain project phases.

In retrospect, the advent of the widespread use of project management in the late 1950s was a decided step forward. Under this concept, a single manager was responsible for the totality of actions required to complete a project. The manager's responsibility cut across the various phases required for the project completion, as well as the various functional groups involved in the task. Prior to employment of this approach, several functional managers were responsible for portions of a project; overall coordination took a back seat. The success of the single-manager concept is well known and has been greatly facilitated (perhaps made possible) by the concurrent availability of certain planning and control techniques, notably the Program Evaluation and Review Technique and the Critical Path Method (PERT/CPM).

Network-based management techniques such as PERT/CPM enabled the project manager to plan diverse undertakings on an itegrated basis, as well as to coordinate project tasks and to control work accomplishment. A major advantage of the networking techniques was that they assisted the manager in assessing two critical interrelationships: (1) the interface between various elements and work groups involved in the project and (2) the relation between current status and likely status at some future point. Thus, the manager was able to plan and institute corrective actions on a more timely and effective basis.

However, as we know, PERT-type systems are most effective in controlling the one-time, developmental, design-oriented stages of a project as opposed to the more reptetitive actions required during later project phases. Also, PERT systems, while of decided value in managing time-related concerns, proved to be cumbersome, at best, in managing costs (PERT/Cost) and of little direct benefit in achieving performance-related variables or in managing risks inherent in project-type undertakings. For these reasons, other techniques have typically been employed in conjunction with PERT systems, some in parallel and some in sequence. For example, Line of Balance–type systems (LOB) monitor production and deliveries, the Graphical Evaluation and Review Technique (GERT) adds to PERT the ability to deal explicitly with uncertainties in flow through the network, and newer approaches such as the Venture Evaluation and Review Technique (VERT) expand GERT to include performance, as well as time and cost, factors. (However, VERT's major advantage is still primarily in the concept and in the design and development stages of a project, although VERT can be applied to any decision involving chance.)

The point is that the project manager is faced with an array of systems and techniques that may be used to manage the time, cost, performance, and risk parameters of a project during its life cycle; the much-heralded "total management system" of a few years back has never materialized. If not properly integrated, this array of somewhat ad hoc systems may cause problems, inefficiencies, overlaps, and voids, particularly for the relatively inexperienced management team. Project managers need a conceptual framework to serve as a "road map" through the morass of information and decision needs, parameters, phases, techniques, and processing requirements.

THE PROJECT LIFE CYCLE

The normal life cycle of a product progresses through several somewhat distinct but partially overlapping stages or phases. For example, most new products begin as concepts that are evaluated in various ways, progress to a design and development phase where the concept is transformed into a product, and then move to a phase where a quantity of the product or item is produced for sale or use. Use of the item constitutes an operational phase, which terminates when the product or item is rendered obsolete, is discontinued, or is phased out in some manner. The relative duration of these stages may vary greatly, depending on the product in question (for example, high-volume, relatively long production and operational phases may occur for consumer products; a moderate-volume, short operational phase for military hardware; a moderate-volume, long operational phase for items such as commercial airliners; and a low-volume operational phase for items such as nuclear reactors). In consumer product parlance, most products go through an "invest/grow" stage followed by a "harvest/divest" stage. Treating the stages of a product's life cycle as phases of a project results in the four basic steps described below.

Concept Phase.   During this phase potentially feasible and profitable candidate products are investigated. Included is an assessment of (1) the likelihood that competing candidates will operate as desired when designed and manufactured, (2) the likelihood that a candidate will meet the real or perceived needs of the ultimate consumers or users, (3) the extent of likely demand, and (4) the likelihood that the product can be designed, developed, produced, and operated as desired within time and cost constraints, and given acceptable levels of risk. At completion of this phase, schedule, cost, and performance estimates and characteristics, plus an identification of the inherent risks, should exist for the product/item. In addition, a specific plan for completion of the remaining phases should exist.

Design and Development Phase.   Upon selection of a concept, detailed design and development activities may begin. These activities include engineering work resulting in the building of test models, or prototypes, for design revision and refinement. The output of this phase should be a set of detailed drawings and specifications (including data pertaining to expected reliability of the product) that are sufficient for manufacture of the product as designed and specified.

Production Phase.   While the production phase normally follows design and development, the two may overlap—at increased risk—if time is a critical parameter. In any event, this phase includes manufacture and on-time delivery of specified quantities of an item with acceptable quality, operational performance, and unit cost. Learning curve effects and necessary engineering changes occur as this phase progresses.

Operational Phase.   This phase typically overlaps with production and begins with the deployment or delivery of an item to the ultimate user. Successful operation includes possession of the desired performance characteristics by the user, with acceptable reliability and maintainability at acceptable costs. This phase is concluded when the product or item is rendered obsolete by changing user/customer needs, by existence of more desirable alternative products/items, or by failure of the item to perform satisfactorily. Disposal of the item completes its life cycle.

While these phases represent the normal life cycle of a successful product or item, few items are project managed throughout all phases. Transition may occur to product management or conventional management at any time and heighten the need for a conceptual framework that can facilitate and ease the "handoff" from the project management group.

PARAMETERS OF CONCERN TO MANAGEMENT

The manager must make many key decisions over the life cycle of an item. These decisions are largely represented by four key, interrelated parametric groupings: time, cost, performance, and risk. The major information needs for management of each of these four parameters in each of the life-cycle phases are shown in table 2.1. The parameters themselves are defined below.

Time. The time parameter initially includes the need to estimate accurately the length of time required and available to accomplish the various activities comprising the early phases of a project. The objective is the development of a realistic and attainable schedule. Once the project is underway, attention is focused on meeting scheduled dates and milestones. In the production phase the time parameter shifts to concern with setting and meeting production rates and delivery schedules, and in the operational phase, to concern with logistics.

Costs. Funding availability and levels are key resource inputs in the early stages of planning. While the level of resource availability constrains what can be accomplished during each phase, a major output of each phase is the further refinement of cost and resource estimates for subsequent phases. Holding costs and resources within approved levels is one of the manager's main responsibilities.

Performance. The product or item must possess an acceptable level of operational capability; otherwise, effort and resource expenditures are wasted. Design activities must result in specifications that permit prototypes and later units to meet test criteria; production items must also perform adequately and must conform to design specifications. In addition, speci-

Table 2.1. Life-Cycle Management Information Needs

Project Parameters	Information Needs			
	Concept Phase	Design & Development Phase	Production Phase	Operational Phase
Time	Completion and quantity guidance; time required for each concept	Milestone accomplishment; estimates-to-complete	Output rates; delivery schedules	Supply & logistical data
Cost	Budget guidance; "should cost" for each concept	Development costs; cost-to-complete	Unit cost	Cost of operation
Performance	Desired operational characteristics; likely performance for each concept	Design achievement; performance of prototypes	Conformance to specifications	Reliability and maintainability
Risk	Probability of achieving time, cost, and performance desired for each concept	Probability of successful accomplishment of above	Probability of successful accomplishment of above	Evaluation of potential successor concepts

SOURCE: Data from Digman and Green (1981).

fied reliability must be proven, and, since breakdowns will occur and service be required, the item must be maintainable under realistic conditions and procedures. Drawings and documentation describing each of these areas must be completed.

Risk.   Since uncertainties exist and unfavorable results may occur, risks are inherent in the very nature of new product/item developments. Acceptable levels of risk must be determined for the many factors comprising the time, cost, and performance parameters. Also, continuous monitoring of the likelihood of achieving unacceptable values of time, cost, and performance must occur.

MANAGEMENT DECISIONS AND INFORMATION REQUIREMENTS

It is difficult to categorize the decisions made by the project manager, even when one classifies them by parameters or phases. At the risk of oversimplification, however, one may conclude that the most difficult planning decisions occur in the concept stage and involve the evaluation and selection of competing candidates. Usually, performance characteristics are the driving factor at this point, and estimates of time, cost, and risk are developed for desired levels of performance. The implicit and/or explicit relative importance of the parameters comes into play as required trade-offs are made between them during the concept stage.

At this point an objective or goal has been created, and a detailed plan (usually based on a network) is developed to accomplish the goal. Design, development, and testing activities begin, and the focus of the project manager shifts to control and corrective action, which remains a major focus from this point forward. It is not the only focus, however; progressively more detailed planning for the production and operational phases are major outputs of the design and development phase.

The management information required for adequate planning and control of each of the project parameters during each life-cycle phase offers a framework for management. This framework provides a "road map" for the manager in charge of the item during any of the phases and enables the manager to focus on critical areas as well as to appreciate what precedes and follows the current phase. For project managers with multiphase responsibility, this perspective is critical; they can make sure that the information requirements of future phases of the project are planned and systems developed and implemented to provide information as required when needed.

PROJECT INFORMATION PROCESSING

Project management, in order to be successful, must depend on a well-designed information storage and retrieval system, timely reports, systems and procedures for revision and updating, and efficient computer programs for processing data. Projects evolve during the life-cycle phases, as do project information needs; an information system must be able to accommodate changing information requirements and project deviations from the original plans. The information system must also produce reports geared to the key parameters and presented in a format that managers may easily use.

Project management puts a premium on three major categories of decision making: planning, scheduling, and control. Project management typically requires different information and/or different application programs for the various project parameters and stages of project completion described earlier. Figure 2.1 presents the relationship between a Management Information System (MIS) and the major managerial requirements for decision making. The MIS data base serves as the primary focal point, since requisite data are needed for the various application programs. (Of course,

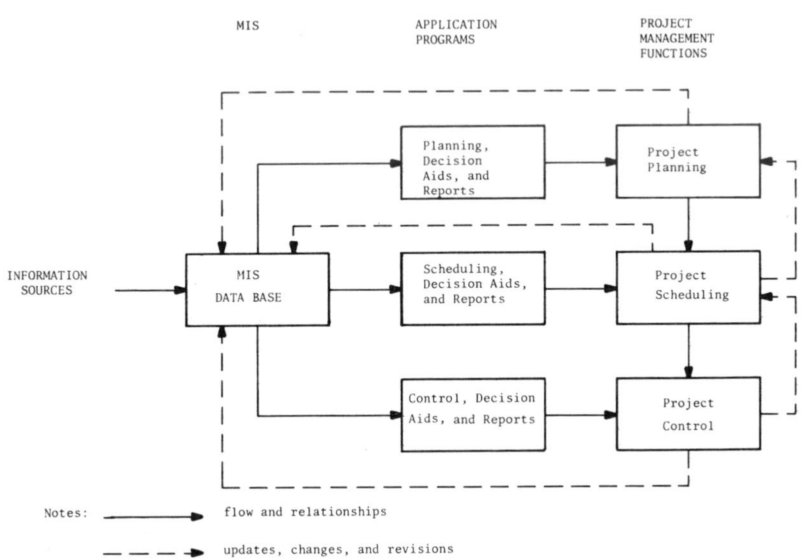

FIGURE 2.1.   Information-Processing Support for Project Management

it is also important to have the appropriate application programs for specific project problems.) Management needs to recognize that the MIS support system must be properly designed; the MIS design must accommodate the various parameters, stages, and potential modifications of projects.

Even historical information may serve a very useful purpose in the planning of future projects with similar objectives, components, problems, operations, labor requirements, and so on. However, collection of data without effective usage could prove very expensive and could lead to system inefficiencies. It would seem absurd for a project to duplicate the information collection and storage of the normal functional systems within the regular operations of the organization. Care should be taken to specify how data will be managed and who will be responsible for data maintenance.

Project management, by definition, implies centralization of project responsibility apart from the normal organizational hierarchy. Thus, an organization may face the practical problem of supporting one or more distinct projects using borrowed organizational resources (labor, facilities, equipment, MIS, and so on) and controlling costs to specified budgets. The MIS should, therefore, be designed to support a number of projects as well as to serve routine organizational requirements. Management cannot delegate responsibility for effective design of MIS procedures, processes, requirements, and usage. This caution applies to the purchase of software systems for project management but does not imply that one should avoid the purchase or development of applications programs. However, these programs must meet the information needs of the manager during all project stages. Most likely, a collection of software packages will be necessary.

Managers should be very skeptical of purchasing computer programs that are touted as capable of performing "project management." As suggested by this book's general theme of modularity, the data processing requirements of decision support for project management are varied and depend on the information needs of a given project problem. Therefore, a project may have a number of different processing needs and capabilities that may result in a number of reports. However, organizations need not spend time and money to develop programs or processing methods that are already available. Whatever the source, computer programs and reports must meet the needs of management and should be applicable to the different decision circumstances (described in table 2.1). Above all, management must not be constrained or limited by the availability of computer programs; such a circumstance would be analogous to a machine's dictating decisions to managers rather than helping managers to make decisions.

EVALUATING ALTERNATIVE VENTURE CONCEPTS

In a broad sense, the approach used to evaluate the alternative candidate products or items during the concept phase of a project involves the evaluation of alternative ventures. For example, similar considerations are involved in evaluating which (if any) competing concepts to develop, which products to market, which markets to enter, which companies to acquire, which strategy to pursue, and so on. In each of these cases the same approach is utilized to evaluate the alternative courses of action, or candidate ventures. All that is required is that the manager/analyst be able to specify desired and likely performance characteristics in relation to time and cost resources consumed in creating results (performance). Results, or performance, are the pertinent outcomes or characteristics important to the manager making the decision. Included would be profits, market share, cash flow, product characteristics (such as size, speed, reliability, output, accuracy, or whatever the manager deems important), sales volume, and productivity. All the manager/analyst must do is to specify desired quantitative results, the variability likely in the results, the time and cost values associated with the results, and the major influences on the results. Also, a decision and possible outcome network must be constructed to represent the flow of decisions, actions, chance, and other influences on the process. Such a network requires nothing more than a structuring and quantification of the decision process the manager normally follows. The result of the simulation runs will enable the manager/analyst to see the complete array of possible outcomes and their likelihood. Armed with this information, the manager can make a more realistic decision about which, if any, of the competing ventures to select and pursue.

While we have discussed project and venture alternatives, operational situations can also be analyzed by the same procedure. For instance, plant locations, facility design and layout, equipment purchases, distribution logistics, and other types of operations management problems are amenable to this type of analysis.

CONCLUSIONS

Networking techniques for management use became popular with the advent of PERT/CPM applications, largely in project-related environments. Earlier, network applications had been used by analysts primarily on operational problems. While the PERT/CPM systems have definite shortcomings, stochastic approaches have broadened the applicability of networking to approximate more closely the parameters and decision processes utilized by managers. We will discuss network models in more detail in chapter 3.

3 INTRODUCTION TO NETWORK MODELS

In this chapter we examine, compare, and evaluate the major networking techniques useful in management applications. First we briefly review the management science concepts of simulation and modeling and recap the various types of models and major approaches to model construction. Next we trace the development of important network models for management use and briefly describe each model. We conclude with a comparative evaluation of the major techniques.

MANAGEMENT SCIENCE MODELS

Management science (MS) may be defined as the application of the scientific method to provide decisionmakers with a quantitative basis for decisions regarding complex situations under their control. Management science involves quantitatively evaluating alternative courses of action that decisionmakers can take and then providing this evaluation to decisionmakers so that they can choose the alternative that will be in the best interest of the organization as a whole.

Distinguishing Characteristics

Management science has four distinguishing characteristics. First, MS is broad in scope, concerned with the organization as a whole rather than with suborganization units. In reality, problems of production are interrelated with, for example, problems of funds, control, personnel, and marketing. MS attempts to define these interrelationships and to solve problems in such a way that the solution is optimal for the organization as a whole. Furthermore, MS is applicable to organizations having definable objectives, where the attainment of these objectives is subject to constraints and where alternative ways of pursuing the objectives exist.

Second, MS emphasizes the utilization of analytic techniques for solving individual problems rather than emphasizing the application of general models. No two problems are exactly alike; therefore, MS attempts to develop a model to fit the problem rather than to modify the problem to fit a model. In the development of models, MS leans heavily on mathemetics and other basic sciences. However, if it is to be successful, MS cannot be limited by the scope and method of these basic sciences. It is limited only by human reasoning ability, not by the nature and content of certain disciplines.

Third, MS uses a "team" effort in dealing with problems. While it may often be possible for one person to do MS work, many problems are of such breadth that the team approach frequently results in more effective and efficient methods of obtaining a solution. The team should bring together persons from various disciplines and backgrounds and persons with different experiences.

Fourth, MS must be concerned with the practical management of an organization. When scientists tackle a real-world problem, they tend to seek solutions in the light of their own background and experiences. But they must present their solutions in the language of the real world. In addition to the use of the scientific method, MS requires that definite conclusions be drawn and that these conclusions be communicated to the decisionmaker.

Management science is not (and does not claim to be) a magic formula that can be manipulated to yield the ultimate, unique solution; it is merely a logical, systematic approach to providing a rational basis for a decisionmaker's choice.

The Scientific Method

Management science follows a systematic approach to find solutions to problems. Specifically, it applies the scientific method, which is simply a logical, systematic approach for conducting research or solving problems.

The scientific method consists of the following sequence of steps: (1) observation, (2) problem formulation, (3) model construction, (4) model solution (5) model and solution test, (6) establishment of controls, and (7) implementation.

Observation. The first step in the scientific method involves ongoing analysis of the organization and its external environment to discover opportunity/problem areas. Opportunities and problems may be highlighted by information from existing reports but may result from informal and judgmental input as well. (The latter circumstance applies particularly in the case of opportunities.) While reports and feedback may indicate the existence of problem areas, the recognition of opportunities will likely be less formal or structured. Both problems and opportunities, however, indicate an area of potential or need requiring further analysis.

Problem Formulation. The second step in the scientific method is formulation of the problem. This phase is often slighted because the analysts (or researchers) are anxious to proceed with the solution phase of the research. It is desirable to employ a systematic procedure in formulating the problem and to allocate adequate time for problem formulation. A suggested procedure is described below:

1. Specifically define the problem. Before a problem can be formulated, analysts should have a general understanding of what the problem is; the nature and content of the problem should be stated as explicitly as possible.
2. Establish objectives. In order for a problem to exist, an organization must want something that it does not have—that is, some objective has not been obtained to the satisfaction of those concerned. The objective is the goal to be obtained through the study or research effort.
3. Determine the constraints. The problem exists in an environment of limited resources such as manpower, machines, material, money, or time. The solution to the problem must "live within" the constraints set by the availability of resources.
4. Determine the key variables. The variables are the factors subject to change—that is, they do not remain constant over time. Variables are of two types: controllable and uncontrollable. The levels (values) of the controllable variables can be set by the decisionmaker; the levels of uncontrollable variables are influenced by factors outside the decisionmaker's control.
5. Determine a measure of effectiveness. Effectiveness of a course of action may be measured in terms of cost, profit, units, time, and so

forth. The measure of effectiveness is used to evaluate and compare a variety of possible alternative courses of action (solutions) so that the decisionmaker can decide what to do on the basis of quantitative data.

Model Construction.   The third step in the scientific method is construction of the model. A model is simply a representation of the real world. Models fall into three general categories: iconic models, analogue models, and symbolic models.

The *iconic model* visually or pictorially represents certain aspects of a system. It looks like what it represents. The model, however, is usually "scaled up or down"—for example, the model airplane is an iconic model of a real airplane; a chemist constructs iconic models of molecules.

The *analogue model* employs one set of properties to represent some other set of properties possessed by the system under study. A very simple analogue is a graph. In graphs we use distance to represent such properties as time, number, percent, and weight.

The *symbolic model* employs symbols (usually mathematical in character) to describe properties of the system under study. Mathematical models are symbolic models; the pertinent factors of a problem are related in abstraction. A simple example is the equation for the area of a circle: $A = \pi r^2$. The value of the symbolic model is that it can be more easily manipulated.

The model of the real-world system used in management science is manipulated according to certain rules, and the consequences of various policy decisions are predicted on the basis of model solutions. As long as the model represents the real world, it is possible to predict the consequences of decisions from the model without actually having to manipulate the real-world system. The object of manipulating the model is to determine an acceptable solution to the problem.

The general form of an operations research model is

$$E = f(c, u),$$

where E is effectiveness, c represents the variables that can be controlled (independent variables), and u represents the variables that cannot be controlled (exogenous variables). In other words, effectiveness is a function of the controllable and uncontrollable variables.

The value of a model is not necessarily its increased detail, but the fact that it contains the "essence of reality." It must incorporate the variables and relationships important to the problem at hand. In fact, increased detail may actually serve to obscure key relationships in certain instances and make a less detailed model more desirable. In any event, the level of detail is dependent upon judgmental and practical concerns.

Model Solution.   The fourth step in the scientific method is obtaining a solution to the model. In the case of an optimization model, the solution represents a set of values for the controllable variables that yields the optimum result while satisfying the constraints. The "solution" to descriptive models may represent the array of possible outcomes and their likelihood.

Solutions may be based on deterministic or probabilistic methods and obtained by analytic or simulation means. A deterministic solution represents assumed conditions of certainty, whereas probabilistic solutions represent the more realistic inclusion of variability in data and relationships incorporated in the model. An analytic solution represents results obtained by mathematical means, whereas simulation solutions describe the results of replicating the operation of the system (model). Simulation solutions are usually employed where probabilistic models and descriptive (versus optimization) outputs are more appropriate. Network models may be either deterministic or probabilistic.

Model and Solution Test.   The fifth step in the scientific method is testing the model. A model is, at best, only a partial representation of reality. The adequacy of a model can be tested by determining how well it can predict the effect of changes in the real system. However, a solution to the model is not necessarily a solution to the problem. The model may be tested by the use of historical data (a retrospective test) or by a trial run or pretest.

Model construction and model testing are not necessarily separated in the actual solution of a problem. In practice, as the study is being made, information that can be used to evaluate the model is continuously becoming available to the analysts. As they test the model, they may discover inadequacies. It is then necessary to modify the model and to retest.

Establishment of Controls.   The sixth step in the scientific method is establishing controls. A solution derived from a model remains optimal or descriptive only as long as no significant changes occur in the system for which the solution was developed. Because systems usually do not remain stable over indefinite periods of time, a control system is required so that significant real-world changes can be detected and required adjustments in the solution and model can be made. Before controls can be established, the method of operation should be defined. The definition should include a description of inputs to the operation, operating procedures, objectives, and outputs. This information can then be used to set up a data feedback system as part of implementation.

Implementation.   The final step in the scientific method is implementation of the solution. In theory, implementation is simply the translation of the solution into actions and procedures, but in reality it is decidedly more complex. Questions of timing, personnel, acceptance by superiors and subordinates, and so forth, make the job of implementation no cut-and-dried affair. As a minimum, efficient implementation, in which control over the new operation is retained, may require training and indoctrinating personnel and setting up checkpoints and a feedback system. Properly trained and oriented to the new objective, the persons most concerned with the operation will be less likely to balk at any new methods. If the control of the decisionmaker is not established and defined, the research and the prior effort expended will most assuredly be wasted. The checkpoints and feedback control system will serve as indicators of any necessary changes in the operation, the model, the implementation process, the components, or the organizations. They will also indicate alterations in any of the facets that may be adverse to the objective of the operation.

If the model and its solution have been properly tested and the trial run effectively analyzed, the implementation step ideally would not require one to institute controls. However, in reality a single trial run or test is too often an inadequate indicator of how a model will perform in the real world. Therefore, an established feedback control system is a necessity if a working operation is to result.

HISTORY OF NETWORK MODELS

Network models represent one application of the scientific method/management science approach to the solution of management problems. The use of network modeling techniques has potential for wide application. While numerous network techniques exist, the manager is interested only in techniques that provide a planning and/or control capability. Furthermore, the project or venture manager's interest is primarily in network-based techniques that assist in managing the time, cost, performance, and risk parameters over a venture's life cycle. Thus, a review of the history and major variations of the key network methods would be helpful in illustrating the capabilities of the techniques and might enable the manager to decide which technique(s) would be most relevant, given various parameters and project/venture stages. Figure 3.1 illustrates the evolution of network models.

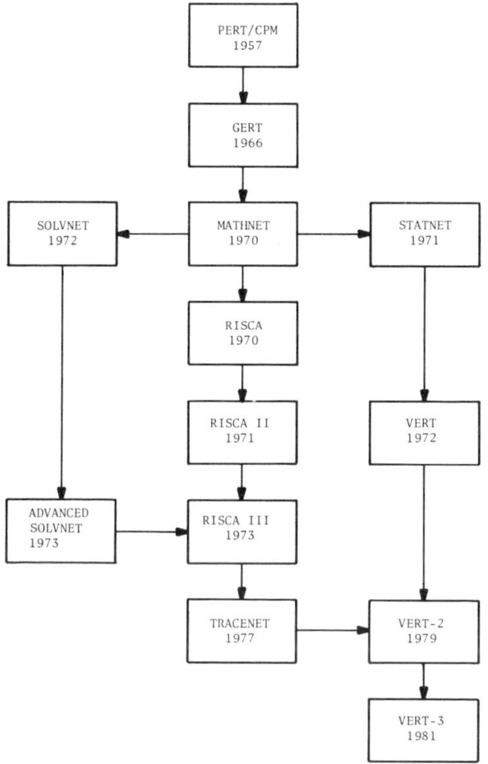

FIGURE 3.1. Evolution of Network Models

Development of PERT and CPM

The U.S. Navy's successful application of PERT to the development of the Polaris Fleet Ballistic Missile System in 1958 generated a very large body of network analysis methodology. PERT, as originally designed, was totally time oriented and did not directly consider cost, availability of resources, resultant performance, or risk. Furthermore, time, as entered in the three time estimates of normal, optimistic, and pessimistic for the beta distribution, has drawn considerable comment in the literature. Later modifications of the technique were designed to deal with the important cost and reliability parameters for project applications. PERT/Cost added resource cost to the PERT/Time schedule. However, cost was given a tagalong, nondecision status. PERT/Reliability was an extension similar to PERT/Cost and designed for inclusion of the reliability variables.

CPM (Critical Path Method) was developed independently at about the same time as PERT for a construction project at DuPont. This method was instituted by a subcontractor, Catalytic Construction Company, and has undergone several refinements. For example, attempts have been made to incorporate the cost parameter, as was done in PERT.

PERT and CPM programming and control methods have been used quite extensively in research and development scheduling, construction planning, and resource allocation models. Pritsker and Happ (1966) attributed the increasing use of network analysis to the ease with which operational systems could be modeled in network form. They also attributed the growth of network analysis to (1) the ability of network analysis to model complex systems by compounding simple systems, (2) the need for a communications mechanism to discuss the operational system in terms of its significant features, (3) the need for a means for specifying the data requirements for analysis of the system, and (4) the need for a starting point for analysis and scheduling of the operational system.

Following the initial and continuing wide use of PERT and CPM, questions were soon raised regarding additional capabilities that would enhance this type of analysis. Some of these questions were: (1) How can one better allocate resources among activities to shorten the project, reduce risk, and reduce cost? (2) How can performance requirements be reduced in critical or pacing activities without greatly reducing the benefits of the program? (3) Are there noncritical activities where costs or risks can be reduced with no increase in the length of the overall program? (4) Is there a better network that would lead to a better combination of program time, cost, and performance values?

Addition of Stochastic Capabilities

The first significant development in expanding the capability of network models to incorporate more stochastic flexibility was the "decision box" introduced by Eisner (1962). Elmaghraby (1974) added additional logic and algebra to network models, and a number of authors have used flowgraphs to represent and analyze probability systems. The next major development was Pritsker's GERT, or Graphical Evaluation and Review Technique (see Pritsker and Happ 1966). GERT is basically an analytical procedure that combines the disciplines of flowgraph theory, moment generating functions, and PERT to obtain solutions to stochastic problems. GERT derives both the probability that a node will be realized and the conditional moment generating function of the elapsed time required to traverse between any two nodes.

Another method for evaluating stochastic design models was proposed by Hespos and Strassmann (1965) when they introduced the concept of the stochastic decision tree. They applied Monte Carlo and GPSS simulation techniques to conduct a risk analysis on investment decisions. Next, Crowstone and Thompson (1967) extended Eisner's "decision box," or decision alternatives, into a CPM network. Their method became what is known as Decision CPM.

Risk Analysis and the Development of Network Technology

With the advent of the large cost overruns and schedule slippages on many major development projects in the defense sector of the U.S. economy in the late sixties and early seventies, military managers realized the need for risk analysis. However, the essential difficulty in applying the techniques available at that time lay in the aggregation of a total risk profile. Stochastic networks are characterized by their events (nodes) and activities (arcs). When all activities leading to an event must be completed prior to the event's taking place, as in PERT, the logic diagram portrayal of the situation corresponds to the intersection, or AND logic. Obviously, requiring *all* activities to be complete before subsequent activities can occur is unrealistically restrictive for decision-related networks. In addition, most real situations the manager faces involve degrees of completion rather than just the yes/no designation. The aggregation of a total risk profile by PERT would require the construction of a network of only AND nodes with deterministic output, where individual-activity risk profiles were obtained from beta or triangular distributions. However, the aggregation of a total risk profile by GERT would specify a case where all nodes have OR input and probabilistic output logic. OR logic specifies that only one activity leading to an event need be completed prior to the event's taking place. (For models that do not fall into either of these two categories, the easiest way to accumulate a total risk profile is through network simulation.)

In view of the lack of generalized tools and the heavy emphasis of Defense Secretary Packard on risk analysis, the U.S. Army let a contract to Mathematica of Princeton, New Jersey, to develop a course on risk analysis (see "Improvement in Weapon Systems Acquisition," 1969). While developing this course, Mathematica pioneered what proved to be a significant change in network analysis by developing a computer program named

MATHNET (Mathematica 1970). The program initially was intended as a teaching aid; however, since it was the only viable tool available for risk analysis, its use soon spread throughout the army. Because MATHNET was structured rather hurriedly and thus was not thoroughly debugged and tested on real problems, it proved to have some computational mistakes. Thus, a number of computer programs that were corrected and expanded versions of MATHNET evolved within the army—for example, RISCA (Risk Information System Cost Analysis), STATNET, and a more advanced step called SOLVNET (see *A Course of Instruction in Risk Analysis,* 1970; Moeller 1971; Percy 1973). These three simulation networking tools present a significant addition to the adaptability available to the development manager. These tools have AND and OR input logic teamed up with ALL and probability output logic. Also, they have some nodes with unit logic that tie a specific input arc to a specific output arc. These nodes selectively transfer flow from the input arcs to the output arcs via time and preference considerations only. These node logic combinations allow for modeling much closer to the real world with less abstraction than was required by PERT or GERT. Additionally, activity processing times can be entered as a normal, uniform, or triangular distribution. Activity times can also be entered in histogram form in STATNET and SOLVNET. Additionally, these two network tools have critical path capabilities. SOLVNET has the capability to structure some time dependencies. The expanded logic capabilities, coupled with the stochastic input capabilities and the simulation treatment given the network by MATHNET and its three offshoots, enabled the development of time-centered risk profiles.

Development of VERT

While the army's techniques constituted a significant advance in networking technology for management, they lacked additional features that seemed highly desirable. For instance, the MATHNET group was time centered like PERT; cost was given a bookkeeping, second-class, tagalong, nondecision status. Additionally, performance was omitted from any direct numeric analysis and considered only in gross alternatives structured into the network. As Archibald (1976) pointed out, "There exists a definite need to evaluate the total project in all three areas simultaneously, on an integrated basis." Also, not enough node logics were available; particularly needed was one enabling the invocation of the time, cost, and performance constraints imposed by management on all new ventures. These deficiencies culminated in the development of the Venture Evaluation and Review Technique (VERT) for military project applications (see Moeller 1972). Since that time the technique has been significantly expanded in terms of capability and application potential to many nonproject, private-sector ventures, as described in detail in part II. Selected nonmilitary applications are described in part III.

The initial version of VERT gave the analyst the ability to model decisions within the network in terms of time and/or cost and/or performance considerations rather than being constrained to time alone. At last, performance could be entered in the network in numerical terms rather than as gross alternatives. VERT thus gave the usual three dimensions used universally to discuss a project (i.e., time, cost, and performance) the same status and treatment level. The ability of a network to represent the real world lies largely in the flexibility of its nodes. VERT made a quantum jump in this area with the introduction of six new types of node logics. But perhaps the most significant innovation was the introduction of VERT's mathematical relationships. VERT is able to establish a mathematical relationship between any given arc's time and/or cost and/or performance and any other arc's and/or node's time and/or cost and/or performance. Thus, any two points within the network can be tied together by a mathematical relationship selected by the user out of the array of mathematical relationships available in VERT. Additionally, these same mathematical relationships can be used to establish relationships among the time, cost, and performance variables of a given arc. VERT is designed to be very open-ended and comprehensive when establishing relationships among network parameters.

VERT-2 and VERT-3

VERT-2, completed in 1979, added many new features and refinements to basic VERT. The latest version, VERT-3, is a completely new program with still more advanced features and improvements in coding. VERT-3 corrects certain malfunctions and defects that remained in VERT-2, and it is accompanied by greatly expanded, more detailed and illustrative documentation.

One example of the new features of VERT-3 is the significant improvement in mathematical relationships. VERT-3 can model without any loss of realism any development/venture problem that does not involve mathematics beyond what is listed in the mathematical relationships as given in chapter 6. The mathematical relationship capability has grown to the point where flows can be isolated, started, stopped, or altered in nearly any conceivable way. For example, separate flows can be established and maintained for each of the pertinent performance characteristics desired if one wants to join these characteristics together in one single value. Cost can be given a similar treatment; however, going beyond three dimensions generally proves difficult for the manager.

More nodes have been added to VERT-3, and many of the nodes in the initial version have been altered or condensed to facilitate modeling and the achievement of realism. To facilitate budgeting, VERT-3 accumulates cost into a histogram form between user-selected time intervals such as years or months. VERT-3 now is able to print out slack times in histogram form. It can also tell the user the optimal path in addition to the critical path and thus help the manager determine where best to expend resources. VERT-3 carries two types of cost: path cost and overall cost. Their names are indicative of the type of information they produce. If a mathematical relationship happened to create a dependency situation on an arc where, for example, cost was dependent on time and time on performance, the user was required by the initial version of VERT to specify the order for processing its time, cost, and performance inputs in order to keep from violating this dependency; VERT-3 now automatically determines these dependencies. To facilitate data handling, VERT-3 has an automated storage and retrieval capability. The initial version of VERT frequently ran out of core when loading large problems; a core-saving, data storage pointer system exists in VERT-3. VERT-3 has been structured to be both very usable and sufficiently powerful to handle any problem not requiring high-level math.

KEY TECHNIQUE DESCRIPTIONS

This section provides a thumbnail sketch of selected deterministic and stochastic networking techniques that have wide acceptance. The material is based on an article by Digman and Green (1981).

Deterministic Techniques

PERT/CPM. The Program Evaluation and Review Technique (PERT) and the Critical Path Method (CPM) represent a generic form of network planning and control techniques. Granted, differences exist between PERT and CPM—principally, PERT is event oriented and CPM is activity oriented —but the similarities far outweigh the differences. In fact, many of the more recent computer programs will accommodate either approach; for example, the user may opt for traditional PERT's three time estimates or CPM's single estimate for activity times (see *Project Management System 360,* 1968). The PERT/CPM approach is described below:

1. Determine the tasks (activities) required for project completion and associated descriptors of project progress (events).
2. Construct a deterministic precedence chart (network) of the activities and events required for project completion. Include time estimates for each of the activities plus required dates for completion of any of the key activities or events (milestones).
3. Perform the network calculations that yield initial values of activity or event slack and criticality.
4. Reallocate resources among activities as required or desired. The result should be a schedule of activities and events that efficiently utilizes resources and, to the extent possible, meets schedule constraints.
5. Begin work accomplishment and control progress through use of actual activity times and updated time estimates for future activities.

In summary, PERT/CPM systems have proven their worth in managing nonrepetitive projects with a number of activities that must be efficiently scheduled and coordinated. A major shortcoming of the systems is the inability to handle decision-type events where, for example, one of several possible activities is selected; in PERT/CPM, *each* activity in the network must be accomplished (see Wiest and Levy 1977).

PERT/Cost. PERT/Cost was an attempt to include the cost parameter in the PERT/CPM system. Beginning with a project tree diagram, or work breakdown structure (WBS), the project was divided into progressively more detailed components, assemblies, parts, or functions. At the most detailed level of breakout, cost categories called work packages were developed; they consisted of one or more network activities. Costs were estimated at this level and could be summarized at higher levels of the WBS and by functional cost category. After work commenced, actual costs were to be accumulated, remaining costs reestimated, and estimates-to-complete developed for control purposes. The network portion of PERT/Cost remained the same as in basic PERT/CPM (see *DOD/NASA Guide,* 1962).

In practice, the project management effort involved in operating the PERT/Cost system was found to be too great. The amount of updating, reestimation, and changes caused by technical and performance problems, as well as the need to estimate "actual" costs to meet report cutoff dates, caused the system to fail. Today project management techniques allow costs to be assigned at higher levels of indenture or by activity. The result is, in effect, a modified PERT/CPM system with an associated cost capability (see, for example, *Project Management System 360,* 1968).

Line of Balance.   Line of Balance (LOB) is a management-oriented charting tool for collecting and presenting information relating to time and accomplishment during production. It shows the progress, status, timing, and phasing of the interrelated activities of the production process; these data provide management with a means of comparing actual with planned performance. In addition, management receives timely information concerning the critical areas where the process is, or will be, behind schedule. LOB differs from other network techniques in that it is utilized mainly in the production process, from the point when incoming or raw materials arrive to the shipment of the end product. It is basically a means of integrating and monitoring the flow of materials, components, and subcomponents into manufacturing in accordance with delivery requirements (see Schoderbek and Digman 1967*a*).

As originally conceived, LOB was a graphical technique. As such, it had little capability to accommodate changing production cycle times caused by learning. In addition, complex production sequences caused problems, and the use of common parts in several subassemblies required separate analysis. Today the graphical technique has been replaced and the LOB type of analysis can be provided by Materials Requirements Planning (MRP) systems.

Stochastic Techniques

PERT, CPM, PERT/Cost, and LOB are deterministic network models —that is, they do not permit probabilistic occurrence of activities. Each activity in the network must occur, and all precedence relationships must be specified by the persons preparing the network. In the real world, however, risk and uncertainty exist, and certain chance events may occur in the course of a project—for example, a certain test may be successful, unsuccessful, or partly successful, and each outcome has a certain likelihood. When the future is "fuzzy," or uncertain, PERT/CPM is not able to accommodate this type of stochastic outcome.

Stochastic networking techniques broaden the model's applicability to include probabilistic nodes (events)—that is, each branch (activity) emanating from a node has associated with it the probability that it will be taken during any given simulation trial. (In PERT/CPM networks, this probability must equal 1; each branch *must* be taken.)

GERT.   Perhaps the best-known and most widely used stochastic networking technique is the Graphical Evaluation and Review Technique (GERT). GERT networks may contain probabilistic branching, deterministic branching, or some combination of the two. Nodes are considered to have an input and an output side, which specify how they interrelate to incoming and outgoing actions (branches). The four types of input logic specify how many incoming activities are required for first and subsequent releases of the node, and the two types of output logic determine the type of branching—deterministic or probabilistic—emanating from the node. GERT has the capability to include repetitive, or recurring, activities as well as cost information.

One of the main advantages of the probabilistic network (in addition to its added realism) is its ability to simulate project outcomes. Since the network can incorporate probabilistic occurrences—both favorable and unfavorable—it is possible to develop a distribution of likely outcomes through repeated runs of the network. This capability is particularly valuable in the planning stages of a project and gives the manager a picture of the probability of success and the risk of failure, as well as indicating which activities are critical to these outcomes. Furthermore, managers might desire to simulate the project impact of varying policies with respect to production loads, materials availability, work scheduling, and so on. GERT would appear to be an excellent vehicle for such analysis. GERT may also include cost analysis by enabling fixed, variable, and total cost to be assigned to each activity, with events serving as cost centers. Halting of activities, partial completion of precedent activities with release of the next event's activities (i.e., production of a product with components missing or partially completed), and activities that are resource-constrained (i.e., lacking sufficient time, labor, materials, or equipment) may be treated by some of the many members of the GERT family (see Moore and Clayton 1976).

It should be noted, however, that cost is afforded a tagalong, nondecision status in GERT. That is, cost is treated as a secondary variable, given certain times for an arc in the network. With GERT, cost is always a dependent variable and thus precludes the manager from, for example, manipulating the budget and determining the impact on the time schedule. This condition is a serious shortcoming in managerial applications because costs and budgets can be of primary rather than secondary importance.

VERT.   The Venture Evaluation and Review Technique (VERT) is, like GERT, a stochastic networking technique. It is similar in basic concept to

GERT but specifically treats time, cost, and performance parameters for nodes and activities (arcs).

Numerical values for each activity's time, cost, and performance parameters may be assigned in terms of (1) one of fourteen standard statistical distributions embedded in the VERT-3 model, (2) a histogram, or (3) a mathematical relationship, provided the time and/or cost and/or performance of this activity is dependent on other nodes and/or arcs that are to be processed (completed) prior to this arc. Performance can be modeled in terms of any meaningful unit of measure—for example, quantities produced, return on investment, or a dimensionless index that combines the many required diverse characteristics needed to define fully the resultant output of the resource expenditure.

Arcs and nodes are similar in that both have time, cost, and performance attributes. Arcs have a primary and cumulative set of time, cost, and performance values associated with them, while nodes have only the cumulative set. The primary set represents the time expended, the cost incurred, and the performance generated in the completion of the specific activity this arc represents. The cumulative set represents the total time expended, cost incurred, and composite performance generated to process all the arcs encountered along the path of the network flow in the completion of the processing of the arc or node in question.

VERT has two types of nodes, which start, stop, or channel the network flow. The most commonly used type is split-node logic; it has separate input and output operations. The second, more specialized, and less frequently used type of node has a single-unit logic that covers both input and output operations simultaneously. Four basic input logics are available for the split-logic nodes. Six basic split-node output logics are available to distribute the network flow to the appropriate output arc(s). The cumulative time, cost, and performance values computed for the active output arcs consist of the sum of the time, cost, and performance values derived for those arcs plus the time, cost, and performance carried by the arc's input node.

After the problem situation has been adequately modeled via VERT's arcs and nodes to achieve the desired level of realism, the problem is ready to simulate. VERT simulation involves the creation of a network flow that traverses the network from initial node(s) to create one trial solution of the problem being modeled. This simulation process is repeated as required to create a sufficiently large sample of possible outcomes. Node completion time, cost, and performance values may be obtained in any of eight desired forms. VERT optionally produces a critical/optimum path index for nodes and arcs. The critical path is the path through the network with the longest completion time, highest cost, lowest performance, or least desirable weighted combination of these factors, based on user-developed weights.

While VERT had direct application to project management, VERT-3 has further application to the general area of strategic decision analysis. To date VERT has been successfully utilized in the assessment of the risks involved in new ventures and projects, in the estimation of future capital requirements, in control monitoring, and in the overall evaluation of ongoing projects, programs, and systems. VERT has been helpful in cases where management must make decisions without complete or adequate information about the available alternatives (see Moeller and Digman 1978).

COMPARISON OF TECHNIQUES

Table 3.1 provides a summary comparison of five key, network-based management techniques evaluated against nine criteria important to managers. Table 3.2 provides a more thorough treatment of each of the five techniques versus the evaluation criteria. Table data, as well as the following discussion, are based on Digman and Green (1981). The criteria are defined below:

1. *Project phase applicability:* Phases during the normal product/item life cycle for which the technique offers significant applicability;
2. *Parametric focus:* Choice of the parametric areas for which the technique offers planning and/or control information for decision purposes;
3. *Preparation requirements:* The amount of time, effort, and resources required to implement the technique, including network preparation, estimating, and other input preparation;
4. *Data base:* Size of data base required to operate the system;
5. *System operating cost:* The costs of utilizing the technique, including processing time, analytical effort, and so on;
6. *Comprehensiveness:* Breadth of applicability of the technique to varying types of activities;
7. *Flexibility:* Ability to handle varying types of situations occurring during the project life cycle;
8. *Ease of update:* The amount of time, effort, and resources required to produce subsequent outputs reflecting changing conditions;
9. *Focus reporting:* Ability of the technique to highlight areas or situations of critical importance to the manager.

Table 3.1. Evaluation of Network-Based Management Techniques

Criterion	PERT/CPM	PERT/Cost	LOB	GERT	VERT-3
Project phase applicability	Design & development	Design & development	Production	Concept, design & development, production	Concept, design & development, production, operations
Parametric focus	Time	Time, add-on cost	Time	Time, add-on cost, risk	Time, cost, performance, risk
Preparation requirements	Moderate	Appreciable	Simple	Can be appreciable	Appreciable
Data base	Moderate	Large	Low	Can be large	Large
System operating cost	Moderate	High	Very low	Moderate to high	High
Comprehensiveness	One-time activities only	One-time activities only	Repetitive activities only	One-time plus repetitive	One-time plus repetitive
Flexibility	Deterministic situations only	Deterministic situations only	Deterministic situations only	Stochastic orientation	Stochastic orientation
Ease of update	Moderate	Burdensome	Easy	Appreciable requirement	Appreciable requirement
Focus reporting	Very good	Excellent	Very good	Outstanding	Outstanding

SOURCE: Data from Digman and Green (1981).

Table 3.2. Evaluation of Major Network-Based Project Management Techniques

Criterion	PERT/CPM	PERT/Cost	LOB	GERT	VERT-3
Project phase applicability	Prime application to one-time projects, largely design & development phase	Same as basic PERT/CPM	Production only	Most value in concept phase; usable in all phases to some degree	Same comment as for GERT
Parametric focus	Time oriented; treats performance & costs as objectives, constraints, or by-products	Adds cost planning & control feature to basic PERT/CPM	Time oriented (schedule & quantity)	Time oriented with add-on cost feature; analyzes time & cost risks	Fully treats time, cost, & various performance measures; analyzes risks in all three
Preparation requirements	Network & time estimating is significant, but is planning that should be done anyway	Cost estimating by work package or activity is significant addition to PERT preparation, but should be done anyway	Main requirement is production flow chart and cycle times	Networking requires special familiarity; multitude of features adds complexity	Same comment as for GERT; use of optional features increases requirement

(cont.)

Table 3.2. (Cont.)

Criterion	PERT/CPM	PERT/Cost	LOB	GERT	VERT-3
Data base	By-product of preparation, as is quality of data	Cost accumulation & control requires extensive data input	Requires only times & quantities at selected control points	Direct function of the number of features selected; large, if technique fully utilized	Same comment as for GERT
System operating cost	Programs easy to use; main cost is effort of preparation & updating	Canned programs expand basic PERT; largest cost by far is preparation & updating	Practically nonexistent, especially if part of MRP or other system	Function of data base; simulation requires multiple runs; input preparation appreciable	Same comment as for GERT; use of additional features increases cost
Comprehensiveness	Limited to time parameter & nonrepetitive activities	Limited to time & cost parameters for nonrepetitive activities	Limited to repetitive situations only	Accommodates most types of activities	Accommodates most activities plus numerous input modes
Flexibility	Handles deterministic situations only; no ability to accommodate decision alternatives or chance events	Deterministic only; same as PERT	Deterministic situations only; fixed production cycle required; learning curve presents problems	Accommodates stochastic & deterministic activities	Significantly more optional input & output features than GERT; has budgeting capability

Ease of update	Relatively simple; requires discipline to insure that future activities are reevaluated; actual times present little problem	Theoretically simple but major effort in practice; estimating "actual" cost data and constant changes are problems	Requires only a physical count of cumulative production	Appreciable, but value of technique in planning rather than control	Same comment as for GERT
Focus reporting	Highlighting of critical activities & problem areas is strong point; forecasts status at completion	Adds to PERT the ability to trace cost problems to the source	Highlights potential delivery schedule problem areas	Risk analysis focuses attention on what is likely to occur and its probability	Analyzes & highlights outcomes in time, costs, & performance; can be used in nonproject strategic planning

SOURCE: Data from Digman and Green (1981).

PERT-type systems possess the advantages of serving as integrated planning and control systems, but their limitations reduce their applicability to one-time, deterministic activities largely in the design and development phase; the performance parameter is not treated. LOB is exclusively a production phase technique that is simple and economical to use but has rather severe limitations. In practice, LOB-type information can be generated by Materials Requirement Planning (MRP) systems. The stochastic, decision-oriented, and simulation features of GERT and VERT provide significantly more powerful analysis than other techniques, particularly in planning (their forte is not as control systems). They allow management to assess potential outcomes much more fully; VERT-3 with its impressive array of input and output logics, fourteen statistical distributions, histogram capability, and output displays, performs such assessments particularly well. VERT-3 accommodates all of the basic parameters when performance is broadly defined to accept any meaningful unit of measure, such as return on investment or dimensionless indices of combined objectives, as well as individual performance characteristics. The technique can work within specified yearly expenditure levels and has particular application to the evaluation of strategic decision alternatives for firms in addition to the more familiar project applications.

CONCLUSIONS

The development of network techniques for management applications is firmly rooted in management science and in the scientific method. A period of significant evolution began with the deterministic, one-parameter PERT/CPM and Line of Balance systems. A quantum-jump advance occurred with the development of GERT, the first and best-known stochastic approach, which overcame the most significant limitations of the deterministic models. From this breakthrough, further evolutionary progress resulted in the 1970s and culminated in VERT, the first technique able to integrate on an equal basis the parameters of prime importance to management: time, cost, and performance, as well as associated risks. The latest version of this simulation technique, VERT-3, is described in detail in part III.

II FUNDAMENTALS OF VERT

Having explored the functions and needs of managers and the role of quantitative techniques, including a comparison of certain management-oriented networking techniques, we are now ready to examine the powerful Venture Evaluation and Review Technique (VERT) in more detail. Chapter 4 provides an overview of VERT-3, including its basic operations and features. Chapters 5 through 7 deal with, respectively, VERT-3's principles of network construction and logic, its input preparation requirements, and the various outputs the program provides the manager or analyst. Chapter 8 provides an overview of the program's computer mechanics.

Selected applications that more fully illustrate the actual use of the VERT-3 technique are included in part III.

4 DESCRIPTION OF THE VENTURE EVALUATION AND REVIEW TECHNIQUE

MANAGEMENT OVERVIEW

As we discussed in part I, one of the most pervasive and recurring situations encountered by management is the necessity of making decisions with incomplete or inadequate information about the available alternatives. These "decisions under risk" usually relate to three general categories: cost, schedule, and performance (production levels, returns on investments, and so on). To assist the manager in making decisions under risk, many techniques have been developed and used in recent years; linear programming, decision trees, and various modeling techniques are some examples. One of the most popular approaches for modeling complex problems has been simulation; the easy accessibility to large-scale, general-purpose computers has made this technique a valuable tool for the manager.

VERT-3, an acronym for Venture Evaluation and Review Technique, third generation, is a computerized, mathematical, simulation-based network technique. It was designed to assess systematically the risks involved in the undertaking of a new venture and in the resource planning, control monitoring, and overall evaluation of ongoing projects, programs and systems. The features structured in VERT-3 allow for expeditious modeling of very complex decisions that heretofore were beyond the scope of existing techniques. Modeling is accomplished with a small set of easily comprehensible operators that readily facilitates the structuring of a symbolic, pictorial network layout of the system under study. VERT-3 is an adaptive tool that allows the scope and level of abstraction to be easily managed by the analyst. Thus, modeling can be accomplished on a one-for-one basis, where one real-world event and activity is represented symbolically in the VERT-3 network as one corresponding event and activity. Modeling can also be accomplished on a compressive basis, where a multitude of real-world events and activities are compressed into the symbolic representation of a few events and activities in the VERT-3 network. However, a compressive model frequently omits important considerations and thus is afflicted with "tunnel vision."

The real power afforded by VERT-3, which is not found in other tools, lies in its ability to handle with complete, unrestricted generality the large-scale management decision problems in which the data and the possible outcomes are typically vested with imprecision and uncertainty. VERT-3 is able to overcome the weakness of decision trees without sacrificing the generality they afford. Decisions can be made and modeled within the network via time and/or performance considerations. The logic structured in VERT-3 allows for local optimization at critical decision points or milestones as well as for overall network optimization.

VERT-3 also has the facility to determine the critical path as well as its opposite, the optimal path. Since time, cost, and performance share the same status level, these critical and optimal paths can be found as a function of the time and/or cost and/or performance generated in the network. The automated data base feature of VERT-3 greatly facilitates the performance of sensitivity analyses, so that "what if" strategy questions can readily be answered.

A final characteristic that makes VERT-3 a desirable tool is the output options available. The program provides distributions depicting the frequency at which certain paths through the network were followed. It also provides distributions showing the times, costs, and performance involved in traversing the different paths.

One need not be a computer programmer or systems analyst to utilize VERT-3 effectively. An individual familiar with basic mathematics and statistics can use VERT-3 productively after several hours of study. Complete mastery of all the model's capabilities would probably require additional effort; however, such proficiency would be necessary only in the simulation of very complex or unique situations. Actually, even in the most complex situations the modeling can be simplified to a great degree by breaking down the network into subnetworks. The results of these subnetworks can then be used as inputs to a higher-level, more general network that ties all the subnetworks together. This approach usually saves time and reduces the number of errors.

THE VERT-3 MODEL

Understanding the VERT concept requires a discussion of the basic network symbols, network construction, input and output logic, statistical distributions and mathematical relationships, and outputs (including simulation results, risk analysis, and critical/optimum path analysis).

Basic Network Symbols

Two basic symbols, each encompassing minor variations, are used to structure the network model. Nodes (squares) are used to represent milestones or decision points, and arcs (lines) are used to represent activities that are basically characterized by three parameters. These parameters are: (1) the time consumed, (2) the cost incurred, and (3) the performance generated in completing the activity. Network, in the VERT context, denotes a pictorial, schematic flow device in which the nodes (decision points) channel or gate the flow into arcs (activities) that carry the flow from an input node to an output node. These nodes and arcs are laid out and coupled together in a working pictorial drawing network to model symbolically the system under analysis. Flow through the network represents the actual completion of the activities and milestones that the flow has traversed.

While a network is being graphically formed, numerical values for each activity's time, cost, and performance parameters may be assigned in terms of (1) one of the standard statistical distributions embedded in the VERT model, (2) a histogram, or (3) a mathematical relationship, provided the time and/or cost and/or performance of this activity is dependent upon other nodes and/or arcs that are to be processed (completed) prior to this arc. These values must be entered in a consistent manner throughout the network. Performance can be modeled in terms of any meaningful unit of measure, such as levels of quantities produced, return on investment, or a dimensionless index that combines the many required diverse characteristics needed to define fully the resultant output of the capital expenditure.

Arcs and nodes are similar in that both have time, cost, and performance attributes. Arcs have a primary and a cumulative set of time, cost, and performance values associated with them, while nodes have only a cumulative set. The primary set represents the time expended, the cost incurred, and the performance generated in the completion of the specific activity this arc represents. The cumulative set represents the total time expended, cost incurred, and composite performance generated to process all the arcs encountered along the path of the network flow in the completion of the processing of the arc or node in question.

VERT has two types of nodes, which start, stop, or channel the network flow. The most commonly used type is split-node logic; it has separate input and output operations. The second, more specialized, and less frequently used type of node has a single-unit logic that covers both input and output operations simultaneously.

Input Logic

Four basic input logics are available for the split-logic nodes.

1. *INITIAL input logic.* INITIAL input logic serves as a starting point for the network flow. Multiple initial nodes may be used. All initial nodes are assigned the same time, cost, and performance values by the user.

2. *AND input logic.* AND input logic requires that *all* the input arcs to the node be successfully completed before the combined input network flow is transferred to the output logic for the appropriate distribution among the output arcs. The time computed for the nodes bearing AND input logic is the maximum cumulative time of all the input arcs. Cost and performance are computed as the sum of all the cumulative cost and composite performance values of all input arcs.

3. *PARTIAL AND input logic.* PARTIAL AND input logic is similar to AND input logic, except that it requires a minimum of one input arc to be successfully completed before allowing flow to continue through this node. However, this logic will wait for all the input arcs to come in or to be eliminated from the network before processing. The same computations are used for calculating the node time, cost, and performance values for this input logic as are used for the AND logic.

4. *OR input logic.* OR input logic requires a minimum of just one input arc to be successfully completed before allowing the flow to continue through this node. This logic, unlike PARTIAL AND logic, will not wait for all the input arcs to come in or to be eliminated from the network before the flow is processed. Therefore, as soon as an input arc is successfully completed, the flow will be sent on to the output logic for processing. The time and performance assigned to the node are the cumulative time and performance values carried by the first successful input arc to be processed. Cost is computed as the sum of all the cumulative costs of all the active input arcs.

Output Logic

The task of the following split-node output logics is to distribute the network flow out to the appropriate output arc(s). Furthermore, the cumulative time, cost, and performance values computed for the active output arcs consist of the sum of the time, cost, and performance values derived for those arcs plus the time, cost, and performance carried by the arc's input node.

1. *TERMINAL output logic.* TERMINAL output logic serves as an end point of the network. It is a sink for network flow(s). Terminal nodes can be given a class designation that allows for optimization within a class as a function of the time and/or cost and/or performance values carried by the active terminal nodes. However, nodes within a class are excluded from competing for the status of optimal terminal node when a terminal node of a higher (more important) class is active.

2. *ALL output logic.* ALL output logic simultaneously initiates the processing of all the output arcs emanating from this node.

3. *MONTE CARLO output logic.* MONTE CARLO output logic initiates the processing of one, and only one, output arc per simulation iteration by the use of the Monte Carlo method. Thus, the output arcs are initiated randomly in accordance with user-developed probability weights that are assigned to these output arcs. The sum of these weights must be equal to 1.0.

4. *FILTER 1 output logic.* FILTER 1 output logic initiates one output arc or a multiple number of output arcs, depending on the joint or singular satisfaction of $T + /C + /P$ (time and/or cost and/or performance) constraints placed on this node's output arcs. These constraints consist of upper and lower $T + /C + /P$ boundaries; therefore, if this node's $T + /C + /P$ lies within the $T + /C + /P$ constraint placed on a given output arc, that arc will be processed. Otherwise, the arc will be eliminated from further consideration for this iteration. $N - 1$ of the N output arcs must have constraints placed on them, and the Nth output arc must be free of constraints. (It will be processed only when none of the constrained arcs can be processed.) Boundaries for the constrained output arcs can be overlapping, continuous, or noncontinuous (i.e., characterized by gaps). Also, the constraints need not be uniformly applied (i.e., a cost and performance constraint may be used on output arc number one with only a time constraint on output arc number two and a time and performance constraint on output arc number three).

5. *FILTER 2 output logic.* FILTER 2 output logic is the same as FILTER 1 except for two factors: First, only one constraint, rather than one to three constraints, can be placed on the constraint-bearing output arcs. This constraint consists of an upper and a lower boundary on the number of input arcs successfully processed. Second, only PAND input logic may be used with FILTER 2 output logic.

6. *FILTER 3 output logic.* FILTER 3 output logic has the same $N - 1$ constrained, as well as one unconstrained, output arc configuration as the other FILTER logics; however, the constraints for FILTER 3 are not boundaries. Rather, they consist of the name(s) of other, previously processed arcs. These constraining arcs are prefixed with a plus ($+$) or a minus ($-$) sign. If a plus sign is attached to the constraining arc's name, this arc must have been successfully processed before the output arc being constrained can be initiated for processing. Each constraint-carrying output arc may have any number of constraints from one up to the total number of arcs in the network minus 1.

Unit Logic

In addition to the split-logic nodes, the following four nodes have unit logic. They are less frequently used; they typically have more specialized applications.

1. *COMPARE logic* initializes its output arcs as a function of the cumulative time and/or cost and/or performance values carried by its input arcs.
2. *PREFERRED logic* initializes its output arcs as a function of user-expressed preferences.
3. *QUEUE logic* simulates a queuing environment where one or more servers process incoming network flows as a function of the time and/or cost and/or performance characteristics specified on this node's output arcs.
4. *SORT logic* takes the incoming network flows from the input arcs and sequentially routes them to output arcs one, two, three, and so on as a function of user-entered time and/or cost and/or performance weights.

Network Construction

At this point the network diagram representing the flow between the arcs and nodes may be constructed. Network construction, like any technique, requires familarity and judgment on the part of the user. The following oversimplified example illustrates how a basic VERT network might be constructed.

The problem consists of a motorist faced with a flat tire. The motorist, driving a borrowed car, is not sure whether the car trunk contains a jack or a spare tire, nor whether the spare tire contains sufficient air pressure. If a jack and a spare with sufficient air pressure are present, the motorist can proceed with the task of changing the tire. The task requires the following sequential actions: unload jack and spare, jack up car, remove flat and mount spare, let down car, load tire and jack into the trunk. If these actions are taken, the task will be successfully completed. However, probability exists that the car will have no jack, that the car will have no spare, and that a spare (if present) will have insufficient air pressure to be functional. Also, variability exists in regard to the performance times of each of the required activities.

For the purposes of network construction, the task can be modeled as shown in figure 4.1. The figure shows the basic arcs (activities) and nodes (events) representing the tire-changing problem. The node labeled "START" ($N1$) is made up of INITIAL input logic, which designates a starting point for the network, and ALL output logic, which indicates that each arc emanating from that node is processed. Nodes $N2$ and $N3$ have MONTE CARLO output logic, which indicates that only one of the arcs emanating from each of the two nodes will be processed, based upon the relative probabilities assigned to the outgoing arcs. For example, node $N3$ is concerned with the status of the spare tire and reflects the following probabilities:

Spare okay	75
No spare	05
Spare flat	20

As the VERT simulation of this problem is run, the MONTE CARLO output logic would randomly route the output to one of the three possible arcs in accordance with its probability of occurrence. Similar logic would hold for node $N2$.

If both the jack and spare are in working order, flow is routed to node $N4$, whose input logic, AND, requires *both* inputs for further processing to

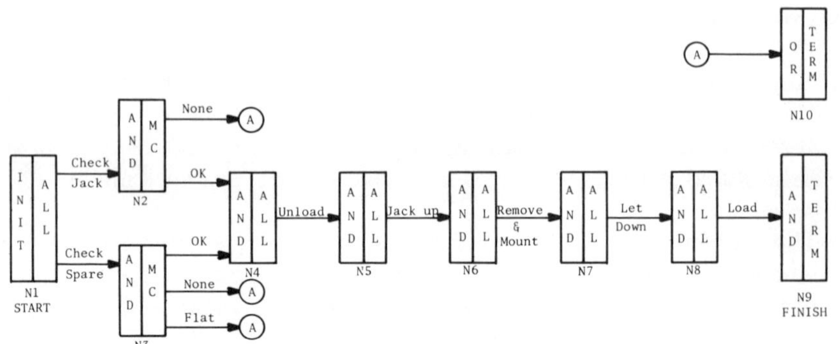

FIGURE 4.1.   Network Construction of Tire-Changing Problem

continue. Flow will then proceed to node $N9$, FINISH, which represents successful termination of the task. If the outputs from $N2$ or $N3$ are not in working order, (that is, no jack, no spare, or flat spare), the flow is routed (designated by A) to node $N10$, which represents unsuccessful completion. Since $N10$ has OR input logic, any one of the three unfavorable circumstances mentioned will trigger this event. By simulating the problem a number of times, the percentage of successful outcomes versus unsuccessful outcomes can be compared and thus can allow for a risk analysis.

By adding variable times to each of the activities represented by the arcs, a distribution of completion times can be generated. For example, assuming the motorist is on a tight time schedule, one could then determine the motorist's likelihood of being on time for the next appointment.

Statistical Distributions and Mathematical Relationships

The fourteen statistical distributions embedded in VERT-3 that may be used in part or in total to model activities are: (1) Constant, (2) Uniform, (3) Triangular, (4) Normal, (5) Lognormal, (6) Gamma, (7) Weibull, (8) Erlang (exponential), (9) Chi-square, (10) Beta, (11) Poisson, (12) Pascal (geometric), (13) Binomial, and (14) Hypergeometric. In addition, VERT-3 allows a histogram input capability for other possible distributions.

Activities may also be modeled as functions of other activities (arcs) or decision points (nodes) in the network via use of fifty mathematical relationships, as described in chapter 6. Table 4.1 illustrates a sample of five of the mathematical relationships, which may also be used to model an activity's time, cost, and/or performance.

Table 4.1. Sample Mathematical Relationships Used
in Network Modeling

Code Number	Transformation	Restrictions
1	$X \cdot Y \cdot Z = R$	
2	$(X \cdot Y)/Z = R$	$Z \neq 0.0$
5	$X + Y + Z = R$	
7	$X - Y - Z = R$	
15	$X \cdot (\mathrm{LOG}_{10}(Y \cdot Z)) = R$	$Y \cdot Z > 0.0$

Structuring a mathematical relationship within a VERT network consists essentially of three steps:

1. Long or complicated mathematical relationships need to be broken down into a series of three-variable unit transformations, as shown in table 4.1
2. Values for each of the three variables (X, Y, and Z) in each single-unit transformation must be defined. These values can be retrieved from (a) previously processed arcs or nodes, (b) constants entered directly in these transformations, or (c) one of the previously processed transformations computed in the current series of transformations used to generate a $T/C/P$ (time or cost or performance) value for the arc under consideration.
3. Results of each of the unit transformations needed to develop a value for an arc's $T/C/P$ can be summed into the overall $T/C/P$ value generated for the arc under consideration or can be omitted. When the resulting value of a unit transformation is omitted, this transformation is used as an intermediate step for calculating the value of a long or complicated mathematical relationship.

VERT Simulation

After the problem situation has been adequately modeled via VERT's arcs and nodes to achieve the desired level of realism, the problem is ready to be simulated. VERT simulation is the creation of a network flow that traverses the network from initial node(s) to terminal node(s) to create one trial solution of the problem being modeled. This simulation process is repeated as many times as the user requests in order to create a sufficiently large sample of possible outcomes. Node completion time, cost, and performance information may be obtained in the following forms: (1) relative frequency distribution, (2) cumulative frequency distribution (ogive), (3) mean observation, (4) standard error (standard deviation of the sample), (5) coefficient of variation, (6) mode, (7) median, (8) Beta 2 measure of kurtosis, and (9) Pearsonian measure of skewness. This information can be displayed for the requested internal nodes and intervals between internal nodes as well as for all terminal nodes. Additionally, all terminal nodes' time, cost, and performance data are combined to give a composite terminal node time, cost and performance printout. The overall network cost incurred and the overall network performance gained between selected time intervals (for example, yearly time intervals) as requested by the user can also be exhibited in the above form. Information of this type is very useful for constructing budgets for future periods of expenditure or for comparing investment alternatives.

Risk Analysis

VERT prints out a bar graph of the optimum terminal node index. The project risk is usually ascertained through use of this printout. A decision/risk analysis network generally has one or several terminal nodes collect successful project completions and one or several terminal nodes collect unsuccessful project completions. Realization of these various terminal nodes compared to the total number of iterations yields a probabilistic indication of project success or failure. In the event that more than one terminal node can be realized at the same time, the optimum terminal node chosen is the one with the lowest completion time, lowest cost, highest performance, or best weighted combination of these factors, per user-specified weights. Entering negative terminal node selection weights will produce an opposite effect.

Critical/Optimum Path

VERT optionally produces a critical/optimum path index for nodes and arcs. The critical path is the least desirable path through the network—that is, the path with the longest completion time, highest cost, lowest performance, or least desirable weighted combination of these factors, per user-developed weights. Entering negative critical/optimum path weights will cause the optimum path—that is, the most favorable path—to be chosen.

Often the budget director can smooth capital flows by trading some re-sources between activities on the cost-traced critical path and the cost-traced optimal path. VERT also allows optional suppression of critical/optimum paths originating from user-selected terminal nodes. This feature facilitates the finding of trouble-producing activities. Since different stochastic paths can be realized in the process of simulating the network, the critical/optimum path tends to change from iteration to iteration. The program computes the portion of time each arc and node is on the critical/optimum path and lists this information in a bar chart display. Time, cost, and performance correlations and plots are printed, upon request, for each terminal node; thus, determination of relationships among these variables is possible.

THE VERT PROCESS

The process of setting up and conducting a VERT analysis can be summarized as five basic steps that, like any networking technique, may require several iterations before the model accurately depicts the real-world situation (see Thomas 1977).

1.   The first effort is to define the decision/risk situation to be analyzed and to specify the objectives of the analysis (e.g., in the hypothetical tire-changing problem mentioned earlier, the risk situation included the likelihood of no jack, no spare, or a flat spare). The decisionmaker should approve the situation description developed during this step before proceeding to the following steps.

2.   The situation defined during step one is now depicted in a generalized flow network. This visual depiction of the risk situations helps to further refine the description developed in the first step. Errors made in verbal descriptions of the situation often become evident at this point.

3.   Concurrent with step two, the data necessary to describe the activities and decision processes should be collected. This data is then organized into some form of a probability distribution or is described by a mathematical equation.

4.   The next effort is to transform the network developed in step two into a network that will facilitate programming. This transformation involves showing the decisions and activities by the use of the logic operators of the VERT program and the conditions developed in step three. There are many possible solutions to this transformed network, and every user will probably accomplish this transformation differently. The important point

is that the transformation accurately describes the decision process of the original network (e.g., the transformed network for the tire-changing problem is shown in figure 4.1). Chapter 5 contains a detailed discussion of the program operations used. After the transformation has been effected, the information is entered in a computer program and the simulation is run. The number of iterations, or runs of the program, is determined by the desired confidence level of the results.

5. The final step is the analysis of the simulation results. The outputs from the simulation will be in the form of distributions indicating the frequency of occurrence of the different outcomes or the distribution of costs and time involved. For instance, a cost distribution would show the average cost and the range of values that might occur. Many other statistics describing the simulation results can be obtained as output from the program. Analysis of all the output information will provide the decisionmaker with an improved understanding of the risks involved in the decision situation. Another benefit can be realized once step five has been reached: It now becomes quite easy to rerun the model using variances of the original conditions. This form of sensitivity analysis will provide the decisionmaker with answers to many "what if" questions—for instance, "what if the market survey phase of a potential new product averages two weeks longer than originally expected?" In this case a simple input data change could be made and the entire simulation rerun in a very short period of time. The new outputs would provide the decisionmaker with good indications of the effects of this hypothetical contingency.

CONCLUSIONS

The initial version of VERT gave the analyst the capability to model decisions within the network in terms of time and/or cost and/or performance considerations rather than being constrained by time alone. That is, performance could be entered in the network in numerical terms rather than as gross alternatives. VERT thus gave the usual three dimensions used universally to describe a project (i.e., time, cost, and performance) equal status and treatment level. The ability to represent the "real world" within a network lies largely in the flexibility of the network's nodes; VERT introduced six new types of node logics. But perhaps the most significant innovation was the introduction of its mathematical relationships, which gave VERT the capability to establish a mathematical relationship between any given arc's time and/or cost and/or performance and any other arc and/or node's time and/or cost and/or performance. Thus, any two points within the network could be tied together by a mathematical relationship selected by the user out of the array of mathematical relationships available in VERT. Additionally, these same mathematical relationships could be used to establish relationships among the time, cost, and performance variables of a given arc.

With VERT, unlike any other technique, each of the three time, cost, and performance parameters can be treated equally rather than as dependent variables. Perhaps the major significance of this feature lies in the financial area. With conventional techniques, cost is a function of other variables; with VERT the manager/analyst can treat cost as the "driving" variable and observe the results on time and/or performance. For instance, management can set an overall project/venture budget as well as yearly budgets. Using VERT, the manager can then see the results (performance) that can be achieved given these constraints, as well as likely activity and project/venture time requirements. In summary, the VERT model is flexible, powerful, and in most cases not overly difficult to use.

5 NETWORK CONSTRUCTION AND LOGIC

As we pointed out in chapter 4, the first requirement of the VERT-3 process is to define the decision situation to be analyzed and to specify its objectives. These specifications are usually accomplished by the executive or manager responsible for the decision and are subsequently given to the analyst for detailed network preparation and analysis. While managers are keenly interested in the results of the analysis, they are usually not involved directly in the stages that intervene between definition of the situation and results of the network analysis. This chapter describes the network construction principles and the logic involved in turning the manager's input into meaningful outputs for further analysis and decision making.

In order to construct the network, the analyst must be familiar with the arc and node logic (operands) of the network analysis technique. A detailed discussion of the operands is more meaningful once the analyst comprehends the network construction syntax. Since we discussed input/output logic briefly in chapter 4, in this chapter we focus our attention first on network construction and then on a detailed operand discussion.

NETWORK CONSTRUCTION AND PARAMETERS

The VERT-3 simulation model uses a generalized networking approach to structure the risk situation to be analyzed. *Network,* in the VERT-3 context, denotes a pictorial, schematic flow device in which the nodes channel or gate the flow into arc(s) that carry the flow from an input node to an output node. These nodes and arcs are structured and coupled together in a working drawing to model symbolically the development of a system. Flow through the network represents the completion or execution of the portion of the system that the flow has traversed. These flows are usually characterized by the three most universally accepted parameters used to discuss a project: time, cost, and performance. However, these flows can also carry just one parameter, such as direct cost, indirect cost, or certain performance factors—for example, weight, speed, or return on investment. As will be-

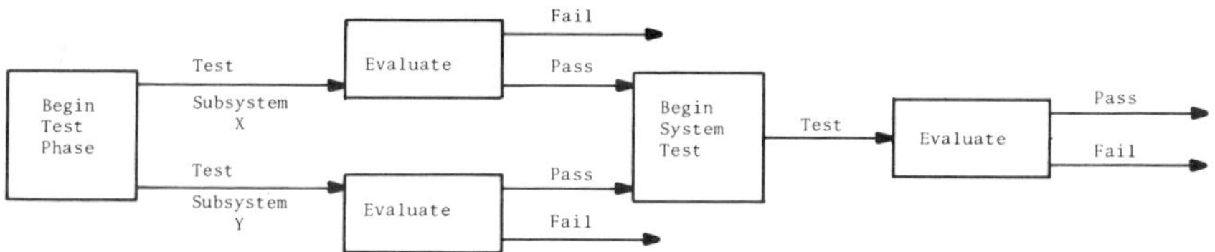

FIGURE 5.1. Test-Phase Flow Diagram

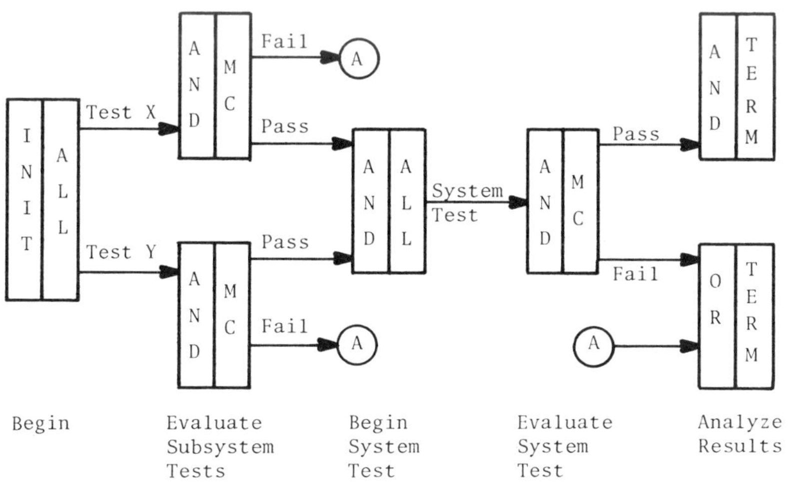

FIGURE 5.2. VERT-3 Test Network

come apparent on mastering VERT-3, the mathematical relationships capability enables one to create and isolate many separate flows, and this capability allows modeling well beyond the basic three dimensions. However, going beyond three dimensions generally proves difficult for the manager.

Figure 5.1 shows a very general flow network depicting the testing phase of a new-product development problem. The manager and analyst have determined the major activities and events and their flow. The next step is for the analyst to prepare the VERT-3 version of the network, as shown in figure 5.2. This network incorporates the probabilities that failure may occur at any one of the test arcs and channels network flow accordingly. This network indicates that both subsystem tests must be passed satisfactorily for flow to continue through system test, which also must be passed for a successful outcome to occur. Failure at any point will route flow to the failure terminal node.

General Concepts

As we stated in chapter 4, arcs and nodes are the basic symbolic operators used to express the unique aspects of the system being modeled. Arcs perform a primary function of representing project activities by using four basic parameters to describe each activity modeled. These parameters are: (1) the probability of successfully completing the activity, (2) the time consumed, (3) the cost incurred, and (4) the performance generated. Arcs also have a secondary function: carrying the network flow from its input node to its output node. When an arc is used in this latter capacity only, it is most often referred to as a *transportation arc*. For some network problems, it is desirable to enter some time and/or cost and/or performance data in the network without this data's going directly into the network flow. This special data input task is accomplished by what is known as a *free arc*. Free arcs are not wired into the network with input and output nodes like the conventional arc previously defined. Free arcs do not have input and output nodes, nor do they have a probability of being successfully completed. They are always assumed to be successfully completed. However, the rest of the input data capabilities associated with the conventional arcs are resident in the free arcs. Conventional arcs within the network flow and other free arcs external to the network flow can reference free arcs when one is structuring mathematical relationships. Free arcs are very useful for entering the many diverse characteristics used to describe performance. After entering these performance characteristics, one can collect data within the network by using mathematical relationships to pull the many diverse characteristics together in one or several meaningful indexes or performance flows.

Nodes gate or channel the network flow they receive from input arc(s) to specific output arc(s) based on the embedded node logic. Nodes generally represent decision points; however, they sometimes represent no particular decision point but rather aid in structuring the model's logic by accumulating or dispersing network flows.

The degree or extent to which a project needs to be segmented into activities and events is determined by the available data and the results desired. Some managers prefer to estimate parameters for entire modules or high-level work packages rather than to estimate parameters for the smaller elemental items in those larger units. Problem size sometimes has a bearing on the way the network is structured. If a problem is large, it is advisable to construct lower-level networks (subnetworks) of major modules. The histogram input capability structured in VERT-3 expedites stochastic substitution of results from lower-level subnetworks into a higher-level network. However, the main task in constructing a VERT-3 model is to structure as much realism in the model as possible with a minimum of abstraction. The realism should be achieved by structuring in the network all the activities (arcs) that require processing (having a flow through them) before a given activity can be processed (have a network flow through this given arc). Most important, the given activity should be a unit of work or a task that can be estimated (in terms of the time required for completion, the cost incurred, and the performance developed) with reasonable accuracy. The precision afforded by the VERT-3 approach will be entirely lost if the unit activities are gross aggregations of units of work or tasks, or if the unit activities are such abstractions of the real world that the estimation of the time, cost, and performance parameters for these unit activities becomes a guessing game.

Parametric Values

Nodes and arcs are similar in that both have time, cost, and performance attributes. Arcs have a primary and a cumulative set of time, cost, and performance values associated with them, while nodes have only a cumulative set. The primary set represents the time expended, the cost incurred, and the performance generated to complete the specific activity this arc represents. The cumulative set represents the total time expended, cost incurred, and performance generated in processing all the arcs encountered along the path of the network flow in order to complete the processing of a given arc or node.

An activity's primary time, cost, and performance can be jointly or singularly modeled as a mathematical relationship (a deterministic equation) with other arcs or nodes in the network and as a random variable. This dual capability enables one to model the residual along with the mathemetical relationship portion of a regression equation. VERT-3 has fifty transformations (see chapter 6), a property that assists in the task of structuring mathematical relationships. An arc's primary $T + /C + /P$ (time and/or cost and/or performance) can be modeled as a function of any previously processed $T + /C + /P$ of any node or arc, including the arc being processed. Thus, an arc's cost can be made a function of the time expended on this activity. Again, fourteen statistical distributions have been embedded in VERT-3 to facilitate the modeling of random variables; other distributions may be entered as histograms.

While sketching out a network, one should assign numerical values for each activity's time, cost, and performance. These values must be measured in consistent units throughout the network. For example, time cannot be entered in terms of weeks in one section of the network and in terms of years elsewhere. Likewise, cost must be measured in identical units of tens, hundreds, or thousands of dollars throughout the network. Performance can be entered in terms of any meaningful unit of measure. It could be expressed as a dimensionless index that combines the many required baseline characteristics such as horsepower, weight, mobility, reliability, availability, range, maintainability, speed, jump, and so on. One method of accomplishing this task is to use the values derived in the design requirements document as a base for normalizing the current estimates of these baseline performance characteristics. The requirement values $(R1, R2, \ldots, Rn)$ are divided into the current estimates $(E1, E2, \ldots, En)$. Further, to give more emphasis to specific performance characteristics, weights $(W1, W2, \ldots, Wn$, where $W1 + W2 + \cdots + Wn = 1)$ may be assigned to each performance characteristic. These weights are then multiplied by the normalized estimates and entered into the network: $[(W1)\ (E1)/R1, (W2)\ (E2)/R2, \ldots, (Wn)\ (En)/Rn]$. If these estimates were exactly equal to the requirements, the value generated for the overall network's performance would be unity.

Table 5.1 illustrates the type of relationships that may exist between the parameters of the test-and-evaluation problem described in figures 5.1 and 5.2. Specified are the types of distributions applicable to each test, as well as their characteristics. For test X, the time required to conduct the test is judged to be normally distributed, with a mean of 1.0 months and a standard deviation of 0.5 months. Previous data have shown that the test results (performance) parameter is triangularly distributed, with a mean of 200 units between a minimum of 150 and a maximum of 300. The cost ($10,000 per month) is a function of the time required to run the test. Thus, the test time and performance are stochastic, while cost is deterministic. Inspection of the test Y functions reveals that the manager/analyst's judgment and

Table 5.1. Arc-Parameter Function in Test Problem

Arc	Parameter	Distribution	Mean	Standard Deviation	Minimum	Maximum
Test X	Time	Normal	1.0 months	0.5 months	150 units	300 units
	Performance	Triangular	200 units			
	Cost	None	$10,000 \times$ time			
Test Y	Time	Normal	1.5 months	0.5 months	200 units	300 units
	Performance	Erlang	250 units			
	Cost	None	$10,000 \times$ time			
System test	Time	Exponential	2.0 months		0.0 months	4.0 months
	Performance	Binomial	180 units		100 units	300 units
	Cost	None	$15,000 \times$ time			

data describe the parametric relationships unique to this test; similarly, the system test relationships describe that particular test.

In general, managers/analysts can be as specific or as general as they desire in describing the arcs' parametric functions and in reflecting their interrelationships both within and among the various arcs and nodes of the network. The examples in part III describe these features more fully.

NETWORK LOGIC

This section describes in more detail the node logic summarized in chapter 4. As was pointed out, VERT-3 has two types of nodes, which start, stop, or channel the network flow. The most commonly used type of node is the one with split-node logic. It has separate input and output logic that invokes specific types of input and output operations. The other node type has a single-unit logic that covers both input and output operations simultaneously.

Split-Node Logic

There are four basic input logics and six basic output logics available for the split-logic nodes. Each is described in the following subsections.

Input Logic.   The four input logics available for the analyst are INITIAL, AND, PARTIAL AND, and OR.

1.   *INITIAL input logic.* INITIAL input logic serves as a starting point for the network flow. Multiple initial nodes may be used. All initial nodes are assigned the same time, cost, and performance values by the user.

(Before defining the three remaining input logics, it should be noted that when the input arcs have a probability of successful completion of less than 1.0, one or more of these input arcs may be failures. When these failure conditions prevail, it may be necessary, as specifically defined for each input logic below, to short circuit the node's output logic and to send the flow out on the escape arc. Rather than letting them hang up on the node, the escape arc acts as a relief path for the escape of failure flows.)

2.   *AND input logic.*   AND input logic requires all its input arcs to be successfully completed before the combined input network flow is transferred over to the output logic for the appropriate distribution among the output arcs. If at least one of the input arcs is a failure, the network flow will be sent out the escape arc. The time computed for the nodes bearing AND input logic is computed as the maximum cumulative time of all the input arcs. Cost and performance are computed as the sum of all the cumulative cost and performance values of all the input arcs. However, if the node is a failure (i.e., if the escape arc is used), the value for performance is set to equal 0.0, while the time and cost computation remain as previously defined.

   3.   *PARTIAL AND input logic.* PARTIAL AND input logic is nearly the same as the AND input logic except that it requires a minimum of one input arc to be successfully completed before allowing flow to continue through this node. However, this logic will wait for all the input arcs to come in or to be eliminated from the network before processing. If all the active input arcs are failures, the network flow will be sent out the escape arc. The same computations are used for calculating this node's time, cost, and performance values as are used for the AND logic, including the times when the node is a failure.

   4.   *OR input logic.* OR input logic is quite similar to the PARTIAL AND logic. OR also requires a minimum of just one input arc to be successfully completed before allowing the flow to continue through this node. This logic will not wait for all the input arcs to come in or to be eliminated from the network before the flow is processed. As soon as an input arc is successfully completed, the flow will be sent on to the output logic for processing. If all the active input arcs are failures, the network flow will be sent out the escape arc. The time and performance values assigned to this node are the cumulative time and performance values carried by the first successfully processed input arc. Cost is computed as the sum of all the cumulative costs of all the active input arcs. If the node is a failure (i.e., if the escape arc is used), then the value given to performance for this node is set to equal 0.0, while the time and cost computations remain as previously described. Arcs flowing directly and indirectly into a node having OR input logic may, at the user's discretion, be pruned from the network, providing that an arc's input node has a larger completion time than the node bearing the OR logic (see chapter 6).

   Arcs emanating from nodes with split-node logic will be eliminated from further consideration as network flow carriers when the input logic cannot be executed. This elimination will occur for the PAND and OR input logic when all the input arcs are not carrying a flow (i.e., when these input arcs have been logically eliminated). However, whenever any of the input arcs for the AND input logic are not carrying flows, all of this node's output arcs will be logically eliminated from further consideration as network flow carriers. It should be observed that AND input logic will impede a flow within the network whenever its input arcs are in a combined state (some carrying and some not carrying flows). All other node logics in VERT-3 are passive in the sense that they will not impede the flow; they will always pass it on.

Output Logic.   The task of the following split-node output logics is to distribute the network flow out to the appropriate output arc(s). However, if some of the input arcs have the potential of failing (i.e., a probability of successful completion of less than 1.0), then the last output arc entered in the computer will be assumed to be the escape arc, which is used as described in the input logic definitions. The output logic, except for the filter output logics described below, will ignore the escape arc. The escape arc used for the filters is assumed to be the same one required of the input logic.

   1.   *TERMINAL output logic.* TERMINAL output logic serves as an end point for the network. It is a sink for network flow(s). Terminal nodes can be given a class designation (see chapter 6), which allows for optimization within a class as a function of the time and/or cost and/or performance values carried by the active terminal nodes. However, nodes within a class are excluded from competing for the status of optimal terminal node when a terminal node of a higher (more important) class is active.

   2.   *ALL output logic.* ALL output logic simultaneously initiates the processing of all the output arcs.

   3.   *MONTE CARLO output logic.* MONTE CARLO output logic initiates the processing of one, and only one, output arc per simulation iteration by the use of the Monte Carlo method. This means that the output arcs are initiated randomly by user-developed probability weights that are placed on the output arcs. The sum of these weights must be equal to 1.0. As an added feature, multiple sets of probability weights may be entered for the purpose of conditionally and randomly initiating the output arc. These sets must be separated by $T/C/P$ (time or cost or performance) boundaries. N separate sets of probability weights are separated by $N - 1$ nondecreasing $T/C/P$ boundaries. These boundaries create $T/C/P$ regions where each of these sets apply. Region selection is based on the $T/C/P$ computed for this node. For example, if the $T/C/P$ computed for this node is less than $T/C/P$ boundary number one, then region number one is applicable and probability set number one will be used. Likewise, if this node's $T/C/P$ lies between $T/C/P$ boundaries one and two, probability set number two will be used. This process continues until, if this node's $T/C/P$ lies beyond the $(N - 1)$st $T/C/P$ boundary, the probability set residing in the Nth region will be used. If $T/C/P$ conditioning is not required, $T/C/P$ boundaries are not needed, and only probability set number one needs to be entered.

4. *FILTER 1 output logic.* FILTER 1 output logic can initiate one or a multiple number of output arcs, depending on the joint or singular satisfaction of the $T + /C + /P$ (time and/or cost and/or performance) constraints placed on this node's output arcs. These constraints consist of upper and lower $T + /C + /P$ boundaries. If this node's $T + /C + /P$ lies within the $T + /C + /P$ constraint boundaries placed on a given output arc, that arc will be processed. Otherwise the arc will be eliminated from further consideration for this iteration. $N - 1$ of the N output arcs must have constraints placed on them. The Nth output arc must be free of constraints. It will be processed only when none of the constrained arcs can be processed. FILTER 1 has an optional feature called the *subtraction feature,* which enables one temporarily to alter this node's $T + /C + /P$ prior to reviewing the output arcs' constraints. This alteration consists of temporarily subtracting, by absolute arithmetic, the $T + /C + /P$ of a designated, previously processed node from this node's $T + /C + /P$. After the constraints have been reviewed, this node's original $T + /C + /P$ values are restored. Boundaries for the constrained output arcs can be overlapping, continuous, or noncontinuous (i.e., characterized by gaps). Also, the constraints need not be uniformly applied (i.e., a cost and performance constraint may be used on output arc number one with only a time constraint on output arc number two and a time and performance constraint on output arc number three).

5. *FILTER 2 output logic.* FILTER 2 output logic is the same as FILTER 1 except for the following three factors: (1) Only one constraint, rather than one to three constraints, can be placed on the constraint-bearing output arcs; this constraint consists of an upper and a lower boundary on the number of input arcs successfully processed. (2) Only PAND input logic may be used with FILTER 2 output logic. (3) FILTER 2 does not have the subtraction feature.

6. *FILTER 3 output logic.* FILTER 3 output logic has the same $N - 1$ constrained, as well as one unconstrained, output arc configuration as the other FILTER logics. The constraints for FILTER 3 are not boundaries; rather, they consist of the name(s) of other, previously processed arcs. These constraint arcs are prefixed with a plus $(+)$ or a minus $(-)$ sign. If a plus sign is attached to the constraining arc's name, this arc must have been successfully processed before the output arc being constrained can be initiated. If a minus sign is attached to the constraining arc's name, this arc must have failed to be successfully processed or eliminated from the network before the output arc being constrained can be initiated for processing. Each output arc may have any number of constraints up to the total number of arcs in the network minus 1 (the given arc carrying the constraints).

The cumulative time, cost, and performance values assigned to an initiated output arc from a node having split-node logic consist of the sum of the primary time, cost, and performance values generated for the specific activity that arc represents plus the time, cost, and performance values computed for that arc's input node.

Unit-Logic Nodes

Four nodes have unit logic rather than the separate input and output logic of the preceding nodes. Two of the nodes, COMPARE and PREFERRED, have N input arcs, each mating with one of N output arcs to allow direct transmission of the network flow from a given input arc to a given output arc. Additionally, one uncoupled output arc must be present. This arc will be initiated when input arc–processing conditions are such that the node logic prevents initiation of any of the mated output arcs. This number of output arcs requested for processing is indicated in the VERT network diagram by a number immediately following the node caption (e.g., COMPARE + 3). This number is preceded by a plus (+) or a minus (−) sign that indicates whether the processing state is a demand (+) or a desired (−) condition. The demand condition requires that the output arc–processing requests be completely filled; otherwise the escape arc will be processed. For instance, if a demand request for the processing of three output arcs has been made, at least three input arcs must be processed successfully to prevent the escape output arc from being the only output arc processed. The desired condition will allow processing of from one to all of the output arcs requested for processing, or of any subset thereof down to one output arc, depending on the number of input arcs successfully processed. When processing is done by the desired condition, the escape output arc will be processed only when none of the input arcs has been successfully processed. The escape arc may be omitted when all the input arcs have a probability of successful completion of 1.0 and when the desired condition is used or when the demand condition is used with only one output arc requested for processing. Output arcs not selected for processing are eliminated from the network for the present iteration. In the event that there are more successfully processed input arcs than there are output arc–processing requests under either the demand or the desired condition, the following logic embedded in each node will be used to select the optimal set of output arcs:

1. *COMPARE node logic.* COMPARE logic selects the optimal output arc set for processing by weights entered for time, cost, and performance. If positive weights are entered, the optimal set consists of those output arcs whose corresponding input arcs have the best weighted combination of minimum cumulative time and cost and maximum cumulative performance. If negative weights are entered, the opposite effect will occur. Negative and positive weights cannot be used in the same application. The time value assigned to this node is the maximum cumulative time required by the most time-consuming arc in the optimum input arc set if time is used as the only decision criterion. If another criterion is used, the node time is computed as the maximum cumulative time of all the processed input arcs. The cost for this node is computed as the sum of the cumulative costs carried by all the processed input arcs. Performance is computed as the average of the cumulative performance carried by all the successfully processed input arcs.

2. *PREFERRED node logic.* PREFERRED logic gives preference to the first input-output arc combination over the second, to the second over the third, and so on. Thus, the criterion for selection is preference. The only thing that will prevent output arc number one from being initialized when operating under the desired condition is that its corresponding input arc was not successfully completed. Cost and performance calculations are the same as the calculations used for the COMPARE logic. Time is computed as the maximum cumulative time carried by the most time-consuming arc in the preferred input arc set.

(For the preceding two nodes, the cumulative cost and performance values assigned to the output arcs are computed as the sum of the primary cost and performance values derived for those arcs plus the cumulative cost and performance values generated for the linked input arcs. The cumulative time value assigned to the output arcs processed under the demand condition is calculated as the sum of the node time and the primary time generated on these arcs. When processing occurs under the desired condition, the cumulative time value assigned to a linked output arc is generally computed as the sum of the cumulative time generated for the corresponding linked input arc and the primary time generated for this output arc. Exceptions to this rule occur when one is using cost or performance weights while using the COMPARE logic and when one is using the PREFERRED logic in the situation where one attempts to process output arcs further down the preferred list than the initial candidate set. In these instances some of the output arcs may have to wait for the processing of input arcs. The escape arc's cumulative time and cost values are computed as the sum of the time and cost values derived for the input node while the value for cumulative performance is set equal to the primary performance generated on this arc.)

Two other nodes also have unit logic. They are similar to the COMPARE and PREFERRED logics in structure, but are quite different in the way they operate on the network flow.

3. *QUEUE node logic.* QUEUE node has the same physical layout as the COMPARE and PREFERRED nodes; it has N input arcs coupled with N output arcs, plus an additional, uncoupled output arc. This arc will be initiated only in the generally unlikely event that the active input arcs are unsuccessfully processed. The primary function of the QUEUE node is to transfer network flows in a queuing manner from an input arc to its mating output arc. As the network flows from the live input arcs arrive, they are queued up and sequentially processed by the server(s). The number of servers is indicated on the network diagram by the number immediately following the node caption (e.g., QUEUE 2). The program assumes that the output arcs carry the time required, the cost incurred, and the performance rendered by the server in processing the flow carried by the mating input arc. The cumulative time computed for a given output arc is calculated as the sum of the following: (a) the cumulative time carried on this arc's mating input arc, (b) the time the flow had to wait in the queue before being served, and (c) the time required by the server to process this flow (the primary time generated on this arc). The cumulative cost and performance for this same output arc are completed in the same way as the cumulative time except that there is no factor number two (i.e., there is no cost or performance generated for waiting in the queue). The time calculated for this node is computed as the maximum cumulative time observed over all the output arcs. The cost calculated for this node is computed as the sum of all the cumulative costs carried by the output arcs. This node's performance is computed in the same manner as the cost except that the total cumulative performance summed over the active output arcs is divided by the number of active output arcs and thus yields an average performance value. Since the escape arc is used in a failure situation, the primary time, cost, and performance generated on it do not relate to the server processing inbound network flows, as they do on the other output arcs. Rather, this arc should be viewed as a point from which to proceed in a new program direction. The computations used to derive the cumulative time, cost, and performance values reflect this point of view as follows: (a) Cumulative time = maximum time observed over all the active input arcs + the primary time generated on this arc. (b) Cumulative cost = the sum of the cumulative cost over all the active input arcs + the primary cost generated on this arc. (*c*) Cumulative performance = the primary performance generated on this arc.

4. *SORT node logic.* The SORT node has the same physical layout as the COMPARE, PREFERRED, and QUEUE nodes: All have N input arcs paired with N output arcs, plus an additional output arc. This arc will be initiated only in the event that all the active input arcs are unsuccessfully processed. The purpose of this node is to transfer flows from input arcs to output arcs by sorting through the use of time and/or cost and/or performance sort weights. If time is given a weight of 1.0 while cost and performance are given weights of 0.0, the flow from the input arc arriving at this node first would be sent out on output arc number one. Likewise, the flow from the input arc arriving at this node second would be sent out on output arc number two, and so on. If the cost weight was equal to 1.0 while the time and performance weights were given the value of 0.0, then the flow coming in on the input arc having the smallest cost would be sent out on output arc number one, and so on. If the performance weight was set equal to 1.0 while the time and cost weights were given the value of 0.0, then the flow coming in on the input arc having the largest performance value would be sent out on output arc number one, and so on. If a mixture of positive weights occurs (e.g., sort time weight = 0.4, cost = 0.3, and performance = 0.3), the flow of the input arc with the best weighted combination of the minimum cumulative time and cost and maximum performance will be sent out on output arc number one. The flow of the input arc having the next best weighted combination will be sent out on output arc number two, and so on. Entering negative weights will produce the opposite effect. Negative and positive weights cannot be used in the same application.

6 INPUT PREPARATION

Upon description of the problem, preparation of the network, and determination of parametric values and relationships, the analyst is ready to prepare the data for computer processing. Loading a network problem in the VERT-3 computer program requires the use of three distinct data modules that specify the various options, features, and outputs the analyst desires to employ and that enter the basic arc and node data. The three modules are the control and problem options module, the arcs module, and the nodes module. This chapter first provides an overview of the purpose and requirements of each module and then gives a detailed description of each module's input requirements.

INPUT MODULES OVERVIEW

Taken together, the three input modules describe the type of problem being processed, the various input and output options desired or selected, how the data is to be processed, all data pertaining to the arcs and nodes of the network, and interrelationships among the arcs and nodes.

74

Control and Problem Options Module

The first module, the control and problem options module, specifies the type of problem being processed and the various program options the analyst has selected for inclusion in the analysis. The module consists of the six types of cards (or 80/80 card images in the case of CRT input) described in the paragraphs that follow.

Control Options Card.   This card indicates which of the program options are to be used. If an option is indicated as being utilized on this card, the input data for that option is entered in the appropriate card(s) in the control and problem options module. If the option is not selected, the card(s) for that option is (are) not required. The following information is specified on the control options card:

- Whether a problem identification card is used to further describe the problem;
- Type of input option—that is, whether the input represents a complete new problem, temporary changes in an existing problem, or permanent changes in an existing problem;
- Types of output options desired, including listings of arc and node activity, a core utilization report, a summary of results for each iteration, the optimum terminal node index, the critical/optimum path index, and various descriptors of network performance (descriptors include minimum, mean, and maximum values, the relative frequency distribution, cumulative distribution, standard error, coefficient of variation, mode, Beta 2 measure of kurtosis, Pearsonian measure of skewness, and median; descriptors apply to node and arc slack times, cost-performance time intervals, internal nodes, intervals between internal nodes, terminal nodes, and the composite terminal node);
- How to value partially complete, uninitiated, and common activities for nodes with OR and COMPARE logic and for incomplete and common network paths;
- Whether to print out the arcs and nodes active during a given iteration;
- Desired correlation and plot combinations for terminal nodes;
- Whether to print cost incurred and/or performance achieved by time interval (e.g., yearly);
- Specified minimum and maximum values for printing of results;

- Type of output desired for composite terminal node;
- Initial values for random numbers;
- Number of iterations desired;
- Inflation rate;
- Discount rate;
- Conversions for other than yearly time intervals;
- Relative time/cost/performance weights for calculation of optimum terminal node and critical/optimum path;
- Initial time, cost, and performance values.

Problem Identification Options Card. This card is used if the analyst desires to include descriptive information regarding the problem.

Full Print Trip Option Card. This card identifies which arcs and nodes are to be printed out if active during the problem iteration.

Correlation Computation and Plot Option Card. This card is used to request correlations and plot combinations between any of the following terminal node values: time, path cost, overall cost, and performance.

Cost-Performance Time Intervals Option Card. This card specifies the printing of cost and/or performance histograms by desired time period. For example, expected cost by year can be printed out for the project or venture.

Composite Terminal Node Histogram Input Options Card. This card specifies the minimum and maximum values for terminal node histograms for time, path cost, overall cost, and performance values.

Arcs Module

The arcs module is used to enter data, provide processing instructions, and specify outputs desired for the arcs of the network. Up to eighteen types of cards may be required, depending upon the program features selected. The types of cards employed are described below.

Master Arc Card. This card is used to enter the arc's code name, description, input node (predecessor), output node (successor), probability of completion, and whether or not a histogram of the arc's slack is desired.

Statistical Distribution Satellite Arc Cards. Three of these cards are used to indicate the type of distribution and input parameters for the arc's time, cost, and performance values.

Histogram Satellite Arc Cards. Four of these cards are used to input histogram data to generate the arc's time, cost, performance, and slack values.

Math-Related Satellite Arc Cards. Three of these cards are used to carry the mathematical relationship(s) used to determine the arc's time, cost, and performance values.

Up to seven additional types of cards can be used to carry information for generating the appropriate output values for the arc's input node (i.e., its predecessor). Three types of cards—FILTER 1, 2, and 3 satellite arc cards—are used to specify upper and lower boundaries of time, cost, and/or performance constraints, upper and lower boundaries on input arc constraints, and the names of input arc constraints corresponding to the output logic (FILTER) of the arc's input node. A fourth type of card—the Monte Carlo satellite arc card—is used to input the arc's probability of MONTE CARLO initiation (if that logic describes the arc's predecessor node). Three additional Monte Carlo cards are used if the arc's probability of initiation is a function of the time, cost, and/or performance of its input node. Therefore, this latter feature enables the arc's initiation to be a conditional probability of values existing at the arc's input node.

Nodes Module

Like the arcs module, the nodes module is used to enter information describing the nodes of the network, including types of logic for each node, desired outputs, and parametric weights for calculation of optimal index values. The four types of cards that comprise the nodes module are described below.

Master Node Card. This card is used to enter the node's code name, node description, type of input logic, type of output logic, whether to print time, cost, and/or performance histograms for the node, and weights assigned to time, cost, and performance parameters in order to select the optimum input arc if COMPARE logic is specified.

Histogram Satellite Node Card.   This card is used to enter minimum and maximum values if histograms are desired for time, path cost, overall cost, and/or performance for the node.

Subtract Satellite Node Card.   This card is used only if the subtract feature of FILTER 1 output logic is specified for the node.

Slack Histogram Satellite Node Card.   This card is used to specify minimum and maximum slack time values if a slack histogram is desired for the node.

DATA INPUT PREPARATION

The following section describes in detail the function and input requirements for each of the modules. Included is a card-by-card description of the data preparation requirements and options.

I.   Control And Problem Options Module

A.   Control Options Card

Column 1, FORMAT I1. Problem Identification Card Option.
Entering a "1" in this column requires a problem identification card to be inserted after this control card.
When "0" is entered or this field is left blank, the
problem identification card must be omitted.

Column 2, FORMAT I1. Type of Input Option. This program
has three input options available for data management.
Option one requires placing a blank or zero in this field.
Under this option the program assumes that a complete,
stand-alone problem is being read that will be placed on
the master file as the new master problem. This new
master problem will replace an old master problem if one
was previously held on this file (IWF1 is the name for
the master file in the FORTRAN program). Option two
requires a "1" to be entered in this field. Under this
option the program assumes that a few temporary changes
to the master problem are desired. These changes are
temporarily merged into the master problem and then
simulated in that state. After simulation these changes
are abandoned and the problem on the master file remains
as it was prior to the simulation. Option three requires
a "2" to be entered in this field. Under this option,
the program assumes that a few permanent changes to the
master problem are desired. Prior to the simulation
these changes are permanently merged into the problem
stored on the master file.
     When utilizing options two or three, making a change
in either an arc's or a node's input data requires
resubmitting all the input cards needed to define that
arc or node. Any arc or node not already on the master
file may be submitted as a change. Arcs or nodes currently on the master file may be deleted by submitting a
card with the arc or node name in columns 1-8 and
"- - - -" (four minus signs) in columns 9-12.
     When options two or three are used, only the change
cards will be listed. To obtain a complete, full listing
of the reconstituted problem, enter a "3" in place of the
"1" for using option two and enter a "4" in place of the

"2" for employing option three.

   Options two and three require as a minimum of input the control card, an ENDARC card, and an ENDNODE card.

<u>Column 3, FORMAT I1. Type of Output Option</u>. The following optional lists are available from VERT-3 in addition to a special listing of the control and identification cards and an 80/80 listing of all the remaining input cards. This special listing is automatically produced every time a problem is processed.

   1. A listing of the two major storage arrays, ASTORE and NSTORE (see chapter 8);
   2. A listing after each iteration of all the flow-carrying arcs and nodes;
   3. A core storage utilization report that shows how well each of the internal storage arrays have been used;
   4. A one-line summary listing of the results obtained after each iteration;
   5. A listing of the optimum terminal node index and arcs' and nodes' critical/optimum path index;

(The following output options apply to (1) node and arc slack times, (2) cost-performance time intervals, (3) internal nodes, (4) intervals between internal nodes, (5) terminal nodes and (6) the composite terminal node.)

   6. A one-line listing of the minimum, mean, and maximum values of the preceding;
   7. A one-page listing carrying (a) the relative frequency distribution, (b) the cumulative frequency distribution (ogive), (c) the mean observation, (d) the standard error (sample standard deviation), (e) the coefficient of variation, (f) the mode, (g) the Beta 2 measure of kurtosis, (h) the pearsonian measure of skewness, and (i) the median (for terminal and composite terminal nodes only);
   8. Same as number 7 except that the relative frequency distribution is omitted;
   9. Inclusion of the median in the preceding list for (a) internal nodes, (b) intervals between internal nodes, (c) node and arc slack times and (d) cost-performance time intervals. This inclusion

requires a significant increase in computer
processing time and is the reason for this setup;

These optional lists are grouped together in what are be-
lieved to be useful output sets as follows:

Option	Field Entry	Preceding Lists Used
A	0 or blank	5 and 6
B	1	5 and 8
C	2	4, 5, and 7
D	3	1, 2, and 3
E	4	5, 8, and 9
F	5	4, 5, 7, and 9

Since option D produces a large amount of output, it is
limited to 100 iterations. This option is designed for
debugging. The remaining options are designed to provide
a diversified informational capability for analyzing the
various types of problems solved by VERT-3.

Column 4, FORMAT I1. Cost and Performance Computational
Variations Available. VERT-3 provides additional options
to the way arc and node cost and performance values are
computed (as was defined in chapter 5). This multi-
purpose field carries the options available for assigning
cost and performance values to arcs and nodes subject to
one or more of the following situations: (1) arcs flowing into a node having OR input logic, (2) arcs flowing
into a node having COMPARE logic when the time selection
weight is equal to 100 percent, (3) arcs not completed
prior to the time the winning terminal node was executed
when the weight placed on time for selecting the winning
terminal node is 100 percent, and (4) nodes intercepting
network paths that share arcs in common.
   VERT-3 is structured to value fully or partially the
cost and performance generated on those arcs that are
partially completed. VERT-3 is also structured to prune
or include the cost and performance values of those activities that have not yet started processing. Lastly,
VERT-3 will multiply or singularly sum the cost and performance of common arcs. The various combinations of
options available are as follows:

1. Full value the partially completed activities;
2. Partial value the partially completed activities;

3. Prune the uninitiated activities;
4. Full value the uninitiated activities;
5. Single-count common activities (elongates computer processing time);
6. Multiple count common activities.

Option	Field Entry	Preceding Computations Used
A	0 or blank	1, 4, and 6
B	1	1, 3, and 5
C	2	2, 3, and 5
D	3	1, 4, and 5.

Column 5, FORMAT I1. Full Print Trip Option. Entering a "1" in this column requires a card to be entered, following the problem identification card, that carries the name of arcs and/or nodes. When any of these arcs of nodes are active, the program will list all the arcs or nodes that were active for the given iteration.

Column 6, FORMAT I1. Correlation Computation and Plot Option. Entering a "1" in this column requires a card to be entered, following the full print trip option card, that carries the correlation and plot combinations wanted for terminal nodes.

Column 7, FORMAT I1. Cost-Performance Time Interval Option. Entering a "1," "2," or "3" in this column requires entering cards, following the correlation computation and plot option card that carry the time intervals and possible upper and lower boundaries for the histograms used to plot the cost incurred and/or performance gained during these time intervals. Entering a "1" in this column indicates that cost only is desired, while entering a "2" indicates that performance only is desired. If both cost and performance are desired, a "3" should be entered in this column.

Column 8, FORMAT I1. Composite Terminal Node Minimums and Maximums Options. Entering a "1" in this column requires a card to be entered following the time interval costing option cards. This card should carry the minimums and the maximums used to print the time, path cost, overall cost, and performance for the composite terminal node.

Columns 9-10, FORMAT I2. Enter the applicable numeric
that indicates the output option desired for the composite
terminal node (see the table under the description of the
input for columns 13-14 under "Master Node Card").

Columns 11-19, FORMAT I1. Enter the initial value assigned to the seed of the uniform (0.0 to 1.0) random number
generator. The ending value of the seed is printed out at
the end of each problem. If this field is left blank or
has a "0" entered in it, the seed will be loaded with the
value of 435459. Further, when running a series of problems via a single computer run, the program will carry the
seed forward to subsequent problems, providing this field
is left blank in those subsequent problems. There is
provision in VERT-3 for embedding two generators rather
than just one uniform random number generator. If the
seed is prefixed with a minus (-) sign, the sign will be
stripped off the seed, and generator number two will be
used for the given problem. If the seed is prefixed with
a plus (+) sign or no sign, the seed will be used as is,
and generator number one will be employed for the given
problem.

Columns 20-24, FORMAT I5. Enter the number of iterations
desired for this problem.

Columns 25-28, FORMAT F4.2. Enter the yearly interest
rate used for inflating cost and/or performance values
for specific arcs as called out by the user. This number
should be entered in percentage form. For example, 7.5
percent should be entered in columns 25-28 as "7.5". If
none of the cost and/or performance values of the arcs in
the network being processed require inflating, leave this
field blank.

Columns 29-32, FORMAT F4.2. Enter the yearly interest
rate used to discount cost and/or performance values for
specific arcs as called out by the user. This number
should be entered in percentage form similar to the preceding field. If none of the cost and/or performance
values of the arcs in the network being processed require
discounting, leave this field blank.

NOTE: The inflation and discounting calculations are made
immediately after generating the time, cost, and per-

formance values for a given arc. These values are then
stored in place of the original values and used in all
future mathematical relationships. However, when the
time, cost, and performance values for a given arc are
interrelated, then the original unadjusted cost and/or
performance values are used in the mathematical relationships to calculate values for the dependent variables.

Columns 33-35, FORMAT F3.2. Enter the time factor that
converts the program time to a yearly basis. This
program computes interest calculations on a yearly basis.
This field carries the number of time units existing in
the network time domain in one year. For example, if the
network time is in months, a '12' should be entered in
columns 33-35. This field should be left blank if the
preceding two fields are blank.

NOTE: Values assigned to the following three fields
must all lie within either the closed interval of -1.0
and 0.0 or the closed interval of 0.0 and +1.0. These
fields must not jointly carry positive and negative
values (i.e., field one cannot have a positive entry
while fields two and/or three have negative entries).
Entering positive values in these fields will give rise
to choosing as the optimal terminal node the one with the
least time and cost and the most performance combination.
Entering negative values in these fields will cause the
terminal node with the largest time and cost and the
least performance to be chosen as the optimum terminal
node. For further information regarding winning terminal
node selection, see the description of the terminal output logic (columns 10-12 of the Master Node Card in
section III-A).

Columns 36-38, FORMAT F3.2. Enter the weight assigned
to time when determining the optimum terminal node.

Columns 39-41, FORMAT F3.2. Enter the weight assigned
to cost when determining the optimum terminal node.

Columns 42-44, FORMAT F3.2. Enter the weight assigned
to performance when determining the optimum terminal
node.

NOTE: Values assigned to the following three fields must
all lie within either the closed interval of -1.0 and 0.0
or the closed interval of 0.0 and +1.0. These fields

must not jointly carry positive and negative value (i.e.,
field one cannot have a positive entry while field two
and/or three have negative entries). Entering positive
values in these fields will give rise to choosing the
critical path as the path with the largest time and cost
and the smallest performance. Entering negative values
in these fields will cause the optimum path to be chosen
as the path with the smallest time and cost and the
largest performance.

Columns 45-47, FORMAT F3.2. Enter the weight assigned
to time when determining the critical/optimum path.

Columns 48-50, FORMAT F3.2. Enter the weight assigned
to cost when determining the critical/optimum path.

Columns 51-53, FORMAT F3.2. Enter the weight assigned
to performance when determining the critical/optimum
path.

Columns 54-62, FORMAT F9.0. Enter the time assigned
to all the initial nodes (network startup time).

Columns 63-71, FORMAT F9.0. Enter the cost assigned
to all the initial nodes (project money spent prior
to the start of the network).

Columns 72-80, FORMAT F9.0. Enter the performance
assigned to all the initial nodes (performance
generated prior to network startup).

B. Problem Identification Options Card

Columns 1-80, FORMAT 20A4. Enter a card carrying any
alpha-numeric information deemed helpful in identifying
this problem.

NOTE: The preceding card may be used only when a "1" has
been entered in column 1 of the control card.

C. Full Print Trip Option Card

Columns 1-8, FORMAT 2A4. Enter the name of the first
node or arc that, when active, will yield a full printout
of all the arcs and nodes that were active during this

iteration.

Continue entering arc and node names in fields of
eight columns until all the arcs and/or nodes desired to
cause this full print option to occur have been listed
or a maximum of ten, which will use up the whole card,
has been reached.

NOTE: The preceding card may be used only when a "1" has
been entered in Column 5 of the control card.

D.   Correlation Computation and Plot Option Card

The following codes must be used to request plotting and
computing the correlation coefficient between the following terminal node variables.

Code Number	Variable
1	Time
2	Path Cost
3	Overall Cost
4	Performance

Columns 1-2, FORMAT 2I1. Enter the code numbers for any
two of the above variables for which a correlation
coefficient is desired. A plot will be made of these
variables in order to observe any possible mathematical
relationship between these two variables.

Continue in fields of two columns, requesting plot and
correlation combinations until all the combinations desired have been requested or until a maximum of twelve such
combinations has been requested.

NOTE: The preceding card may be used only when a "1"
has been entered in column 6 of the control card.

E.   Cost-Performance Time Intervals Option Card(s)

The program considers only positive cost and/or performance observation within the designated time interval.
(Negative observations or observations having a value of
zero are ignored.)

Columns 1-10, FORMAT F10.0. Enter the lower boundary of
the time interval.

The last card in this series of cards must have ENDCTPR punched in columns 1-7 of this field, and the rest of the card must be left blank.

<u>Columns 11-20</u>, FORMAT F10.0. Enter the upper boundary of the time interval.

<u>Columns 21-30</u>, FORMAT F10.0. Enter the lower value used to structure the cost histogram.

<u>Columns 31-40</u>, FORMAT F10.0. Enter the upper value used to structure the cost histogram. If this field and the preceding field are left blank or have zeros entered in them, the program will use the minimum and the maximum cost values observed during the simulation to construct this histogram.

<u>Columns 41-50</u>. Same as columns 21-30 except substitute the word "performance" for the word "cost."

<u>Columns 51-60</u>. Same as columns 31-40 except substitute the word "performance" for the word "cost."

Input in the following two fields will activate calculations that will aid management in the budgeting process. VERT-3 assists in ascertaining the funds made available for critical budgeting periods of a development and for the entire life of the development have a very high chance of being adequate. For these calculations, VERT-3 assumes that the first $N - 1$ cost-performance time interval request covers the entire planning horizon in $N - 1$ ascending unique nonoverlapping units of time. The last request of the series being entered, the Nth request, covers the entire planning horizon also, but in just one unit of time. An example of this situation would be entering eleven cost-performance time interval requests, the first covering year one, the second covering year two, the third covering year three, and so on. However, the eleventh request would cover the entire ten years.

To assist in this confidence-level budgeting process, VERT-3 requires the assignment of a desired confidence level to each cost-performance time interval request. The assignment of a unit step for each of the first $N - 1$ cost-performance time interval requests is also required. The unit step will be used to adjust incre-

mentally either upward or downward all of the confidence
levels of the first N - 1 periods to attain the assigned
confidence level of the overall period (the Nth period).
Entry of each of the unit steps enables one to hold the
critical periods at a relatively fixed confidence level,
while the confidence level of the remaining periods can
vary more. The values assigned to the unit steps should
be relatively small because VERT-3 solves this problem
by an iterative method. VERT-3 will incrementally
adjust each period's confidence level by its assigned
unit step, compute the resultant sum of each period's
cost, and compare this sum to the cost associated with
the overall period's confidence level. Once the cost
associated with the overall period's confidence level has
been crossed over, VERT-3 will print the adjusted
confidence levels of the previous iteration. Hence,
the exactness of the solution depends upon the unit step
size and the number of periods (cost-performance time intervals) requested.

Columns 61-70, FORMAT F10.0. Enter the confidence level
described in the preceding two paragraphs. If this field
and the following field are left blank or have zeros
entered in them, VERT-3 will not attempt any budget
confidence calculations.

Columns 71-80, FORMAT F10.0. Enter the unit step size
described above.

NOTE: The preceding card(s) may be used only when a "1,"
"2," or "3" has been entered in column 7 of the control
card.

F.   Composite Terminal Node Minimums and Maximums Option
     Card

Columns 1-10, FORMAT F10.0. Enter the lower boundary
value desired for the time histogram.

Columns 11-20, FORMAT F10.0. Enter the upper boundary
value desired for the time histogram. If this field and
the preceding field are left blank or have zeros entered
in them, the program will use the minimum and maximum time
value observed for this histogram during the simulation to
construct this histogram.

Columns 21-30.   Same as columns 1-10 except substitute
the words "path cost" for the word "time."

Columns 31-40.   Same as columns 11-20 except substitute
the words "path cost" for the word "time."

Columns 41-50.   Same as columns 1-10 except substitute
the words "overall cost" for the word "time."

Columns 51-60.   Same as columns 11-20 except substitute
the words "overall cost" for the word "time."

Columns 61-70.   Same as columns 1-10 except substitute
the word "performance" for the word "time."

Columns 71-80.   Same as columns 11-20 except substitute
the word "performance" for the word "time."

NOTE: The preceding card may be used only when a "1"
has been entered in column 8 of the control card.

II.   Arcs Module

A.   Master Arc Card

Columns 1-8, FORMAT 2A4.   Enter the name of the arc being
modeled.   The last card of this module must have "ENDARC"
punched in columns 1-6, and the rest of the card must
be left blank.

Columns 9-16, FORMAT 2A4.   Enter the arc's input node
name.   If this arc is a "free arc," enter "NOFLOW" in
columns 9-14.

Columns 17-24, FORMAT 2A4.   Enter the arc's output node
name.   If this arc is a "free arc," enter "DATAGEN" in
columns 17-23.

Columns 25-28, FORMAT F4.2.   Enter the probability of successfully completing this arc (activity).   Acceptable entries are 1.0 and all the values between 0.0 and 1.0.
If this arc is a "free arc," the program ignores any
entry and puts a 1.0 in this field.

<u>Column 29</u>, FORMAT A1. Enter the letter "S" in this
column to request a histogram of the slack time present
on this arc; otherwise, leave this field blank. This
histogram will be structured only when a time critical
path (1.0 weight on the critical path time) is requested.
Further, a satellite arc card (see subsection R) that
will input a scale for this histogram may be entered.
If this satellite card is omitted, the program will use
the observed minimum and maximum slack time to structure
its own scale. If this arc is a "free arc," the program
ignores any entry and puts a blank in this field.

<u>Columns 30-80</u>, FORMAT A3, 12A4. Enter the description of
the activity this arc represents.

<u>NOTE</u>: The following nine satellite arc cards (B-J) are
the basic vehicles used to input time, cost, and performance data for each arc in the network. These cards
carry the data needed to define an arc's time, cost, and
performance values.

B.   <u>Time Statistical Distribution Satellite Arc Card</u>

This card carries the input parameters needed to define
in part or in total the time value generated for this arc
via the use of one of the following statistical distributions.

Type of Distribution	Field No. 1	Field No. 2	Field No. 3	Field No. 4	Field No. 5	Field No. 6
Constant	1	Cons				
Uniform	2	Min	Max			
Triangular	3	Min	Max	Most likely		
Normal	4	Min	Max	Mean	Std.Dev.	
Lognormal	5	Min	Max	Mean	Std.Dev.	
Gamma	6	Min	Max	Mean	Std.Dev.	
Weibull[1]	7	Min	Max	Scale Par.	Shape Par.	
Erlang	8	Min	Max	Mean	No Exp. Dev.	
(Exponential)	8	Min	Max	Mean	1	
Chi-Square	9	Min	Max	No Deg. Free.		
Beta[2]	10	Min	Max	A	B	

Poisson[3]	11	Min	Max	L		
Pascal[4]	12	Min	Max	P	K	
(Geometric)[4]	12	Min	Max	P	1	
Binomial[5]	13	Min	Max	P	N	
Hypergeometric[6]	12	Min	Max	P	N	M

[1] The minimum observation is the location parameter.

[2]
$$F(X) = \frac{G(A+B)X^{A-1}(1-X)^{B-1}}{G(A)G(B)}$$
A greater than zero
B greater than zero
G = gamma function

[3]
$$F(X) = E^{-L}\frac{(1)}{XF}X \quad X=0,1,2\ldots$$
L greater than zero
E = natural log base
F = factorial

[4]
$$F(X) = (K+X-1)_{(P)}P^{K}Q^{X} \quad X=0,1,2\ldots$$
$$Q=1-P$$

[5]
$$F(X) = F(X) = (N)_{(X)}P^{X}Q^{N-X} \quad X=0,1,2\ldots N$$
$$Q=1-P$$

[6]
$$F(X) = \frac{(NP)_{(X)}(NQ)_{(M-X)}}{(N)_{(M)}} \quad \begin{array}{l} X=0,1,2\ldots N \\ M-X=0,1,2\ldots NQ \\ Q=1-P \end{array}$$

Columns 1-8, FORMAT 2A4. Enter the name of the arc for which this satellite card is carrying information.

Columns 9-13, FORMAT A4,A1. Enter the satellite type identifier--"DTIME."

Columns 14-15, FORMAT I2. Enter the card sequence number. Only one card is needed to carry all the possible distribution data needed to define time in terms of one of the above statistical distributions. Therefore, enter a "1" in column 15.

Columns 16-25, FORMAT F10.0. Enter the data for field one as defined above.

Columns 26-35, FORMAT F10.0. Enter the data for field
two as defined above.

Columns 36-45, FORMAT F10.0. Enter the data for field
three as defined above.

Columns 46-55, FORMAT F10.0. Enter the data for field
four as defined above.

Columns 56-65, FORMAT F10.0. Enter the data for field
five as defined above.

Columns 66-75, FORMAT F10.0. Enter the data for field
six as defined above.

C. Cost Statistical Distribution Satellite Arc Card

This card carries the input data needed to define in part
or in total the cost value generated for this arc via the
use of one of the previously defined statistical distributions. This card is the same as card B, except enter
"DCOST" in columns 9-13 and enter the appropriate cost
data. Further, if it is desired to inflate and/or
discount the cost generated for this arc, in place of
entering "DCOST" in columns 9-13, enter "DCOST" to inflate
the cost, enter "DCOSD" to discount the cost, or enter
"DCOSB" to both inflate and discount the cost by the
appropriate interest rates entered in the control card.

D. Performance Statistical Distribution Satellite Arc
    Card

This card carries the input data needed to define in part
or in total the performance value generated for this arc
via the use of one of the previously defined statistical
distributions. This card is the same as card B, except
enter "DPERF" in columns 9-13 and enter the appropriate
performance data. Further, if it is desired to inflate
and/or discount the performance generated for this arc,
in place of entering "DPERF" in columns 9-13, enter
"DPERI" to inflate the performance, enter "DPERD" to
discount the performance, or enter "DPERB" to both inflate
and discount the performance by the appropriate interest
rates entered in the control card.

E. <u>Time Histogram Satellite Arc Card(s)</u>

This card carries histogram data used to generate in part
or in total the time value for this arc.

<u>Columns 1-8</u>, FORMAT 2A4. Enter the name of the arc for
which this satellite card is carrying information.

<u>Columns 9-13</u>, FORMAT A4, A1. Enter the satellite type
identifier--"HTIME."

<u>Columns 14-15</u>, FORMAT I2. Enter the card sequence number.
The cards required to accomplish this task must be sequentially numbered. The maximum number of cards allowed
equals the current maximum number of arc allowed (value
of the check variable MARC) divided by 6 or a total number
of 99, whichever is smaller.

<u>Columns 16-25</u>, FORMAT F10.0. Enter the left hand time
boundary of cell number one.

<u>Columns 26-35</u>, FORMAT F10.0. Enter the probability den-
sity of cell number one.

<u>Columns 36-45</u>, FORMAT F10.0. Enter the time boundary sep-
arating probability cells one and two.

<u>Columns 46-55</u>, FORMAT F10.0. Enter the probability den-
sity of cell number two.

<u>Columns 56-65</u>, FORMAT F10.0. Enter the time boundary sep-
arating probability cells two and three.

<u>Columns 66-75</u>, FORMAT F10.0. Enter the probability den-
sity of cell number three.
  Repeat the above sequence of steps for additional cards.
The maximum number of cards allowed has been previously
defined in columns 14-15.

F. <u>Cost Histogram Satellite Arc Card(s)</u>

This card carries histogram data used to generate in part
or in total the value for the cost carried by this arc.
This card is the same as card E except enter "HCOST" in
columns 9-13 and enter the appropriate cost data. Further,

if it is desired to inflate and/or discount the cost
generated for this arc, in place of entering "HCOST" in
columns 9-13, enter "HCOSI" to inflate the cost, enter
"HCOSD" to discount the cost, or enter "HCOSB" to both
inflate and discount the cost by the appropriate interest
rates entered in the control card.

G.   Performance Histogram Satellite Arc Card(s)

This card carries histogram data used to generate in part
or in total the value for the performance carried by this
arc. This card is the same as card E, except enter
"HPERF" in columns 9-13 and enter the appropriate
performance data. Further, if it is desired to inflate
and/or discount the performance generated for this arc,
in place of entering "HPERF" in columns 9-13, enter
"HPERI" to inflate the performance, enter "HPERD" to
discount the performance, or enter "HPERB" to both inflate
and discount the performance by the appropriate interest
rates entered in the control card.

H.   Time Math-Related Satellite Arc Card(s)

This card carries the mathematical relationship(s) used
to create in part or in total the time value for the arc
under consideration. Entering mathematical relation-
ship(s) requires using one or more of the following unit
transformations:

Code No.	Transformation	Restrictions	Notes
1 or 51	X*Y*Z	= R	(Multiply)
2 or 52	(X*Y)/Z	= R Z NE 0.0	(NE=not equal to)
3 or 53	X/(Y*Z)	= R Y*Z NE 0.0	
4 or 54	1/(X*Y*Z*)	= R X*Y*Z* NE 0.0	
5 or 55	X+Y+Z	= R	
6 or 56	X+Y-Z	= R	
7 or 57	X-Y-Z	= R	
8 or 58	-X-Y-Z	= R	
9 or 59	X*(Y+Z)	= R	
10 or 60	X*(Y-Z)	= R	
11 or 61	X/(Y+Z)	= R	
12 or 62	X/(Y-Z)	= R	
13 or 63	$X*(Y)^Z$	= R Y GT 0.0	(GT=Greater than)

```
14 or 64    X*(LOG (LOG
                  E
            (Y*))        E
                              = R  Y*Z GT 0.0  (E=natural log)
15 or 65    X*(LOG (Y*Z))    = R  Y*Z GT 0.0
                   10
16 or 66    X*(SIN(Y*Z))     = R
17 or 67    X*(COS(Y*Z))     = R
18 or 68    X*(ARCTAN
            (Y*Z))           = R
19 or 69    X       GE
             19 or 69

            Y        ---
             19 or 69

            X            = R            (GE=greater than
             20 or 70      20 or 70      or equal to)
```

    (Transformation 19 or 69 must be followed by 20 or 70)

```
20 or 70    X       LT
             19 or 69

            Y        ---
             19 or 69

            Y            = R            (LT=less than)
             20 or 70      20 or 70
21 or 71    X GE Y ------ Z = R
            X LT Y ------ X = R
22 or 72    X GE Y ------ X = R
            X LT Y ------ Z = R
23 or 73    (X*Y)+Z          = R
24 or 74    (X*Y)-Z          = R
25 or 75    (X/Y)+Z          = R  Y NE 0.0
26 or 76    (X/Y)-Z          = R  Y NE 0.0
27 or 77    (X+Y)*Z          = R
28 or 78    (X+YY)/Z         = R  Z NE 0.0
29 or 79    (X-Y)*Z          = R
30 or 80    (X-Y)/Z          = R  Z NE 0.0
31 or 81    X+(Y*Z)          = R
32 or 82    X-(Y*Z)          = R
33 or 83    X+(Y/Z)          = R  Z NE 0.0
34 or 84    X-(Y/Z)          = R  Z NE 0.0
35 or 85    -X-Y+Z           = R
36 or 86    -X+Y+Z           = R
37 or 87    X/Y/Z            = R  Y and Z NE 0.0
38 or 88#   X,Y,Z FUN(TAB1)= R  X,Y,Z within TAB1's bounds
39 or 89#   X,Y,Z FUN(TAB2)= R  X,Y,Z within TAB2's bounds
40 or 90#   X,Y,Z FIM(TAB3)= R  X,Y,Z within TAB3's bounds
41 or 91#   X,Y,Z FUN(TAB4)= R  X,Y,Z within TAB4's bounds
42 or 92#   X,Y,Z FUN(TAB5)= R  X,Y,Z within TAB5's bounds
43 or 93#   X,Y,Z FUN(TAB6)= R  X,Y,Z within TAB6's bounds
44 or 94#   X,Y,Z FUN(TAB7)= R  X,Y,Z within TAB7's bounds
```

```
45 or  95@ X,Y,Z FUN(SUB1)= R
46 or  96@ X,Y,Z FUN(SUB2)= R
47 or  97@ X,Y,Z FUN(SUB3)= R
48 or  98@ X,Y,Z FUN(SUB4)= R
49 or  99@ X,Y,Z FUN(SUB5)= R
50 or 100@ X,Y,Z FUN(SUB6)= R
```

$Transformation numbers 1-50 and 51-100 use floating point computations to initially derive a value for R. However, transformations 51-100 truncate R to an integer, while transformations 1-50 retain R in its floating point form.

#Transformation 38-44 and 88-94 utilize table lookups. X, Y, and Z in each of these transformations are rounded up by 0.5, integerized by truncation, and then used to point to a specific location within the arrays TAB1-TAB7 from where the value of R is retrieved. Subroutines LOADT and DOARC are the only subroutines involved with table lookup operations. These subroutines respectively load and process the table data. The user must write the necessary coding to read the data in subroutine LOADT. Arrays TAB1-TAB7 are carried in the common block labeled/TABLE/, which appears only in subroutine LOADT and DOARC. Also, the values of the minimums and maximums of the arrays TAB1-TAB7 must be loaded in subroutine LOADT in the check variable arrays LXCK(7), MXCK(7), LYCK(7), MYCK(7), LZCK(7), and MZCK(7). For example, suppose the X (the first), Y (the second), and Z (the third) dimensions of the table being entered in TAB5 are 15, 60, and 45 for the maximums and 1, 1, and 1 for the minimums. The coding for each of these check variables is as follows: LXCK(5) = 1, MXCK(5) = 15, LYCK(5) = 1, MYCK(5) = 60, LZCK(5) = 1, and MZCK(5) = 45.

In the event the user desires entering a one - or two-dimensional table, the dimensions of the array used to carry the table must be shaped to carry that table. Also, the retrieval coding in DOARC must drop the IZ and IY or IZ variables. Additionally, the values given the check variables for the unused dimension(s) must be compatible with the values Z, or Z and Y will be carrying in as entered from the transformation cards. Otherwise, the restriction will be violated and an error message will be unnecessarily registered. Tables beyond three dimensions must be broken down into three or fewer dimensions to be entered in VERT-3.

@Transformations 45-50 and 95-100 utilize subroutines
SUB1, SUB2, SUB3, SUB4, SUB5, and SUB6, which must be
coded by the user. These subroutines enable the user to
create transformations that are not available in the
preceding list of transformations. The user must supply
the checks necessary to insure that dividing by zero and
other such computational mistakes are not made.
  Structuring a mathematical relationship within a
VERT-3 network consists essentially of the following
three phases:

1. Long or complicated mathematical relationships
    need to be broken down into a series of three
    variable-unit transformations shown above.
2. Values for each of the three variables (X, Y,
    and Z) in each single-unit transformation must be
    defined. These values can be retrieved from (a)
    previously processed arcs or nodes, (b) constants
    entered in these satellite arc cards, or (c) one
    of the previously processed transformations com-
    puted in the current series of transformations
    used to generate a time value for the current arc
    under consideration. Values calculated for each
    unit time transformation are consecutively,
    temporarily stored in a one-dimensional array.
    This array enables one to retrieve the value
    calculated for a prior transformation for use in
    the current unit transformation. Upon the com-
    pletion of all the unit time transformations for a
    given arc, this temporary storage array is cleared.
    Thus, only the values calculated for previously
    derived unit time transformations developed for the
    current arc under consideration can be referenced.
    When one is retrieving numerical values from a
    previously processed arc or node, the time or cost
    or performance value calculated for the referenced
    node or the primary (not cumulative) time or cost or
    performance value generated for the referenced arc
    is retrieved.
3. Results of each of the unit transformations needed
    to develop a value for an arc's time can be either
    summed into the overall time value generated for the
    arc under consideration or can be omitted. When the
    resulting value of a unit transformation is omitted,
    this transformation is generally being used as an

intermediate step for calculating the value of a
long or complicated mathematical relationship. An
example mathematical relationship follows the
complete description of all three types of mathematical relationships.

Columns 1-8, FORMAT 2A4. Enter the name of the arc for
which this satellite card is carrying information.

Columns 9-13, FORMAT A4, A1. Enter the satellite type
identifier--"RTIME."

Columns 14-15, FORMAT I2. Enter the card sequence number.
The cards required to accomplish this task must be sequentially numbered. The maximum number of cards allowed
equals the current maximum number of arcs allowed (value
of check variable MARC) or the current maximum number of
nodes allowed (value of check variable MNODE) or a total
of 99, whichever is the smallest.

Columns 16-17, FORMAT I2. This field aids phase one of
the transformation process. In this field, enter the
code number of the transformation desired to be used.

Column 18, FORMAT A1. This field is concerned with phase
one of the transformation process. Enter the letter "S"
to sum the resulting value of this transformation into the
time value generated for this arc. Otherwise, enter the
letter "O" to omit it. In the event an arc or a node used
in a given transformation is logically eliminated or is a
failure, backup or alternate transformations may be entered.
This task is accomplished by entering, directly after an
initial transformation, additional backup transformation(s)
carrying the letter "B" in column 18 (or column 48). An
unlimited number of backup transformations may then follow
a given initial transformation. The program will sequentially try processing each one of these backup transformations until it finds one that can be computed. It will
then ignore the rest of the backup transformations. However, if the initial transformation plus all of its back-
ups are infeasible, an error number will be listed and the
simulation will then be terminated. The program assumes
that the sum or omit disposition that applies to the
initial transformation should apply to the backup
transformations.

The following three groups of two fields per group are concerned with phase three, the retrieval phase, of the transformation process. These three groups structure the retrieval of numerical information for the transformation variables X, Y, and Z. The specific layout for the X group and similar layouts for the Y and Z groups are as follows:

<u>Column 19</u>, FORMAT A1. Acceptable entries in this field are the letters "T," "C," "P," "K," or "-" (blank).

<u>Columns 20-27</u>, FORMAT 2A4 or F8.0. If a "T" is entered in column 19, the time value carried by the node entered in this field or the primary time value carried by the arc entered in this field will be loaded into the transformation variable X prior to executing the transformation. Also, entering a "C" or "P" in column 19 will promulgate loading into X the value of the cost or performance carried by the arc or node whose name is entered in this field. If a "K" is entered in column 19, a numerical constant must be entered in this field. The constant will be loaded into the transformation variable X prior to executing the transformation. If column 19 is left blank, it is assumed that the value calculated for a previous transformation in the current series of time transformations will be entered in variable X prior to executing the transformation. The series number of that previous transformation must be entered in this field. For example, if it is desired to load the resulting value of the second unit time transformation into the variable X in the third unit time transformation, "2.0" should be entered in this data field when structuring the third unit time transformation.

<u>Column 28</u>. Same as column 19.

<u>Columns 29-36</u>. Same as columns 20-27, except substitute "Y" for "X."

<u>Column 37</u>. Same as column 19.

<u>Columns 38-45</u>. Same as columns 20-27, except substitute "Z" for "X."

This card is structured to carry two unit time transformations. The fields in the second part of this card equate with the fields in the first part of this card as follows:

Columns of first part		Columns of second part
16-17	=	46-47
18	=	48
19	=	49
20-27	=	50-57
28	=	58
29-36	=	59-66
37	=	67
38-45	=	68-75

NOTE: The second part of this card may optionally be left blank when loading a series of unit time transformations. However, the first part of this card must always be used.

I.   Cost Math-Related Satellite Arc Card(s)
This card(s) carries the mathematical relationship(s) used to create in part or in total the cost value for the arc under consideration. The description of this card is the same as card H, except enter "RCOST" in columns 9-13 and substitute the word "cost" for the word "time." Further, if it is desired to inflate and/or discount the cost generated for this arc, in place of entering "RCOST" in columns 9-13, enter "RCOSI" to inflate the cost, enter "RCOSD" to discount the cost, or enter "RCOSB" to both inflate and discount the cost by the appropriate interest rates entered in the control card.

J.   Performance Math-Related Satellite Arc Card(s)
This card carries the mathematical relationship(s) used to create in part or in total the performance value for the arc under consideration. The description of this card is the same as card H, except enter "RPERF" in the columns 9-13 and substitute the word "performance" for the word "time." Further, if it is desired to inflate and/or discount the performance generated for this arc, in place of entering "RPERF" in columns 9-13, enter "RPERI" to inflate the performance, enter "RPERD" to discount the performance, or enter "RPERB" to both inflate and discount the performance by the appropriate interest rates entered in the control card.

<u>Transformation Example</u>. Suppose the value for the performance of a given arc is related to the time, cost, and performance values generated on this arc and other previously processed arcs and nodes as follows:

$$PA10 = \frac{(PA1 + PA2 + PA3)*(TA1)*(LOG_E (CA1 * CA2)) +}{(PA4 * PA5 * PA6)}$$

$$\frac{(188.6)*(TA10)}{CA10} + (15.8)*(TN1)$$

where:

```
TN1  = the time value for the node named N1.
TA1  = the time value for the arc named A1.
TA10 = the time value for the arc named A10.
CA1  = the cost value for the arc named A1.
CA2  = the cost value for the arc named A2.
CA10 = the cost value for the arc named A10.
PA1  = the performance value for the arc named A1.
PA2  = the performance value for the arc named A2.
PA3  = the performance value for the arc named A3.
PA4  = the performance value for the arc named A4.
PA5  = the performance value for the arc named A5.
PA6  = the performance value for the arc named A6.
PA7  = the performance value for the arc named A7.
PA8  = the performance value for the arc named A8.
PA9  = the performance value for the arc named A9.
PA10 = the performance value for the arc named A10.
```

The following dimensioned card layouts illustrate how the preceding equation is put into card form:

```
A10        RPERF 1 50PA1       PA2        PA3        trans. no. 1
A10        RPERF 2 40PA4       PA5        PA6        trans. no. 2
A10        RPERF 3140TA1       CA1        CA2        trans. no. 3
A10        RPERF 4 is 1.0      2.0        3.0        trans. no. 4
A10        RPERF 5 2SK188.6    TA10       CA10       trans. no. 5
A10        RPERF 6 1SK15.8     TN1        K1.0       trans. no. 6
********+++++**++*+*********+*********+********
********+++++**++*+*********+*********+********
Cols 1-8++++++**++*+  20-27 + 29-36  + 38-45
        +++++**++*+          +         +
        9-13**++*19          28        37
           **++*
        14-15++18
           ++
           ++
        16-17
```

The preceding layouts illustrate that the above equation
can be modeled by using six sequential transformations.
The first three transformations compute the values for
(PA1 + PA2 + PA3), (1/(PA4 * PA5 * PA6)), and ((TA1)*
(LOGE(CA1 * CA2)), respectively. The letter "0" in column
18 of these transformations indicates that the resultant
values of each of these transformations are not summed in
the performance value for arc A10. However, transformation
four is summed into the resulting performance value of
arc A10. It pulls these three previously derived values
together to derive a composite value for the first major
term of the equation. Transformations five and six
compute the values for the second and third terms of the
equation. These values are summed directly into the resulting performance value calculated for arc A10.

NOTE: The following seven satellite arc cards (K-Q) assist
an arc's input node in its logic function. These satellites
carry information used in conducting the output logic
function of a given arc's input node.

K. FILTER NUMBER 1 Satellite Arc Card (this arc's input
node must have FILTER 1 output logic).

Columns 1-8, FORMAT 2A4. Enter the name of the arc for
which this satellite card is carrying information.

Columns 9-13, FORMAT A4, A1. Enter the satellite type
identifier--"FILT1."

Columns 14-15, FORMAT I2. Enter the card sequence
number. Only one card per arc is required to carry all
the information needed for this task. Therefore, enter
a "1" in column 15.

Columns 16-25, FORMAT F10.0. Enter the lower boundary of
the time constraint placed on this arc.

Columns 26-35, FORMAT F10.0. Enter the upper boundary of
the time constraint placed on this arc.

Columns 36-45, FORMAT F10.0. Enter the lower boundary of
the cost constraint placed on this arc.

<u>Columns 46-55</u>, FORMAT F10.0. Enter the upper boundary of the cost constraint placed on this arc.

<u>Columns 56-65</u>, FORMAT F10.0. Enter the lower boundary of the performance constraint placed on this arc.

<u>Columns 66-75</u>, FORMAT F10.0. Enter the upper boundary of the performance constraint placed on this arc.

L. <u>FILTER NUMBER 2 Satellite Arc Card</u> (this arc's input node must have FILTER 2 output logic).

<u>Columns 1-8</u>, FORMAT 2A4. Enter the name of the arc for which this satellite card is carrying information.

<u>Columns 9-13</u>, FORMAT A4, A1. Enter the satellite type identifier--"FILT2."

<u>Columns 14-15</u>, FORMAT I2. Enter the card sequence number. Only one card per arc is required to carry all the information needed for this task. Therefore, enter a "1" in column 15.

<u>Columns 16-25</u>, FORMAT F10.0. Enter the lower boundary of the number of successfully completed input arcs constraint(s) placed on this arc.

<u>Columns 26-35</u>, FORMAT F10.0. Enter the upper boundary of the number of successfully completed input arcs constraint(s) placed on this arc.

M. <u>FILTER NUMBER 3 Satellite Arc Card</u> (this arc's input node must have FILTER 3 output logic).

<u>Columns 1-8</u>, FORMAT 2A4. Enter the name of the arc for which this satellite card is carrying information.

<u>Columns 14-15</u>, FORMAT I2. Enter the card sequence number. The cards required to accomplish this task must be sequentially numbered. The maximum number of cards allowed equals the current maximum number of arcs allowed (value of the check variable MARC) divided by 6 or a total number of 99, whichever is smaller.

<u>Column 16</u>, FORMAT 1X. Leave blank.

Column 17, FORMAT A1. Enter a plus (+) sign if the following constraining arc must have been successfully completed
before the output arc being constrained can be initiated.
Otherwise, enter a minus (-) sign if the constraining arc
must have been unsuccessfully processed or eliminated from
the network before the output arc being constrained can be
initiated.

Column 18-25, FORMAT 2A4. Enter the name of the first
constraining arc.

Column 26. Same as column 16. Leave blank.

Column 27. Same as column 17 (if another constraining arc
is needed).

Column 28-35, FORMAT 2A4. Enter the name of the second
constraining arc.
    Continue this process until all the constraints have
been entered or until the maximum number of these cards
have been entered as previously defined in the card sequence field (columns 14-15).

N.   MONTE CARLO Satellite Arc Card (this arc's input node
must have MONTE CARLO output logic).

Columns 1-8, FORMAT 2A4. Enter the name of the arc for
which this satellite card is carrying information.

Columns 9-13, FORMAT A4, A1. Enter the satellite type
identifier--"M   " in column 9 and leave column 10-13
blank.

Columns 14-15, FORMAT I2. Enter the card sequence number.
Only one card is required to carry all the information
needed for this task per arc. Therefore, enter a "1" in
column 15.

Columns 16-25, FORMAT F10.0. Enter this arc's probability
of being MONTE CARLO initiated.

NOTE: The following three satellite arc cards are an
addendum to card N. These cards will enable the construction of conditional probability situations where the
probability of arc initiation is a function of either the

time, cost, or performance accumulated on this arc's input
node. The layout of these card types consists of a probability element in the first data field, followed by
either a time, cost, or performance boundary in the second
data field, followed by a probability element in the third
data field, followed by another boundary in the fourth data
field, and so on, through the last element, which must be a
probability element. The boundaries must be identical in
numerical value and field placement over all the output
arcs of this arc's input node. The probability elements
within a given field must have a value greater than zero
and less than one. The sum of the probability elements
in each of the probability element data fields must sum to
one when these items are being accumulated over all the output arcs of this arc's input node. The probability field
selected when the network is being processed will be the
one whose lefthand time, cost, or performance boundary is
less than or equal to the value generated for this arc's
input node's time, cost, or performance, and whose right-
hand time, cost, or performance boundary is greater than
the value generated for this arc's input node's time, cost,
or performance.

O. MONTE CARLO Time-Conditioned Satellite Arc Card (this
arc's input node must have MONTE CARLO output logic).

Columns 1-8, FORMAT 2A4. Enter the name of the arc for
which this satellite is carrying information.

Columns 9-13, FORMAT A4, A1. Enter the satellite type
identifier--"MTIME."

Columns 14-15, FORMAT I2. Enter the card sequence number.
The cards required to accomplish this task must be sequentially numbered. The maximum number of cards allowed
equals the current maximum number of arcs allowed (value
of the check variable MARC) divided by 6 or a total number
of 99, whichever is smaller.

Columns 16-25, FORMAT F10.0. Enter this arc's element to
distribution number one.

Columns 26-35, FORMAT F10.0. Enter time boundary number
one.

<u>Columns 36-45</u>, FORMAT F10.0. Enter this arc's element to distribution number two.

<u>Columns 46-55</u>, FORMAT F10.0. Enter time boundary number two.

<u>Columns 56-65</u>, FORMAT F10.0. Enter this arc's element to distribution number three.

<u>Columns 66-75</u>, FORMAT F10.0. Enter time boundary number three.
   Repeat the above steps for any additional cards needed. The maximum number of cards allowed has been previously defined in the card sequence field (columns 14-15).

P. <u>MONTE CARLO Cost-Conditioned Satellite Arc Card</u> (this arc's input node must have MONTE CARLO output logic).
   This description is the same as card O, except enter MCOST in columns 9-13 and substitute the word "cost" for the word "time" in the narrative.

Q. <u>MONTE CARLO Performance-Conditioned Satellite Arc Card</u> (the input node for this arc must have MONTE CARLO output logic).
   This description is the same as card O, except enter MPERE in columns 9-13 and substitute the word "performance" for the word "time" in the narrative.

<u>NOTE</u>: The following satellite arc card is used to aid in reporting the slack time on an arc.

R. <u>Slack Histogram Satellite Arc Card</u>
This card is used to input the minimum and maximum slack time value used to construct arc slack time histograms. This feature is optional. If this card is omitted, the program will use the minimum and maximum values observed during the simulation to construct the histogram cells. If the values generated during the simulation lie outside the minimum and maximum boundaries entered, the program accumulates these values in minimum and/or maximum overflow cells. Thus, outliers are accumulated in these peripheral cells while the interior contents of the histogram are being pictorialized.

<u>Columns 1-8</u>, FORMAT 2A4. Enter the name of the arc for
which this satellite card is carrying information.

<u>Columns 9-13</u>, FORMAT A4, A1. Enter the satellite type
identifier--"SLAK."

<u>Columns 14-15</u>, FORMAT I2. Enter the card sequence number.
Only one card per arc is required to carry all the information needed for this task. Therefore, enter a "1" in
column 15.

<u>Columns 16-25</u>, FORMAT F10.0. Enter the minimum slack time
desired for constructing the slack time histogram.

<u>Columns 26-35</u>, FORMAT F10.0. Enter the maximum slack time
desired for constructing the slack time histogram.

<u>Columns 36-80</u>. Leave blank.

III. Nodes Module

A. <u>Master Node Card</u>

<u>Columns 1-8</u>, FORMAT 2A4. Enter the name of the node being
modeled. The last card of this module must have "ENDNODE"
punched in columns 1-7 and the rest of the card must be
left blank.

<u>Column 9</u>, FORMAT I1. Enter the input logic code number
(defined as follows):

<u>Input Logic Code Number</u>	<u>Type of Input Logic</u>
1	INITIAL
2	AND
3	PARTIAL AND
4	OR

<u>NOTE</u>: The order in which the arc cards are entered in the
computer is critical for nodes having COMPARE, PREFERRED, or
QUEUE logic. The first input arc read in for those nodes
will be linked or mated with the first output arc read in;
likewise, the second input arc read in will be linked or
mated with the second output arc read in, and so on. It
does not matter whether the input arcs or the output arcs
are read in first or whether the input and output arc cards

are intermixed while being read in. The relative order in
which the input and output arcs, by themselves, are read in
is the important factor.

5	COMPARE
6	PREFERRED
7	QUEUE
8	SORT

Columns 10-12, FORMAT I3. Enter the output logic code
number (defined below), the number of servers if QUEUE
logic is used, or the number of output arcs desired to be
initiated if COMPARE or PREFERRED input logic was requested. Under this latter option, a minus (-) sign
should prefix this number if utilization of the desired
condition is wanted. Otherwise, this number will be picked
up as a positive number, and thus the demand condition will
be invoked.

Output Logic Code Number	Type of Output Logic
1	TERMINAL*
2	ALL
3	MONTE CARLO
4	FILTER 1
5	FILTER 2
6	FILTER 3

*If the "1" in column 12 is prefixed with a "1," "2,"
"3,"---,"99" to give a total field entry of "11," "21,"
"31,"---, "991," the terminal node is given a 2nd-, 3rd-,
4th-,--- 100th-class designation. The first-class status
is assigned by leaving columns 10 and 11 blank. The higher
the prefix number, the lower the class. (When choosing the
winning terminal node as described in I - A, see the note
before the description of columns 36-38.) The first-class
terminal nodes take precedence over the second-class
terminal nodes, the second-class terminal nodes take precedence over the third-class terminal nodes, and so on.
Competition is first conducted among the first-class
terminal nodes, providing there is at least one active
first-class terminal node for the given iteration. However, if there are not any active first-class terminal
nodes, then competition is conducted at the second-class
level, or the third-class level, or at the highest class
level where active terminal nodes exist. There are no

class size limitations; however, there cannot be more
class levels than the number of nodes read in minus 1, or
a grand total of 100 (including the zero - first-class
level), whichever is smaller.

Columns 13-14, FORMAT I2. Enter the numeric code given
below for the type of output desired from the next field.

Numeric Code	Time	Path Cost	Overall Cost	Performance
Blank or zero	Yes	Yes	Yes	Yes
1	Yes	No	No	No
2	No	Yes	No	No
3	No	No	Yes	No
4	No	No	No	Yes
5	Yes	Yes	No	No
6	Yes	No	Yes	No
7	Yes	No	No	Yes
8	No	Yes	Yes	No
9	No	Yes	No	Yes
10	No	No	Yes	Yes
11	Yes	Yes	Yes	No
12	Yes	Yes	No	Yes
13	Yes	No	Yes	Yes
14	No	Yes	Yes	Yes
15	No	No	No	No
16	*	*	*	*

*Entering a "16" in this field will cause punching HTIME,
HCOST, and HPERF stochastic histogram satellite arc cards
carrying the histograms. This will facilitate the substitution of the results obtained from a lower-level network into a higher, summary-level network. Generally, a
terminal node in the lower-level network becomes an arc in
the higher-level network.

Columns 15-16, FORMAT I2. This program has the facility
for printing time, cost, and performance histograms for a
limited number of internal nodes. This limit is set by
the programs embedded check variable MHIST. Internal
nodes can be designated as candidates for statistical
printouts by sequentially numbering these nodes in this
field up to and including the value of MHIST. If it is
desired to construct time, cost, and performance histograms
for an interval between two nodes, enter the same number in
this field for the two nodes bridging the interval. Only

two nodes at a time can be used to develop interval histograms.

Terminal node histograms will be listed automatically.
Therefore, this field should be left blank when one desires
the normal histogram listings for the terminal nodes. However, the previous field (columns 13-14) should have an
entry when this node is a terminal node. If a "1" is entered in this field when this node is a terminal node, all
critical/optimum paths terminating in this node will be
suppressed from the critical/optimum path analysis.

NOTE: Values assigned to the following three fields must
all lie within either the closed interval of -1 and 0 or
the closed interval of 0 and +1. These fields must not
jointly carry positive and negative values (field one cannot have a positive entry at the same time that fields two
and/or three have negative entries, and so on). Entering
positive values in these fields will cause the optimum input arc set to be chosen as the one having the least time
and cost and the most performance. Entering negative
values in these fields will cause the optimum input arc
set to be chosen as the one having the largest time and
cost coupled with the least performance.

Columns 17-20, FORMAT F4.3. Enter the weight assigned to
time when choosing the optimum input arc set via COMPARE
logic or when sorting input flows via SORT logic.

Columns 21-24, FORMAT F4.3. Enter the weight assigned to
cost when choosing the optimum input arc set via COMPARE
logic or when sorting input flows via SORT logic.

Columns 25-28, FORMAT F4.3. Enter the weight assigned to
performance when choosing the optimum input arc set via
COMPARE logic or when sorting input flows via SORT logic.

Column 29, FORMAT A1. Enter the letter "S" in this column
to request a histogram of the slack time available at this
node. Otherwise, leave this field blank. This histogram
will be structured only when a time critical path (1.0
weight on the critical path time) is requested. Further, a
satellite node card (see card D) that will input a scale
for this histogram may be entered. If this satellite
card is omitted, the program will use the observed minimum
and maximum slack time to structure its own scale.

Columns 30-80, FORMAT A3, 12A4. Enter the description of the event this node represents.

B. Histogram Satellite Node Card

This card is used to input the minimum and maximum time, path cost, overall cost, and performance values used to construct histograms generated for this node. Therefore, this node must be either a terminal node or an internal node used to gather statistics. This card is optional. If it is omitted, the program will use the minimum and maximum values observed during the simulation for constructing the histogram cells. However, if this card is used and values are generated that exceed the minimum and/or maximum boundaries entered, the program accumulates these values in minimum and/or maximum overflow cells. Thus, outliers can be accumulated in these peripheral cells while one is pictorializing the interior content of the data.

Columns 1-8, FORMAT 2A4. Enter the name of the node for which this satellite node card is carrying information.

Columns 9-12, FORMAT A4. Enter the satellite type identifier--"HIST" (abbreviation for histogram).

Columns 13-20, FORMAT F8.0. Enter the lower boundary value desired for the time histogram.

Columns 21-28, FORMAT F8.0. Enter the upper boundary value desired for the time histogram. If this field and the preceding field are left blank or have zeros entered in them, the program will use the minimum and maximum time values observed for this histogram during the simulation for its construction.

Columns 29-36. Same as columns 13-20, except substitute the words "path cost" for the word "time."

Columns 37-44. Same as columns 21-28, except substitute the words "path cost" for the word "time."

Columns 45-52. Same as columns 13-20, except substitute the words "overall cost" for the word "time."

Columns 61-68. Same as columns 13-20, except substitute the word "performance" for the word "time."

Columns 69-76. Same as columns 21-28, except substitute the word "performance" for the word "time."

C.   Subtract Satellite Node Card

This card is used to input the subtract node. This node must have FILTER NUMBER 1 output logic. If this card is omitted, the subtraction feature (see Chapter 5) will not be utilized when using FILTER NUMBER 1 logic.

Columns 1-8, FORMAT 2A4. Enter the name of the node, for which this satellite node card is carrying information.

Columns 9-12, FORMAT A4. Enter the satellite type identifier--"SUBT" (abbreviation for subtract).

Columns 13-20, FORMAT 2A4. Enter the subtract node name.

D.   Slack Histogram Satellite Node Card

This card is used to input the minimum and maximum slack time values used to construct slack time histograms for this node. This card is optional in the same sense as the histogram satellite node card is (see card B).

Columns 1-8, FORMAT 2A4. Enter the name of the node for which this satellite node card is carrying information.

Columns 9-12, FORMAT A4. Enter the satellite type identifier--"SLAK" (abbreviation for slack).

Columns 13-20, FORMAT F8.0. Enter the minimum slack time desired for the slack time histogram.

Columns 21-28, FORMAT F8.0. Enter the maximum slack time desired for the slack time histogram. If this field and the preceding field are left blank or have zeros entered in them, the program will use the minimum and maximum time values observed during the simulation to construct this histogram.

Columns 29-80. Leave blank.

7 OUTPUTS AND REPORTS

The output options, data, and reports generated by VERT are extensive and are perhaps some of the most important reasons why the technique is so useful in network analysis (see Brown 1975). Of course, since the analyst selects the outputs to assist the manager in the decision/risk analysis, only those outputs pertinent to the problem at hand and meaningful to the manager need be selected.

The available output options are described in this chapter, and several sample outputs are included. Embedded in the program described here are 146 error messages that may provide assistance to the analyst. They are available from the authors on request.

In analyzing the output data, it should be remembered that VERT-3 also includes a cost-pruning option. Arcs in the stream of a network path going into OR logic and COMPARE logic nodes, as well as other arcs, may be only partially completed or may never start processing prior to the completion of a given simulation run. VERT-3 is structured to cost fully or partially activities partially completed and allows cost pruning of activities not started.

OUTPUTS

It should be remembered that the simulation process creates a network flow that traverses the network from initial node(s) to terminal node(s) and results in one trial solution or outcome to the problem being modeled. This simulation process is repeated as many times as the user requests to generate a sufficiently large sample of possible outcomes to aid in the subsequent analysis. The distribution of time, cost, and performance outcomes enables the manager/analyst to evaluate the likelihood of occurrence of various levels of these parameters. Thus, managers can take into account the variability that may occur in terms of confidence intervals; consequently, they are on more solid ground when budgeting for projects or ventures.

For each terminal node, composite terminal node, requested internal nodes, and internal intervals, the program produces the following measures of time, cost, and performance results:

1. Relative frequency distribution;
2. Cumulative frequency distribution (ogive);
3. Mean observation;
4. Standard error (standard deviation of the sample);
5. Coefficient of variation;
6. Mode;
7. Median;
8. Beta 2 measure of kurtosis;
9. Pearsonian measure of skewness.

This information is displayed for requested internal nodes, intervals between internal nodes, and all terminal nodes. Additionally, all terminal nodes' time, cost, and performance data are combined to give a composite terminal node time, cost, and performance printout.

Two sets of cost data are generated for each of the preceding node printouts. The first set, labeled *path cost,* consists of the total cost accumulated in processing all the activities on the path(s) through which the network flow(s) had to come in order to process the node requesting the printout information. The second set, labeled *overall cost,* consists of the path cost plus the cost of all the other activities processed during and prior to the time this node was processed.

Also, slack time on each arc and node per user request is exhibited in the above form. *Slack time* is the excess time available for processing an arc or the additional amount of time that a decision can be delayed before the node appears on the critical path.

The overall network cost incurred and the overall network performance gained between selected time intervals (for example, yearly time intervals) as requested by the user is also exhibited in the above form. Information of this type is very useful for constructing budgets for future periods of expenditure or for comparing investment alternatives.

Additionally, time/cost, time/performance, and performance/cost correlations are graphed for all terminal nodes, including the composite terminal node. Some other forms of output include the following:

1. A listing of major variable storage arrays;
2. A listing of all flow-carrying arcs and nodes realized in each iteration;
3. A one-line summary listing of the results of each iteration;
4. A listing of the mean, minimum, and maximum for the time, path cost, overall cost, and performance for each terminal node and for requested internal nodes;
5. Time, path cost, and performance correlations;
6. A listing of the optimal terminal node index and an accompanying arcs and nodes critical/optimum path index.

VERT-3 prints out a bar graph on the optimum terminal node index. Through use of this printout, the project risk can be ascertained. A decision/risk analysis network takes the usual form of having one or several terminal nodes collect successful project completions and having one or several terminal nodes collect unsuccessful project completions. Realization of these various terminal nodes compared to the total number of iterations gives an indication of project success or failure. In the event that more than one terminal node can be realized at the same time, the optimum terminal node is the one with the lowest completion time, lowest cost, highest performance, or best weighted combination of these factors, per user-developed weights. Entering negative terminal node selection weights will produce an opposite effect.

The program next prints out the critical/optimum path index for nodes and arcs. The critical path is the path through the network with the longest completion time, highest cost, lowest performance, or least desirable weighted combination of these factors, per user-developed weights. Entering negative critical/optimum path weights will cause the optimum path to be chosen. VERT-3 allows optional suppression of critical/optimum paths originating from user-selected terminal nodes. This feature facilitates the finding of trouble-producing activities. Since different stochastic paths can be realized in the process of simulating the network, the critical/optimum path tends to change from iteration to iteration. The program computes the portion of time each arc and node is on the critical/optimum path and lists this information in a bar chart display. Time, cost, and performance correlations and plots are printed upon request for each terminal node. This information allows the possible determination of relationships among these variables.

INTERPRETATION OF STATISTICAL MEASURES

The relative frequency distribution provides a picture of the range and concentration of the time, cost, and performance values observed on a given node. The probability of exceeding certain value levels can be obtained from the cumulative frequency distribution, which results in the ability to infer confidence levels.

The *mean* is the average of all the observations. The sum of the squares of the differences between the observations and the mean value is divided by the number of observations to compute the *variance,* or the *mean square.* The square root of the variance is the *standard deviation,* also known as the *root mean square.* The standard deviation, being in original units, is an absolute measure of dispersion and does not permit comparisons to be made between the dispersion of various distributions that are on different units of measurement. The *coefficient of variation* has been designed for such comparative purposes. Since it is the ratio of the standard deviation to the mean, the coefficient of variation is an abstract measure of dispersion. The greater the dispersion of a distribution, the higher the value of the standard deviation relative to the mean. Hence, the relative dispersion of a number of distributions may be determined simply by comparing the values of their coefficients of variation.

The value in a series of observations that occurs with the greatest frequency is called the *mode.* It is the most meaningful measure of central tendency in the case of strongly skewed or nonsymmetric distributions since it provides the best indication of the point of heaviest concentration. Though a distribution has only one mean and one median (midpoint), it may have several modes, depending upon the number of peaks of concentration. The mode is not affected by extreme values, while the mean is influenced by such values. In a symmetrical distribution, these two measures of central tendency are equal. But if the distribution is skewed, the value of the mean will be strongly influenced in the direction of the skew, while the mode will remain stationary. Hence, the difference between these two measures of central tendency is a measure of the skewness of a distribution. This measure of skewness can be converted into relative terms by dividing it by the standard deviation. As a general rule, a distribution is not considered to be markedly skewed as long as the Pearsonian formula described above yields an absolute value less one.

Kurtosis is a measure of the relative height of a distribution—that is, its peakedness. A distribution is said to be *mesokurtic* if it has so-called normal kurtosis, *platykurtic* if its peak is abnormally flat, and *leptokurtic* if its peak is abnormally sharp. The *Beta 2* measure of kurtosis is defined as the fourth moment about the mean divided by the standard deviation raised to the fourth power. Beta 2 is a relative measure of kurtosis based on the principle that, as the relative height of a distribution increases, the value of the standard deviation decreases relative to its fourth moment. In other words, the more peaked a distribution is, the greater the value of Beta 2. For the standard normal distribution, Beta 2 is equal to 3. Since the normal distribution plays such an important role in statistical theory, this value is taken as the norm. The more platykurtic a distribution is, the further Beta 2 decreases below 3; the more leptokurtic a distribution is, the more Beta 2 will exceed 3.

SAMPLE REPORTS

The output reports shown in figures 7.1 through 7.6 are taken from the alternative energy source problem described for illustrative purposes in chapter 13. The problem involves evaluating three alternative power-generating ventures; the manager is required to select among the three.

The optimum terminal node index bar chart in figure 7.1 indicates that the probability that one of the three ventures under study will be successfully developed is 94.8% (FUWINNER + FIWINNER + COWINNER = 7.1% + 33.6% + 54.1%). It can further be observed that there exists about a 1% chance of failing in the pilot-plant test phase (FAILPILT), a 4% chance of failing in the shock-test phase (FAILSHOC), and virtually no

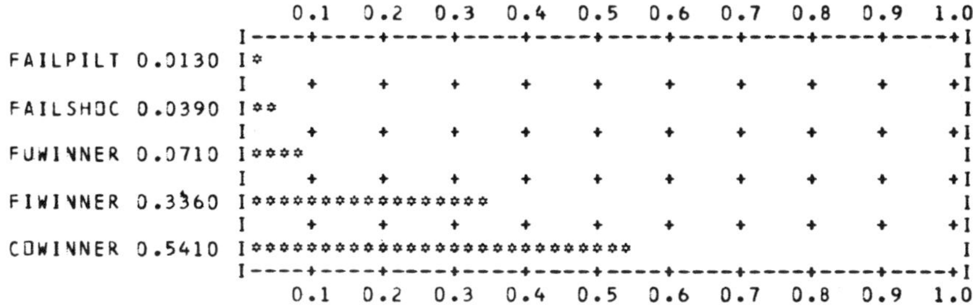

FIGURE 7.1. Optimum Terminal Node Index Chart

chance of failing in the R&D phase. (Since FAILRD did not occur in any of the simulation runs, it does not show up on figure 7.1.)

The cumulative frequency distribution (CFD) of the network time for the composite terminal node (see figure 7.2) indicates that the venture will be

```
       CFD  0.1  0.2  0.3  0.4  0.5  0.6  0.7  0.8  0.9  1.0
11.08   I----I----I----I----I----I----I----I----I----I----I MIN
        I                                                     0.005
12.00   I
        I                                                     0.005
12.23   I
        I                                                     0.006
12.45   I
        I                                                     0.007
12.68   I
        I                                                     0.007
12.91   I
        I                                                     0.007
13.14   I
        I                                                     0.007
13.36   I
        I                                                     0.007
13.59   I
        I                                                     0.007
13.82   I
        I                                                     0.007
14.05   I
        I                                                     0.008
14.27   I
        I                                                     0.008
14.50   I
        I                                                     0.009
14.73   I
        I*                                                    0.020
14.95   I
        I**                                                   0.044
15.18   I
        I*****                                                0.096
15.41   I
        I********                                             0.165
15.64   I
        I**************                                       0.271
15.86   I
        I**********************                               0.432
16.09   I
        I******************************                       0.627
16.32   I
        I**************************************               0.787
16.55   I
        I*********************************************        0.910
16.77   I
        I**************************************************** 0.981
17.00   I
        I*****************************************************1.000
17.53   I----I----I----I----I----I----I----I----I----I----I MAX
        NO OBS------------    1000   STD ERROR-        .6070
        COEF OF VARIATION-    0.04   MEAN------        16.10
        KURTOSIS (BETA 2)-   19.66   MEDIAN----        16.16
        PEARSONIAN SKEW---   16.20   MODE------        .1620
```

FIGURE 7.2.   Network Time for the Composite Terminal Node

completed somewhere within a time span of 11.08 to 17.53 years. There is a clustering of times between 14.95 and 17 years that accounts for approximately 96% of the observations.

```
             CFD  0.1  0.2  0.3  0.4  0.5  0.6  0.7  0.8  0.9  1.0
 223.2        I----I----I----I----I----I----I----I----I----I----I MIN
              I                                                      0.0
 223.2        I
              I**                                                    0.044
 236.2        I
              I*************                                         0.271
 249.2        I
              I*********************                                 0.464
 262.2        I
              I**********************                               0.488
 275.2        I
              I***************************                          0.552
 288.2        I
              I*************************************************     0.762
 301.2        I
              I******************************************************  0.845
 314.2        I
              I*******************************************************  0.849
 327.2        I
              I*******************************************************  0.849
 340.2        I
              I*******************************************************  0.849
 353.2        I
              I*******************************************************  0.849
 366.2        I
              I*******************************************************  0.849
 379.2        I
              I*******************************************************  0.849
 392.2        I
              I*******************************************************  0.849
 405.2        I
              I*******************************************************  0.849
 418.2        I
              I********************************************************  0.856
 431.2        I
              I*********************************************************  0.879
 444.2        I
              I***********************************************************  0.916
 457.2        I
              I************************************************************  0.941
 470.2        I
              I*************************************************************  0.957
 483.2        I
              I**************************************************************  0.985
 496.2        I
              I***************************************************************1.000
 509.2        I
              I***************************************************************1.000
 509.2        I----I----I----I----I----I----I----I----I----I----I MAX
              NO OBS------------    1000   STD ERROR-        74.59
              COEF OF VARIATION-    0.25   MEAN------       297.8
              KURTOSIS (BETA 2)-    4.34   MEDIAN----       279.1
              PEARSONIAN SKEW---  247.15   MODE------        .6794
```

FIGURE 7.3.   Overall Cost for the Composite Terminal Node

The CFD of the overall cost for the composite terminal node indicates that the venture will cost somewhere between $223.2 and $509.2 million (see figure 7.3). Furthermore, it can be observed that two definite areas of concentration exist in this distribution. Approximately 85% of the observations

```
          CFD  0.1  0.2  0.3  0.4  0.5  0.6  0.7  0.8  0.9  1.0
209.4     I----I----I----I----I----I----I----I----I----I----I MIN
          I                                                     0.0
209.4     I
          I*                                                    0.014
213.5     I
          I*                                                    0.014
217.6     I
          I*                                                    0.028
221.7     I
          I****                                                 0.070
225.8     I
          I********                                             0.169
229.9     I
          I**************                                       0.282
234.0     I
          I****************                                     0.352
238.1     I
          I*****************                                    0.408
242.2     I
          I******************                                   0.423
246.2     I
          I******************                                   0.423
250.3     I
          I******************                                   0.423
254.4     I
          I******************                                   0.423
258.5     I
          I*******************                                  0.451
262.6     I
          I*********************                                0.493
266.7     I
          I**********************                               0.521
270.8     I
          I**************************                           0.592
274.9     I
          I*****************************                        0.634
279.0     I
          I************************************                 0.761
283.1     I
          I****************************************             0.845
287.2     I
          I*******************************************          0.930
291.3     I
          I********************************************         0.944
295.4     I
          I*********************************************1.000
299.5     I
          I*********************************************1.000
299.5     I----I----I----I----I----I----I----I----I----I----I MAX
          NO OBS-------------      71   STD ERROR-        26.31
          COEF OF VARIATION-      0.10   MEAN------       259.6
          KURTOSIS (BETA 2)-      1.45   MEDIAN----       267.6
          PEARSONIAN SKEW---    281.72   MODE------        .8403
```

FIGURE 7.4.   Path Cost for Node FUWINNER

lie within the first area of concentration between \$223.2 and \$327.2 million. The remaining 15% of the observations fall between \$418.2 and \$509.2 million. The path cost CFDs for terminal nodes FUWINNER, FIWINNER, and COWINNER (see figures 7.4, 7.5, and 7.6) indicate that the FUWIN-

```
          CFD  0.1  0.2  0.3  0.4  0.5  0.6  0.7  0.8  0.9  1.0
109.5     I----I----I----I----I----I----I----I----I----I----I MIN
          I                                                      0.0
109.5     I
          I                                                      0.003
110.4     I
          I                                                      0.006
111.3     I
          I                                                      0.006
112.2     I
          I                                                      0.009
113.2     I
          I*                                                     0.015
114.1     I
          I***                                                   0.060
115.0     I
          I****                                                  0.080
115.9     I
          I*******                                               0.131
116.8     I
          I**********                                            0.190
117.7     I
          I**************                                        0.292
118.6     I
          I******************                                    0.381
119.5     I
          I*********************                                 0.464
120.4     I
          I***************************                           0.586
121.3     I
          I*********************************                     0.693
122.2     I
          I************************************                  0.756
123.1     I
          I****************************************              0.824
124.0     I
          I********************************************          0.878
124.9     I
          I**********************************************        0.932
125.8     I
          I************************************************      0.970
126.7     I
          I**************************************************    0.982
127.6     I
          I***************************************************   0.988
128.5     I
          I****************************************************1.000
129.5     I
          I****************************************************1.000
129.5     I----I----I----I----I----I----I----I----I----I----I MAX
          NO OBS------------     336   STD ERROR-      3.435
          COEF OF VARIATION-     0.03  MEAN------      120.6
          KURTOSIS (BETA 2)-     2.82  MEDIAN----      120.5
          PEARSONIAN SKEW---   121.05  MODE------       .1334
```

FIGURE 7.5.   Path Cost for Node FIWINNER

NER alternative can be expected to cost considerably more than the
FIWINNER or COWINNER alternatives. The FUWINNER alternative
also has a much wider variance. These facts account for the shape of the
total project cost CFD as shown in figure 7.3.

```
          CFD  0.1   0.2   0.3   0.4   0.5   0.6   0.7   0.8   0.9   1.0
 90.26    I----I----I----I----I----I----I----I----I----I----I MIN
          I                                                           0.0
 90.26    I
          I*                                                          0.017
 91.31    I
          I****                                                       0.072
 92.36    I
          I******                                                     0.124
 93.40    I
          I*********                                                  0.183
 94.45    I
          I**************                                             0.274
 95.49    I
          I*****************                                          0.344
 96.54    I
          I************************                                   0.460
 97.58    I
          I***********************************                        0.579
 98.63    I
          I*****************************************                  0.671
 99.67    I
          I*************************************************          0.784
100.7     I
          I****************************************************       0.861
101.8     I
          I******************************************************     0.902
102.8     I
          I********************************************************   0.933
103.9     I
          I*********************************************************  0.967
104.9     I
          I********************************************************** 0.982
105.9     I
          I********************************************************** 0.989
107.0     I
          I*********************************************************0.993
108.0     I
          I*********************************************************0.994
109.1     I
          I*********************************************************0.996
110.1     I
          I*********************************************************0.998
111.2     I
          I*********************************************************0.998
112.2     I
          I*********************************************************1.000
113.3     I
          I*********************************************************1.000
113.3     I----I----I----I----I----I----I----I----I----I----I MAX
          NO OBS-------------    541   STD ERROR-          3.717
          COEF OF VARIATION-     0.04  MEAN------          97.97
          KURTOSIS (BETA 2)-     3.22  MEDIAN----          97.85
          PEARSONIAN SKEW---    97.65  MODE------          .8636E-01
```

FIGURE 7.6.   Path Cost for Node COWINNER

Figure 7.7 shows the cost likely to be incurred during selected time periods. In this problem, a five-year breakout of costs was desired for the twenty-year project in order that realistic budgets could be prepared for each of the four five-year periods. Many ventures and managers would, per-

```
         CFD  0.1  0.2  0.3  0.4  0.5  0.6  0.7  0.8  0.9  1.0
180.1    I----I----I----I----I----I----I----I----I----I----I MIN
         I                                                    0.005
185.0    I
         I                                                    0.009
186.2    I
         I*                                                   0.021
187.4    I
         I***                                                 0.052
188.5    I
         I*****                                               0.106
189.7    I
         I*********                                           0.175
190.9    I
         I*************                                       0.250
192.1    I
         I******************                                  0.354
193.3    I
         I***********************                             0.450
194.5    I
         I****************************                        0.557
195.6    I
         I*********************************                   0.654
196.8    I
         I************************************                0.742
198.0    I
         I****************************************            0.818
199.2    I
         I*******************************************         0.858
200.4    I
         I*********************************************       0.885
201.5    I
         I***********************************************     0.905
202.7    I
         I************************************************    0.921
203.9    I
         I**************************************************  0.936
205.1    I
         I*************************************************** 0.949
206.3    I
         I**************************************************** 0.967
207.5    I
         I***************************************************** 0.976
208.6    I
         I****************************************************** 0.985
209.8    I
         I******************************************************0.994
211.0    I
         I*******************************************************1.000
212.8    I----I----I----I----I----I----I----I----I----I----I MAX
         NO OBS-----------     1000    STD ERROR-        5.116
         COEF OF VARIATION-    0.03    MEAN------        195.5
         KURTOSIS (BETA 2)-    3.72    MEDIAN----        195.0
         PEARSONIAN SKEW---  195.07    MODE------        .8996E-01
```

FIGURE 7.7.   Positive Cost Incurred between periods 0–5

haps, prefer a shorter cost breakout—possibly yearly. This type of information puts the manager's budget requests on much more realistic, firm ground.

CONCLUSIONS

The large quantity of output information available from VERT-3 makes it possible for the manager to evaluate alternatives more in line with the criteria upon which he or she must make decisions. In addition, it is possible for the analyst to make an in-depth review of the model and sensitivity of the data structured within it. The addition of the coefficient of variation, Beta 2 measure of kurtosis, and Pearson's measure of skewness permits the analyst to compare distributions more precisely. The coefficient of variation is a dimensionless measure of dispersion (standard deviation/mean). Increasing values for the coefficient of variation indicate increasing dispersion. Pearson's measure of skewness ([mean-mode]/standard deviation) provides a comparison of central tendency. Absolute values for Pearson's statistic that are less than one imply that the distribution is not markedly skewed. The third comparison measure of the distribution provided in VERT is its peakedness or kurtosis. The Beta 2 measure of kurtosis (fourth moment about the mean/standard deviation) for the standard normal distribution is three.

VERT also has the capability of present-value analysis. The analyst inputs a time factor and the interest rate for the analysis. Another significant feature of VERT is the accompanying auxiliary programs. A redimensioning program is available which punches new program cards for the VERT source program. This allows the user to easily contract or expand the capability of VERT to fit a smaller core computer or a larger network application. A second program provides the user with a list of the VERT source program. Error checking within the system is greatly facilitated by the 146 error messages.

8 COMPUTER MECHANICS

This chapter describes the general computer processing aspects of using VERT-3, including required computer equipment, major program limitations, functions of the main routine and subroutines, and the program's use of an optional overlay structure approach. General information concerning auxiliary programs used to aid VERT-3 analysis, plus the major data storage arrays, is also included. The complete computer program listing is included in the appendix.

COMPUTER EQUIPMENT

The VERT-3 program was written for the IBM 360/65 computer in FORTRAN IV and is essentially a batch-oriented system. For conversion of the program to another computer, the user may be required to modify certain "read" statements, as well as the random number generator; the program should then readily compile and execute on any moderate-sized computer possessing a FORTRAN IV capability.

The program allows an optional overlay structure (described later in this chapter) that allows more efficient use of core storage for computer systems not having a virtual memory capability. Similar overlay capabilities on other equipment than IBM are also possible. Other required equipment includes the following (or their capability equivalents):

An eighty-column card reader (INPT);
An eighty-column card punch (IPNH);
A 132-character printer (IOUT);
Four peripheral sequential storage devices (IWF1, IWF2, IWF3, and IWF4) among which data are transferred (hence, they should be put on separate disc or tape drives).

MAIN AND SUBROUTINE FUNCTIONS

The program consists of a main routine and twenty-three subroutines. The primary function of the main routine is to sequentially call the major subroutines. The result is the most efficient use of the overlay structure. The twenty-three subroutines are described below.

Subroutine ERROR.   This subroutine lists an error whenever it occurs in an operating subroutine. Errors encountered while data is being read in call for an immediate dumping of the rest of the problem. The dumping is performed in the last part of this subroutine.

Subroutine SEEK.   This subroutine pulls data out of the general arc storage array.

Subroutine LOADF.   This subroutine initializes the program and reads in all the data except when a major error is encountered; in such a case, subroutine ERROR reads and lists the remainder of the problem. The program then reverts to the subroutine LOADF, which reads and stores the next sequential problem being loaded, providing no errors are encountered. LOADF uses four files to load the data: File IWF1 is the master problem file, and IWF2 is an intermediate file whose function consists of merging changes into the master problem while retaining the master file as it was prior to loading these changes. Thus, it is possible to make changes in the master problem without having to submit the entire problem. IWF3 is a transport file that carries a condensed version of the problem to an error check routine within this subroutine. This check consists of looking for alpha data in a numeric field. While this check is being made, the data is loaded on transport file IWF4, which has the main task of carrying the input data from subroutine LOADF to subroutine LOADA.

Subroutine CHECK.   This subroutine assists subroutine LOADF in checking for alpha data in numeric fields.

Subroutine LOADT.   This subroutine loads the tables for table lookup.

Subroutine LOADA.   This subroutine takes the arc data from the transport file IWF4 and loads it into core. The first part of this subroutine stores the addresses of the input and output nodes and checks to see if the arc currently being read in has any satellite arc cards. If any Monte Carlo, filter, or slack histogram satellite arc cards are present, they are the next items stored. If any time and/or cost and/or performance distribution, histogram, or mathematical relationship satellite arc cards are present, they are the last items stored, along with the dependency indicator. The dependency indicator tells the program in what order to process the time, cost, and performance inputs for a given arc.

Subroutine SEREAD.   This subroutine assists subroutine LOADA in reading data off transport file IWF4.

Subroutine LOADNR.   This subroutine loads node data into core storage, lists the arc and the node main storage arrays, and performs its main function of extensively checking all input data for possible errors. Some preliminary error checks have already been made in subroutines LOADA, LOADF, and SEREAD. However, these checks are mainly concerned with the tasks of immediate reading and storing of data. One can best understand the errors checked for in this subroutine by reading the list of error and warning messages generated by this subroutine. The front part of this subroutine reads the node data, the middle part lists the main storage arrays, and the last part checks for errors.

Subroutine DONET.   This subroutine simulates the network, determines the winning terminal node, finds the critical/optimum path, computes values for the slacks, gives a complete listing of all active arcs and nodes, and, lastly, computes internal node statistics.

Subroutine DOARC.   This subroutine generates each arc's primary time, cost, and performance values via stochastic and/or mathematical relationships. DOARC also determines the success or failure status for each arc.

Subroutine SVECT.   This subroutine stores arc cost and performance address vectors in the node data storage array.

Subroutine INITIAL.   This subroutine initializes the output arcs of the node currently being processed. Additionally, the output nodes of these arcs are given an initiated status that allows them to be reviewed by the subroutine DONET as possible candidates for processing.

Subroutine DVECT.   This subroutine has the function of developing a vector of arc addresses that is input to the development of the cumulative cost (first half of this subroutine) or performance (second half of this subroutine) attained by a given arc.

Subroutine HELPDV.   This subroutine assists subroutine DVECT in failing (giving an arc an unsuccessful completion state) and adjusting cost (first half of this subroutine) of arcs whose completion time exceeds their output node completion time.

Subroutine OCOST.   This subroutine calculates overall cost for terminal and internal nodes.

Subroutine RANDOM.   This subroutine generates uniform variates. This subroutine is currently set up for using two generators that operate on IBM 360 hardware. These generators are pseudo–random number generators—that is, they will produce the same sequence of random numbers upon being initialized with the same seed. The seed for these generators is carried in ISEED and is initially set in the control card. When this program is put on another machine, these generators will probably have to be replaced.

Subroutine NORM.   This subroutine develops a standard normal variate (a variable with mean 0 and standard deviation of 1).

Subroutine GAM.   This subroutine generates partial gamma variates.

Subroutine OUTFLO.   This subroutine directs the construction of histograms and bar charts. The first part of this subroutine, along with subroutine HISTO, develops slack histograms for the internal arcs and nodes. Time, cost, and performance histograms are developed for internal nodes, again with subroutine HISTO. The second part of this subroutine, along with the subroutines MS and HISTO, develops terminal node time, cost, and performance histograms. The third part of this subroutine, again along with subroutines MS and HISTO, develops the composite terminal node time, cost, and performance histograms. The subroutine next lists the terminal node realization index, which indicates the overall project risk (success-fail status), and then lists node and arc critical path indexes. Lastly, the subroutine lists the storage utilization analysis.

Subroutine MEDIAN. This subroutine computes the median of the sequence of observations that are sent to it by other subroutines.

Subroutine MS. This subroutine computes the mean, and the standard deviation and fills the histogram cells for the terminal nodes and the composite terminal mode.

Subroutine HISTO. This subroutine constructs all the time, cost, and performance histograms requested. This subroutine also develops cards or card images for terminal node time, cost, and performance histograms. These cards can then be inserted into higher-level networks.

Subroutine CORR. This subroutine calculates the correlation between time and/or cost and/or performance and makes two-dimensional plots for terminal nodes.

In addition to the above subroutines, six more (SUB1–SUB6) may be used to generate special transformations for particular applications at the option of the user. Such transformations are not contained within the program; they must be programmed by the user.

OVERLAY STRUCTURE

An overlay is a feature of a computer system's software that enables one to run a program with only parts of the program, rather than the whole program, in core at one time. It requires some additional time to roll segments of the program in and out of core memory; however, the overlay structure shown in figure 8.1 does not significantly impact the time required to process a network. The overlay feature is useful for computer systems without a virtual memory capability.

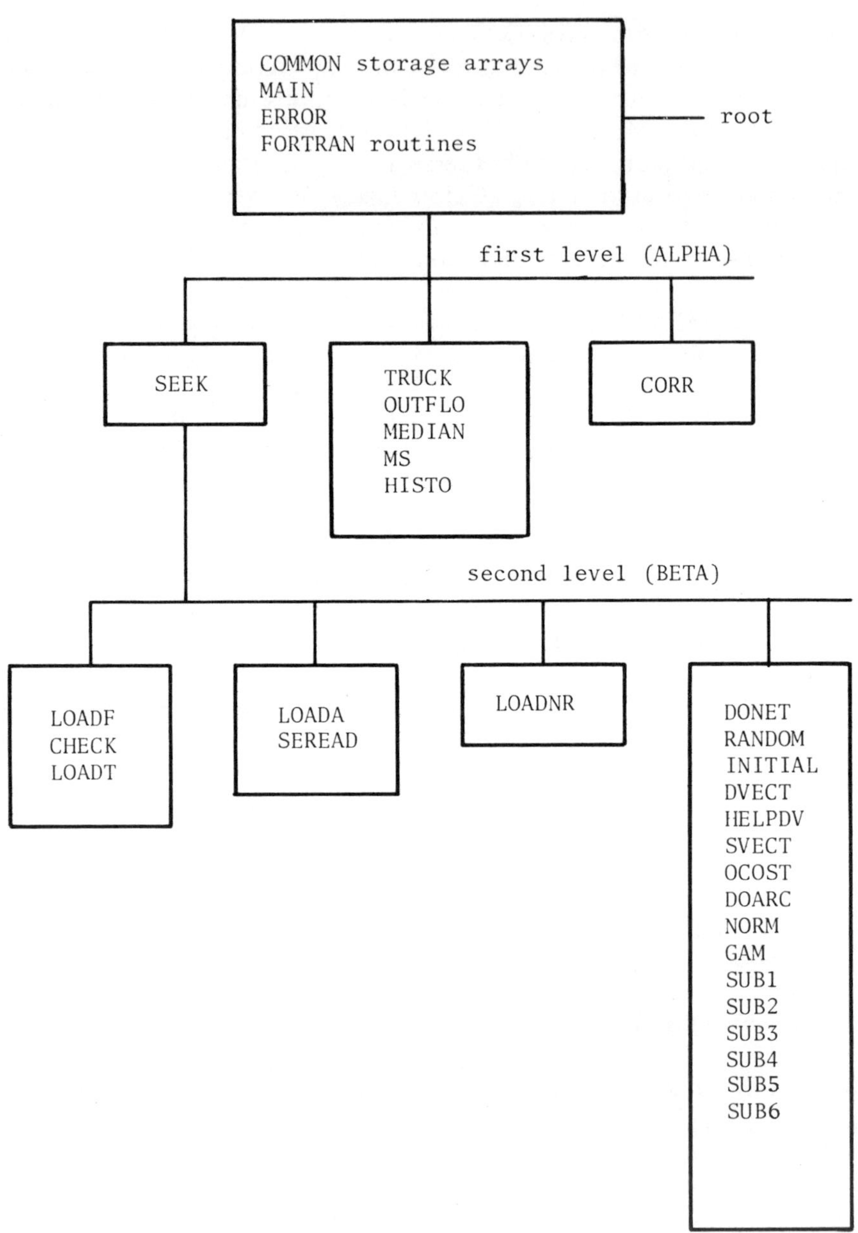

FIGURE 8.1.   VERT-3 Overlay Structure

DATA STORAGE ARRAYS

The function of this program is to simulate a VERT-3 network over many iterations. This function requires almost all the input data to be recalled from storage and to be used for each iteration. From a cost-of-computation point of view, this requirement nearly dictates the use of internal (core) storage rather than the slow-speed external files. Accordingly, most inputs are stored in one-dimensional arrays; a few are stored in two- and three-dimensional arrays. Because of computer core limitations, these data storage arrays may need to be redimensioned to accommodate different sizes of jobs. These dimensions are specified in the various blocks of the COMMON and the DIMENSION statements. A set of variables defined in subroutine LOADF of VERT-3 is used to conduct a check on the quantity of the various types of data entered in VERT-3. These checks insure that the boundaries of the storage arrays are not exceeded; the check will be done correctly only when the value assigned to each check variable is equal to the magnitude of the dimensioned storage area it is checking. The check variables and a set of their assigned values that would result in a requirement for 194 K bytes of storage are listed in table 8.1.

Seven blocks comprise the data storage arrays. Their functions are described below.

TRIALS: This block carries the optimum terminal node addresses, time, path cost, overall cost, and performance observations for each simulation iteration.

Table 8.1.   Check Variables and Assigned Values

Check Variables	*Values Assigned*
MITER	1000
MARC	350
LARC	2800
MNODE	200
LNODE	1800
MTAG	3600
MHIST	20
MTERM	10
MSLACK	20
MPCOST	10

ARCS: This block carries arc-related data and has two types of storage areas. All arcs carry a certain quantity of fixed information, regardless of the task they perform. This information is stored in certain variables comprising this block, as is the arc's variable information.

NODES: This block carries the node-related data. Like the arc's block, it has fixed and variable storage segments.

HIST: If internal node statistics have been requested, the variables in the common block labeled *HIST* store information needed for this task.

SLACK: If slack time statistics have been requested, the variables in the common block labeled *SLACK* store information needed for this task.

CPGAP: If the cost-performance time interval statistics have been requested, the variables in the common block labeled CPGAP are store information needed for this task.

MEDAN: If the median has been requested, the variables in the common block labeled MEDAN are assigned the job of storing the information needed for the calculation of the median.

VERT-3 gives a core storage analysis at the end of the run when output option D has been selected. This analysis will enable one to determine how well the various storage arrays are being utilized for their primary storage functions. Their secondary function of temporarily storing data at data read-in time is not included in this analysis since this task does not generally use much core.

III APPLICATIONS OF VERT-3

In part I we laid the background of network analysis through a discussion of the analytical functions performed by management, the concepts involved in managing significant new projects and ventures, and the major types of network models for management use. In part II we investigated in detail an advanced network technique—the Venture Evaluation and Review Technique, third generation (VERT-3)—through discussion of its general concepts, network construction and logic, input requirements, output reports, and computer mechanics.

It is often difficult to appreciate the advantages and power of a technique, as well as to understand fully its working and features, when dealing in generalities. For this reason we have carefully selected examples that illustrate the application of VERT-3 to a cross section of situations, including project management, multiple-criteria decisions, new-product development decisions, plant or facility location problems, evaluation of alternative energy sources, and analysis of strategic decisions relating to mergers and acquisitions. Each of the six chapters of part III illustrates an application of VERT-3 to one of these management decision problems.

9 PROJECT MANAGEMENT WITH VERT-3

As we saw in chapters 2 and 3, network planning and control techniques have found heavy application in project-type environments over the past twenty-five years. Traditionally, the most widely used networking techniques have been the deterministic PERT/CPM and related approaches, used largely to plan and control well-defined, nonrepetitive endeavors. Stochastic techniques such as GERT and VERT overcome many of the limitations of the deterministic systems by allowing the inclusion of probabilistic events and thus giving the manager or planner a more realistic picture of the likely effects of risk and uncertainty on a project's results, time, and cost. This ability has made the stochastic technique much more valuable to the manager/planner who must evaluate alternative project concepts and approaches.

Since project applications remain one of the major applications of stochastic networking techniques, our first example illustrates the use of VERT-3 in a simplified project setting.

THE Z-CAR PROJECT

The hypothetical example that follows is designed to illustrate an oversimplified project management application of VERT-3 that employs interrelated time, cost, and performance parameters.

An automobile manufacturer is faced with financial problems resulting from a sudden change in consumer preferences. The manufacturer has been advised that a new model in tune with customer tastes must be introduced in short order for the company to survive. The new model must be designed and tested within the extremely short time frame of 20 months and must meet a number of performance targets in order to gain public acceptance. In addition, the company cannot afford to spend more than $30 million on this project. The company and its lenders are vitally interested in learning the likelihood of the project's success *prior* to the commitment of additional resources.

Project Management

The project will consist of three major phases: design, component testing, and system test. In the design phase, three major efforts are required: chassis design, engine design, and drive train design. The chassis design work includes the car's body and related equipment but does not require all design details and refinements to be completed prior to the car's initial system test. The engine design work consists of modifying and resizing existing engines. The drive train design work includes the transmission and running gear appropriate for the size of car and of engine utilized. Project engineers have developed most likely time and cost estimates for each of the major activities involved in the project (see figure 9.1). In many instances performance can be improved if more time is spent on these activities—particularly design work—but increased design time will add to cost. In addition, "crashing" certain activities to reduce their time requirements below a certain level may actually increase costs because of overtime, subcontracting, and so forth. Based upon the most likely estimates of time and cost, the project should be completed in 18 months at a cost of $30.5 million. While the time requirement is within the 20-month time frame, the cost exceeds the limitation by $500,000. Can an acceptable cost/time trade-off be imple-

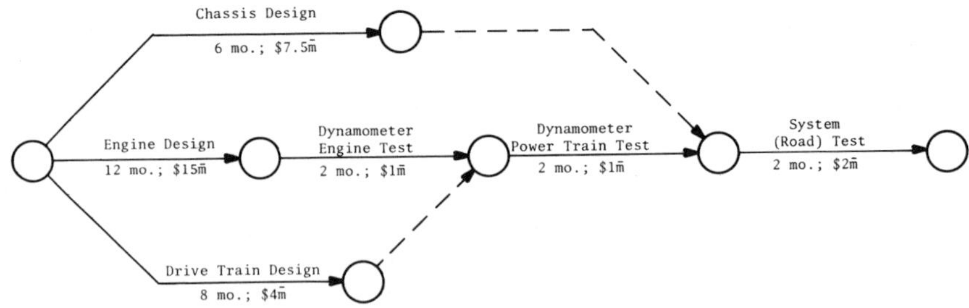

FIGURE 9.1.   Z-Car Development Network

mented? How certain are we with regard to these figures? And, more importantly, what is the likelihood of meeting the performance requirements? If the performance requirements are not met, little consolation will be found in the fact that the 20-month and $30 million limits were not exceeded.

Engineering Relationships

Knowledgeable technical people have developed time, cost, and performance relationships for each of the major activities of the project. Figures 9.2a through 9.2d show the graphical relationships, equations, design goals, and absolute requirements (if any) for important parameters at each major activity.

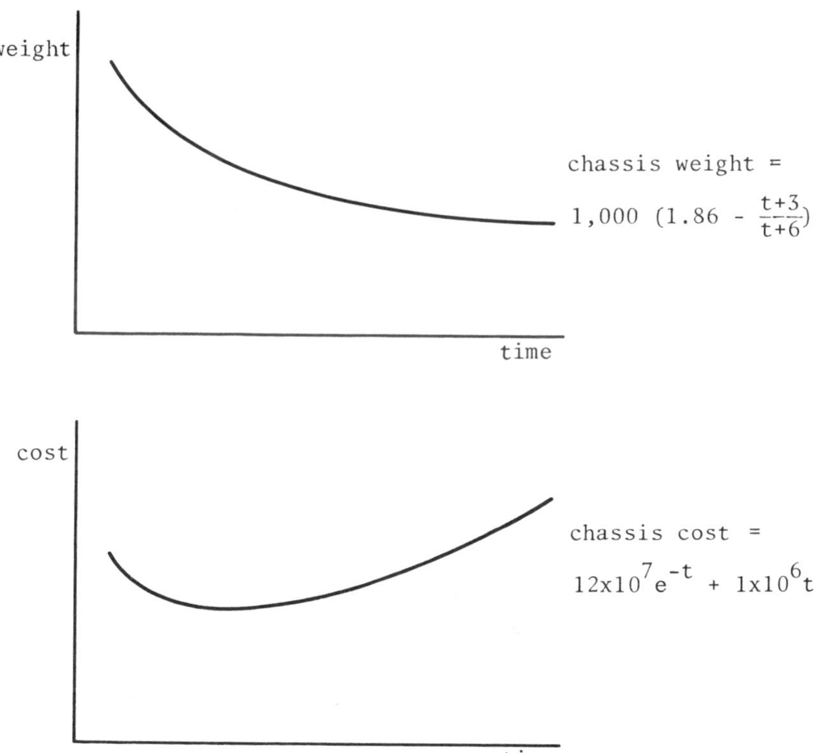

FIGURE 9.2a.   Chassis Design. Weight and cost are each a function of design time in months. There is a weight goal of 1,000 lbs.

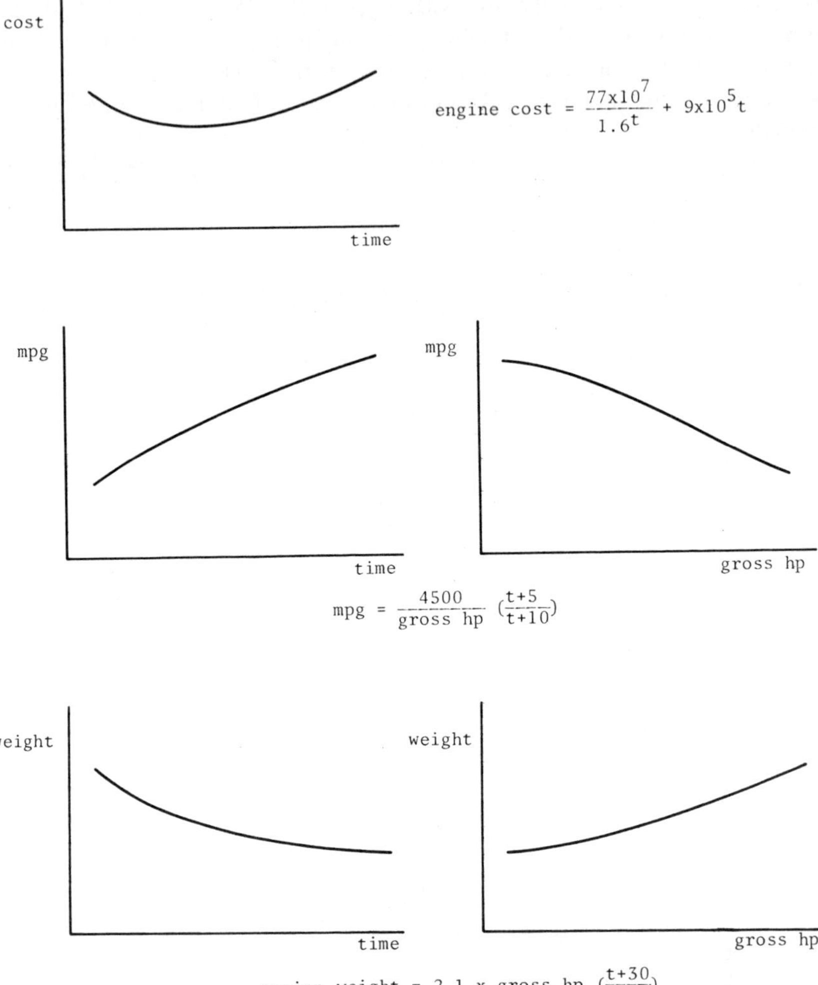

FIGURE 9.2*b*.   Engine Design. Engine weight is a function of time and gross horsepower. A goal of 500 lbs. and 100 gross hp exists. Cost is a function of design time, and economy is a function of design time and gross hp. A goal of 40 mpg exists.

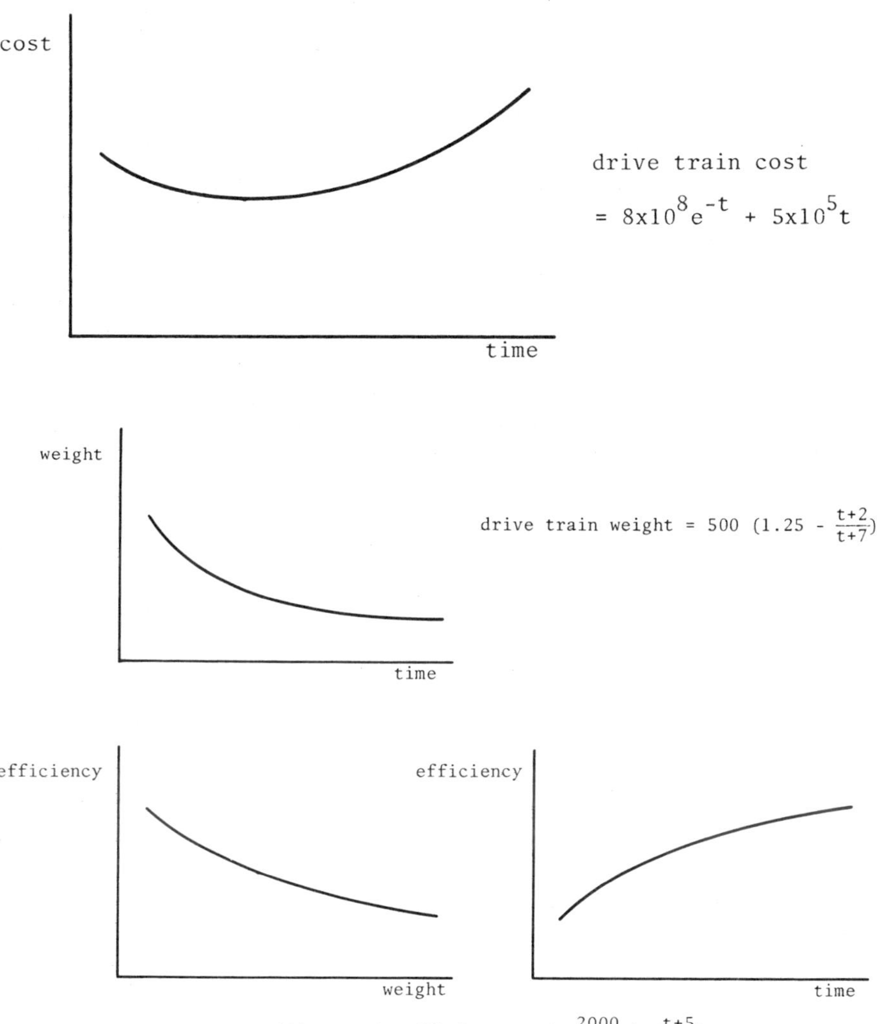

FIGURE 9.2c.   Drive Train Design. Drive train design cost is a function of design time, as is its weight, which has a design goal of 300 lbs. Drive train efficiency is a function of its weight and design time.

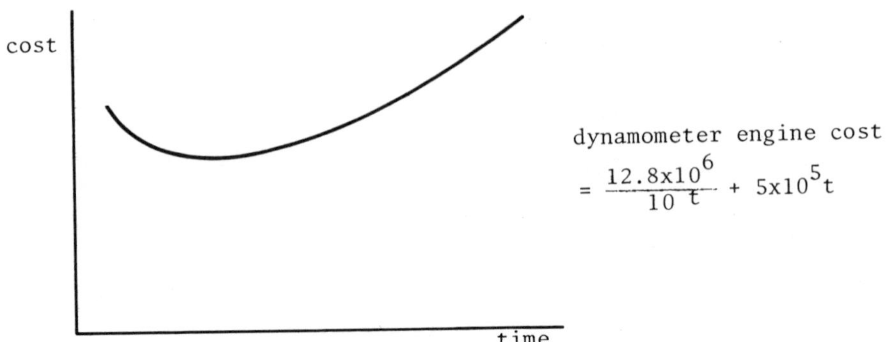

$$= \frac{12.8 \times 10^6}{10\,t} + 5 \times 10^5 t$$

FIGURE 9.2*d*.   Dynamometer Engine and Power Train Test. The cost of the dynamometer engine test is a function of time incurred. Goals of 40 mpg and 100 gross hp exist. The power train consists of the engine plus the drive train. The net horsepower indicated by this test is the product of the engine gross hp times drive train efficiency, and a goal of 95 net hp exists for this test. The cost of this test is a function of time, identical to the dynamometer engine test cost equation.

System (Road) Test.   At this point the entire car (chassis and power train) is tested as a unit under actual road conditions. Top speed is a function of total weight (chassis plus power train) and net horsepower. Acceleration is also a function of these two factors, as is economy. The stability of the vehicle is an index number dependent upon total weight, and the cost of the test is a function of the time consumed. While performance goals exist for speed, acceleration, economy, and stability, required values also exist. Not meeting any one of the required values will result in the project's being judged a failure, just as will exceeding the 20-month time frame and the $30 million budget limitation. The performance goals and requirements are shown below. The relationships are shown in Figure 9.3.

Measure	Goal	Requirement
Speed	75 mph	75 mph
Acceleration	10 sec.	13 sec.
Economy	40 mpg	35 mpg
Stability	0.75	0.65

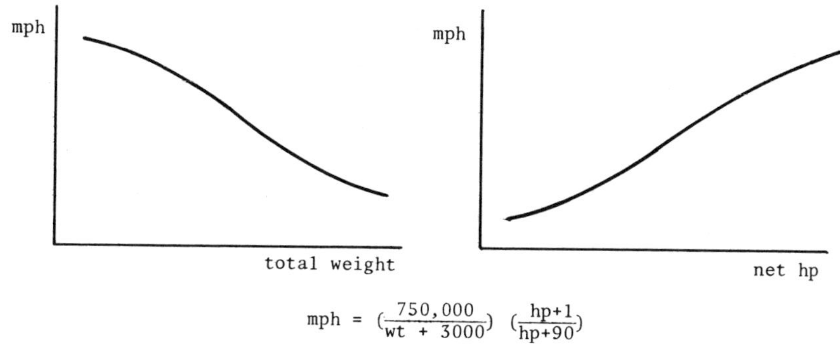

$$mph = \left(\frac{750,000}{wt + 3000}\right) \left(\frac{hp+1}{hp+90}\right)$$

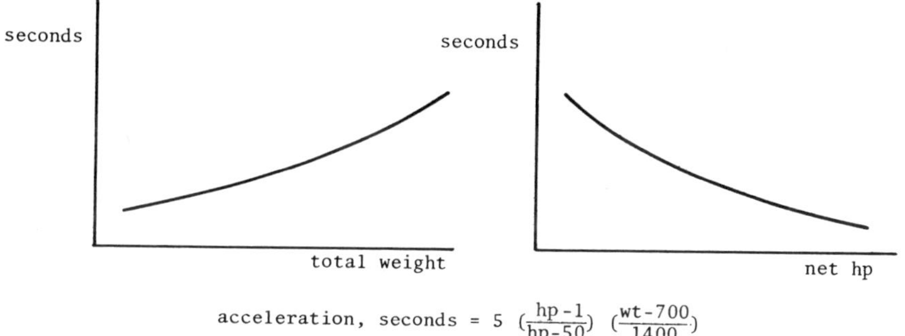

$$\text{acceleration, seconds} = 5 \left(\frac{hp-1}{hp-50}\right) \left(\frac{wt-700}{1400}\right)$$

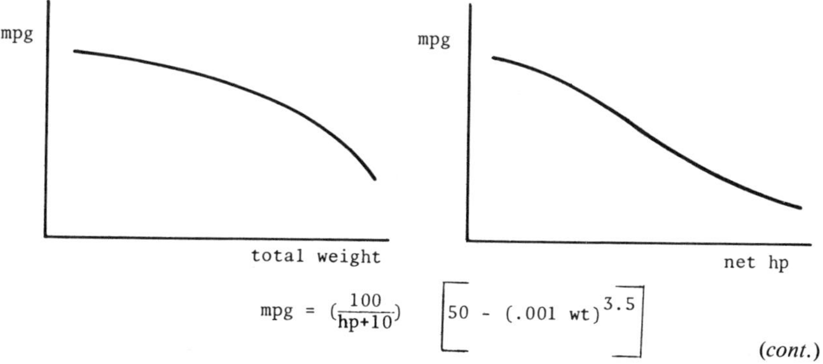

$$mpg = \left(\frac{100}{hp+10}\right) \left[50 - (.001\ wt)^{3.5}\right]$$

(*cont.*)

FIGURE 9.3. System (Road) Test

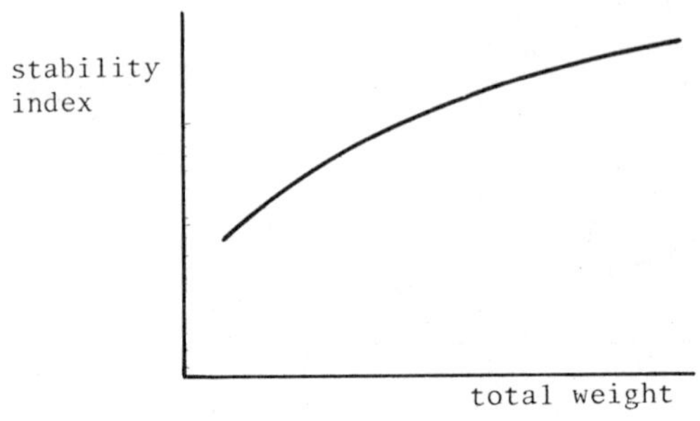

$$\text{index} = \frac{wt-1320}{wt-1200}$$

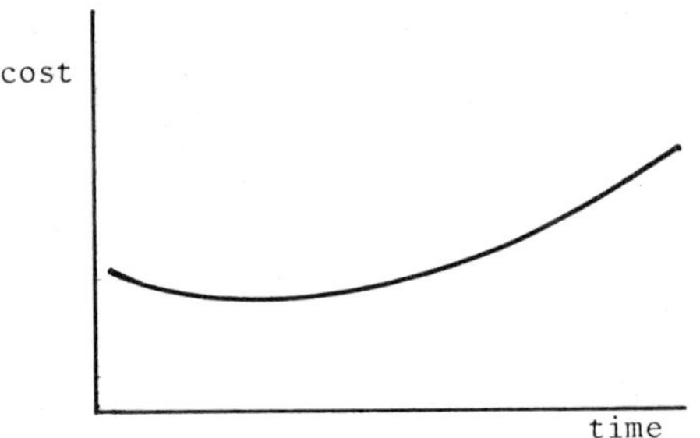

$$\text{cost} = \frac{9x10^6}{6t} + 8x10^5 t$$

FIGURE 9.3.   (Cont.)

VERT-3 MODEL

The flow diagram for the Z-car project is shown in figure 9.4. This diagram represents the logical flow of activities and decisions that result in an ultimate pass/fail set of outcomes. If the project fails any one of the performance, time, or cost criteria, the project will be judged a failure; only if it passes on all counts will it be judged a success. Obviously, the decisionmakers are interested in learning the risk of failure: What is the probability of not meeting all of the requirements? Here the advantage of a stochastic networking technique, (VERT-3, in particular) becomes evident. Not only has the decisionmaker been able to include stochastic information concerning three types of parameters (time, cost, and performance); he or she has been able to include their interrelationships and interactions. Furthermore, the decisionmaker will receive output showing the percentage of time the simulation runs resulted in a successful outcome or in a failure. Thus, the decisionmaker has an advance indication of the "odds" for success and failure.

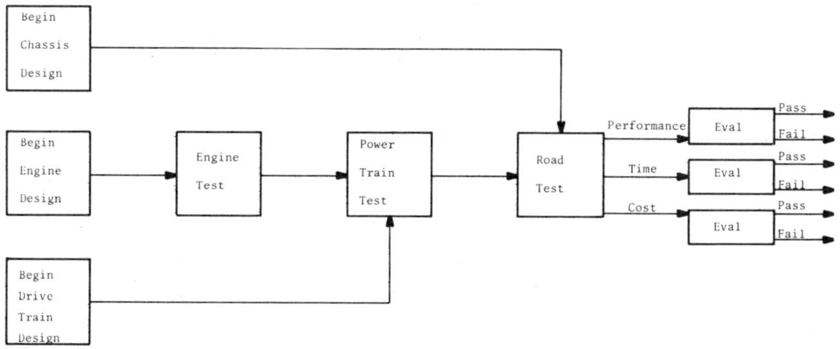

FIGURE 9.4.   Z-Car Project Flow Diagram

VERT-3 Network

The next step is the creation of the **VERT-3** network through use of the appropriate arcs and nodes, including input and output logic, to simulate the desired flow and parametric relationships. The network is shown in figure 9.5. The nodes appearing on the network function as follows:

START: acts as the beginning node for each iteration of the network flow;

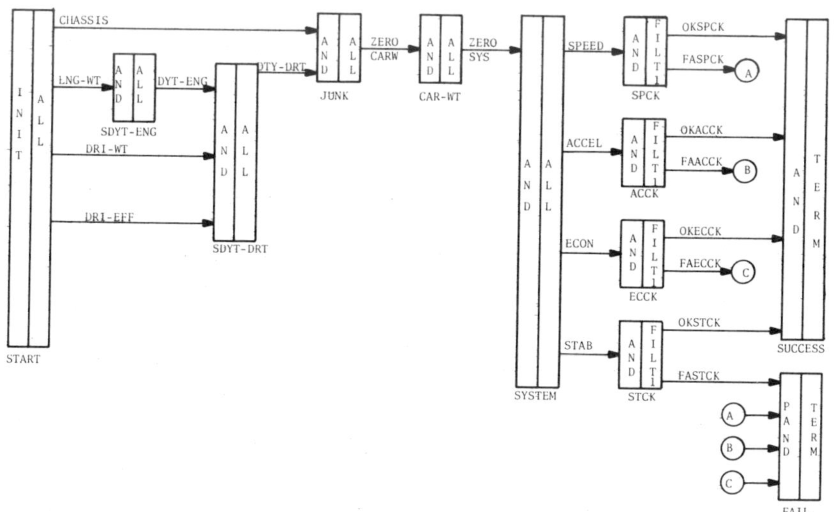

FIGURE 9.5.   Z-Car Vert-3 Network

SYDT-ENG: separates the engine design arc from the engine dynamometer test arc;

SYDT-DRT: initiates dynamometer test of the drive train, which follows engine test and drive train design;

JUNK: carries an accumulation of unrelated performance characteristics (however, the time and cost values carried by this node represent the total incurred to that point by the engine, chassis, and drive train design plus the engine and power train dynamometer tests);

CAR-WT: carries the cumulative time, cost, performance, and total car weight generated on arc ZERO CARW;

SYSTEM: carries the total time and cost incurred on the car development; also starts off the development of speed, acceleration, economy, and stabilization performance calculations;

SPCK: performs a check on the time and cost values accumulated to this point, as well as on results of the speed test; filters out to the failure node (FAIL) flows over 20 months, greater than $30 million, and slower than 75 miles per hour;

ACCK: checks acceleration performance; filters out to the failure node (FAIL) flows whose acceleration time exceeds 13 seconds:

ECCK: checks economy; filters out to the failure node (FAIL) flows with less than 35 miles per gallon;

STCK: checks stability; filters out to the failure node (FAIL) flows with a stability index of less than 0.65;

SUCCESS: catches all completions not routed to the failure node;

FAIL: catches all completions filtered out by any one of the four check nodes.

The arcs appearing on the network function as follows:

GROSSHP: functions as a free arc not explicitly appearing on the network; generates gross horsepower values;

CHASSIS: generates the weight of the chassis plus the time and cost incurred in chassis design;

ENG-WT: generates the weight of the engine plus the time and cost incurred in engine design;

DRI-WT: generates the weight of the drive train plus the time and cost incurred in drive train design;

DRI-EFF: generates the drive train efficiency (note that it uses the drive train weight generated on the previous arc);

DYT-ENG: generates the time and cost incurred in conducting the engine dynamometer test;

DYT-DRT: generates the time and cost incurred in conducting the drive train dynamometer test; generates the net horsepower;

ZEROCARW: zeroes out the unrelated performance characteristics carried on node JUNK and then generates in the performance field the total weight of the car by reaching back to arcs CHASSIS, ENG-WT, and DRI-WT to add together the weight of the chassis, engine, and drive train: Note, however, that this performance value (total car weight) is carried in the secondary (cumulative) field of the performance vector, while the primary performance field carries the numeric value of the performance of node JUNK plus total car weight; when referencing an arc in a math relationship, pick up the primary value, not the secondary or cumulative;

ZEROSYS: zeroes out the car weight in the performance field and generates the time and cost incurred in conducting the systems test;

SPEED: generates the car speed;

ACCEL: generates the car acceleration;

ECON: generates the car economy (miles per gallon);

STAB: generates the car stability;

OKSPCK: receives the acceptable flows from node SPCK;

FASPCK: receives the unacceptable flows from node SPCK;

OKACCK: receives the acceptable flows from node ACCK;

FAACCK: receives the unacceptable flows from node ACCK:

OKECCK: receives the acceptable flows from node ECCK;

FAECCK: receives the unacceptable flows from node ECCK;

OKSTCK: receives the acceptable flows from node STCK;

FASTCK: receives the unacceptable flows from node STCK.

Data Input.   Table 9.1 lists the data input required to process the Z-car problem. Note the sequence of the input data cards as described in chapter 6.

Table 9.1.   Z-Car Input Data

```
PROBLEM IDENTIFICATION CARD OPTION------------------------------        1

TYPE OF INPUT OPTION------------------------------------------        0

TYPE OF OUTPUT OPTION-----------------------------------------        4

COSTING AND PRUNING OPTION------------------------------------        2

FULL PRINT TRIP OPTION----------------------------------------        0

CORRELATION COMPUTATION AND PLOT OPTION--------------------        0

COST-PERFORMANCE TIME INTERVAL OPTION----------------------        0

COMPOSITE TERMINAL NODE MINIMUMS AND MAXIMUMS OPTION-------        0 (cont.)
```

Table 9.1.   (Cont.)

```
NUMBER OF ITERATIONS-------------------------------------------------    500

YEARLY INTEREST RATE USED FOR INFLATION ADJUSTMENTS--------         0.00

YEARLY INTEREST RATE USED FOR PRESENT VALUE DISCOUNTING----         0.00

TIME FACTOR WHICH CONVERTS PROGRAM TIME TO A YEARLY BASE---         0.00

                                          TIME      COST      PERF

TERMINAL NODE SELECTION WEIGHTS           0.00      0.00      1.00

CRITICAL - OPTIMUM PATH WEIGHTS           1.00      0.00      0.00

INITIAL VALUES                            0.00      0.00      0.00

GROSSHP NOFLOW  DATAGEN  1.0 GROSS HORSEPOWER
GROSSHP DPERF 1   2.0         75.0       130.0

CHASSIS START    JUNK       1.0 CHASSIS DESIGN
CHASSIS DTIME 1   4.0         5.25        9.75        7.5         0.75
CHASSIS RCOST 1130K   1.0    K 2.71828TCHASSIS  2SK120000.0K  1000.0      1.0
CHASSIS RCOST 2 1SK1000000.TCHASSIS K   1.0
CHASSIS RPERF 1 50TCHASSIS K 3.0     K   0.0      50TCHASSIS K   6.0    K   0.0
CHASSIS RPERF 2 20   1.0     K 1.0       2.0      10SK  1000.0K  1.86       3.0

ENG-WT  START    SDYT-ENG 1.0 ENGINE DESIGN WEIGHT AND COST CALCULATIONS
ENG-WT  DTIME 1   4.0         8.4        15.6        12.0        1.2
ENG-WT  RCOST 1130K   1.0    K   1.6    TENG-WT    2SK770000.0K  1000.0      1.0
ENG-WT  RCOST 2 1SK900000.0TENG-WT   K   1.0
ENG-WT  RPERF 1 50TENG-WT   K  30.0    K   0.0      50TENG-WT   K   3.0    K   0.0
ENG-WT  RPERF 2 20   1.0     K   1.0       2.0      1SK   2.1    PGROSSHP     3.0

DRI-WT  START    SDYT-DRT 1.0 DRIVE TRAIN WEIGHT AND COST CALCULATIONS
DRI-WT  DTIME 1   4.0         5.6        10.4        8.0         0.8
DRI-WT  RCOST 1 10K  800.0 K  1000.0K   1000.0 10TDRI-WT   K  -1.0    K   1.0
DRI-WT  RCOST 213S   1.0     K 2.71828    2.0
DRI-WT  RPERF 1 50TDRI-WT   K   2.0    K   0.0      50TDRI-WT   K   7.0    K   0.0
DRI-WT  RPERF 2 20   1.0     K   1.0       2.0      10SK  100.0 K  1.25       3.0

DRI-EFF START    SDYT-DRT 1.0 DRIVE TRAIN EFFICIENCY CALCULATIONS
DRI-EFF RPERF 1 50TDRI-WT   K   5.0    K   0.0      50TDRI-WT   K   6.0    K   0.0
DRI-EFF RPERF 2 20   1.0     K   1.0       2.0      50PDRI-WT   K  1700.0K   0.0
DRI-EFF RPERF 3 2SK  2000.0    3.0       4.0

DYT-ENG SDYT-ENGSDYT-DRT 1.0 DYNAMOMETER TEST ENGINE COST CALCULATIONS
DYT-ENG DTIME 1   4.6         1.4         2.6        2.0         0.2
DYT-ENG RCOST 1130K   1.0    K  10.0    TDYT-ENG   2SK 12800.0K 1000.0      1.0
DYT-ENG RCOST 2 1SK500000.0TDYT-ENG K   1.0

DYT-DRT SDYT-DRTJUNK       1.0 DYNAMOMETER TEST ON DRIVE TRAIN NET HP & COST CAL
DYT-DRT DTIME 1   4.0         1.4         2.5        2.0         0.2
DYT-DRT RCOST 1130K   1.0    K  10.0    TDYT-DRT   2SK 12800.0K 1000.0      1.0
DYT-DRT RCOST 2 1SK500000.0TDYT-DRT K   1.0
DYT-DRT RPERF 1 1SPGROSSHP PDRI-EFF K   1.0

ZEROCARWJUNK        CAR-WT   1.0 ZERO OUT AND CALCULATE TOTAL CAR WEIGHT
ZEROCARWRPERF 1 1SPJUNK      K  -1.0    K   1.0      5SPCHASSIS PENG-WT   PDRI-WT

ZEROSYS CAR-WT   SYSTEM   1.0 ZERO OUT CAR WT & COMPUTE SYS TEST TIME & COST
ZEROSYS DTIME 1   4.0         1.4         2.6        2.0         0.2
ZEROSYS RCOST 1130K   1.0    K   6.0    TZEROSYS   2SK  9000.0K  1000.0     1.0
ZEROSYS RCOST 2 1SK  800.0 K  1000.0TZEROSYS
ZEROSYS RPERF 1 1SPCAR-WT   K  -1.0    K   1.0

SPEED     SYSTEM  SPCK      1.0 SYSTEM SPEED CALCULATIONS
SPEED     RPERF 1 50PDYT-DRT K   1.0    K   0.0      50PDYT-DRT K   90.0   K   0.0
SPEED     RPERF 2 20  1.0    K   1.0       2.0      50PCAR-WT   K  3000.0K   0.0
SPEED     RPERF 3 2SK750000.0    3.0       4.0

ACCEL     SYSTEM  ACCK      1.0 SYSTEM ACCELERATION CALCULATIONS
ACCEL     RPERF 1 50PDYT-DRT K  -1.0    K   0.0      50PDYT-DRT K  -50.0   K   0.0
```

Table 9.1.   (Cont.)

```
ACCEL      RPERF 2 20     1.0     K    1.0       2.0     300PCAR-WT   K   700.0 K   1400.0
ACCEL      RPERF 3 1SK    5.0          3.0       4.0

ECON       SYSTEM   ECCK       1.0 SYSTEM ECONOMY (MPG) CALCULATIONS
ECON       RPERF 1110K   100.0 PDYT-DRT K   10.0    20PDYT-DRT K   1.0   K   1000.0
ECON       RPERF 2130K    1.0         2.0    K  3.5   10S   1.0    K   50.0      3.0

STAB       SYSTEM   STCK       1.0 SYSTEM STABILITY CALCULATIONS
STAB       RPERF 1 7OPCAR-WT   K   1320.0K   0.0       7OPCAR-WT   K   1200.0K   0.0
STAB       RPERF 2 2S     1.0     K    1.0       2.0

OKSPCK     SPCK     SUCCESS   1.0 SPEED CHECKS OUT OK
OKSPCK     FILT1 1          0.0        20.0        0.030000000.0      75.099999999.0

OKACCK     ACCK     SUCCESS   1.0 ACCELERATION CHECKS OUT OK
OKACCK     FILT1 1                                   6.099999999.0

OKECCK     ECCK     SUCCESS   1.0 ECONOMY (MPG) CHECKS OUT OK
OKECCK     FILT1 1                                  35.099999999.0

OKSTCK     STCK     SUCCESS   1.0 STABILITY CHECKS OUT OK
OKSTCK     FILT1 1                                   0.65        1.0

FASPCK     SPCK     FAIL     1.0 FAIL SPEED CHECK

FAACCK     ACCK     FAIL     1.0 FAIL ACCELERATION CHECK

FAECCK     ECCK     FAIL     1.0 FAIL ECONOMY (MPG) CHECK

FASTCK     STCK     FAIL     1.0 FAIL STABILITY CHECK

ENDARC

START    1   2                     START NETWORK

SDYT-ENG2   2                      ENGINE DEVELOPMENT COMPLETE, START ENGINE DYNO TEST

SDYT-DRT2   2                      DRIVE TRAIN COMPLETE, START DRIVE TRAIN DYNO TEST

JUNK     2   2                     ACCUMULATION OF PERFORMANCE JUNK

CAR-WT   2   2  4  1               TOTAL CAR WEIGHT

SYSTEM   2   2  6  2               START SYSTEMS CALCULATIONS

SPCK     2   4  4  3               SPEED CHECK

ACCK     2   4  4  4               ACCELERATION CHECK

ECCK     2   4  4  5               ECONOMY (MPG) CHECK

STCK     2   4  4  6               STABILITY CHECK

SUCCESS  2   115                   SUCCESS SINK

FAIL     3   115-l                 FAILURE SINK

ENDNODE
```

Output Reports.   The selected output reports reflecting the 500 simulation
iterations of the input data on the VERT network are shown in figures 9.6
through 9.13. Figure 9.6, depicting the network time for node SYSTEM,
shows that the mean completion time for the project was approximately 17.9
months, slightly below the 18 months predicted by a PERT approach and
significantly below the 20-month time period specified for project fail-

```
          CFD   0.1   0.2   0.3   0.4   0.5   0.6   0.7   0.8   0.9   1.0
14.9195 I----I----I----I----I----I----I----I----I----I----I MIN
        I                                                             0.000
14.9195 I
        I                                                             0.006
15.1918 I
        I*                                                            0.028
15.4640 I
        I**                                                           0.042
15.7363 I
        I****                                                         0.072
16.0086 I
        I******                                                       0.120
16.2809 I
        I********                                                     0.158
16.5532 I
        I**********                                                   0.198
16.8255 I
        I************                                                 0.250
17.0978 I
        I*****************                                            0.336
17.3701 I
        I**********************                                       0.420
17.6424 I
        I***************************                                  0.492
17.9146 I
        I********************************                             0.594
18.1869 I
        I***************************************                      0.682
18.4592 I
        I******************************************                   0.750
18.7315 I
        I**********************************************               0.810
19.0038 I
        I***************************************************          0.874
19.2761 I
        I******************************************************       0.904
19.5484 I
        I*********************************************************     0.934
19.8207 I
        I***********************************************************   0.958
20.0929 I
        I************************************************************* 0.978
20.3652 I
        I**************************************************************0.994
20.6375 I
        I**************************************************************1.000
20.9098 I
        I**************************************************************1.000
20.9098 I----I----I----I----I----I----I----I----I----I----I MAX

          NO OBS------------       500   STD ERROR-            1.2344
          COEF OF VARIATION-      0.07   MEAN------          17.8935
          KURTOSIS (BETA 2)-      2.59   MEDIAN----          17.9345
          PEARSONIAN SKEW---      0.17   MODE------          18.1003
```

FIGURE 9.6.   Network Time for Node SYSTEM

```
               CFD   0.1   0.2   0.3   0.4   0.5   0.6   0.7   0.8   0.9   1.0
23858923.9264  I----I----I----I----I----I----I----I----I----I----I----I MIN
               I                                                          0.000
23858923.9264  I
               I                                                          0.010
24179020.9813  I
               I**                                                        0.032
24499118.0363  I
               I****                                                      0.082
24819215.0913  I
               I********                                                  0.180
25139312.1463  I
               I************                                              0.280
25459409.2013  I
               I******************                                        0.410
25779506.2562  I
               I************************                                  0.518
26099603.3112  I
               I*******************************                           0.636
26419700.3662  I
               I*************************************                     0.730
26739797.4212  I
               I******************************************                0.818
27059894.4762  I
               I**********************************************            0.864
27379991.5312  I
               I***********************************************           0.896
27700088.5861  I
               I************************************************          0.914
28020185.6411  I
               I*************************************************         0.936
28340282.6961  I
               I**************************************************        0.948
28660379.7511  I
               I***************************************************       0.960
28980476.8061  I
               I****************************************************      0.970
29300573.8610  I
               I*****************************************************     0.978
29620670.9160  I
               I******************************************************    0.986
29940767.9710  I
               I*******************************************************   0.994
30260865.0260  I
               I********************************************************  0.998
30580962.0810  I
               I*********************************************************1.000
30901059.1360  I
               I*********************************************************1.000
30901059.1360  I----I----I----I----I----I----I----I----I----I----I----I MAX

          NO OBS------------       500   STD ERROR-      1214722.9356
          COEF OF VARIATION-      0.05   MEAN------     26218741.4828
          KURTOSIS (BETA 2)-      4.42   MEDIAN----     26048202.2264
          PEARSONIAN SKEW---      0.47   MODE------     25644080.5791
```

FIGURE 9.7.   Overall Cost for Node SYSTEM

ure. In fact, there appears to be approximately a 94% probability of completion in 20 months or less, or a 6% chance of failure on account of the time constraint.

```
              CFD   0.1   0.2   0.3   0.4   0.5   0.6   0.7   0.8   0.9   1.0
1538.8763 I----I----I----I----I----I----I----I----I----I----I MIN
          I                                                            0.000
1538.8763 I
          I*                                                           0.014
1560.8422 I
          I**                                                          0.040
1582.8082 I
          I*****                                                       0.100
1604.7742 I
          I*******                                                     0.152
1626.7401 I
          I**********                                                  0.214
1648.7061 I
          I***************                                             0.298
1670.6721 I
          I*******************                                         0.366
1692.6380 I
          I*********************                                       0.406
1714.6040 I
          I************************                                    0.462
1736.5700 I
          I****************************                                0.524
1758.5359 I
          I********************************                            0.604
1780.5019 I
          I************************************                        0.662
1802.4679 I
          I****************************************                    0.734
1824.4338 I
          I*********************************************               0.800
1846.3998 I
          I**********************************************              0.862
1868.3658 I
          I*************************************************           0.908
1890.3317 I
          I***************************************************         0.948
1912.2977 I
          I****************************************************        0.964
1934.2637 I
          I******************************************************      0.986
1956.2297 I
          I*******************************************************0.992
1978.1956 I
          I********************************************************0.996
2000.1616 I
          I*********************************************************1.000
2022.1276 I
          I*********************************************************1.000
2022.1276 I----I----I----I----I----I----I----I----I----I----I MAX


          NO OBS------------       500   STD ERROR-        105.3647
          COEF OF VARIATION-      0.06   MEAN------       1747.0637
          KURTOSIS (BETA 2)-      2.09   MEDIAN----       1751.6683
          PEARSONIAN SKEW---      0.81   MODE------       1661.4232
```

FIGURE 9.8.   Path Performance for Node CAR-WT

```
      CFD   0.1   0.2   0.3   0.4   0.5   0.6   0.7   0.8   0.9   1.0
76.9650 I----I----I----I----I----I----I----I----I----I----I MIN
        I                                                      0.000
76.9650 I
        I*                                                     0.018
77.7329 I
        I***                                                   0.062
78.5008 I
        I*****                                                 0.098
79.2687 I
        I*******                                               0.152
80.0366 I
        I********                                              0.184
80.8045 I
        I**********                                            0.204
81.5724 I
        I************                                          0.238
82.3403 I
        I**************                                        0.278
83.1082 I
        I*****************                                     0.336
83.8761 I
        I*******************                                   0.374
84.6440 I
        I*********************                                 0.418
85.4119 I
        I***********************                               0.462
86.1798 I
        I************************                              0.490
86.9477 I
        I***************************                           0.540
87.7156 I
        I*****************************                         0.584
88.4835 I
        I********************************                      0.642
89.2515 I
        I************************************                  0.706
90.0194 I
        I******************************************            0.790
90.7873 I
        I**********************************************        0.862
91.5552 I
        I**************************************************    0.940
92.3231 I
        I**************************************************** 0.976
93.0910 I
        I****************************************************1.000
93.8589 I
        I****************************************************1.000
93.8589 I----I----I----I----I----I----I----I----I----I----I MAX

        NO OBS-------------      500    STD ERROR-              4.6750
        COEF OF VARIATION-       0.05   MEAN------             86.2976
        KURTOSIS (BETA 2)-       1.84   MEDIAN----             87.0523
        PEARSONIAN SKEW---       0.90   MODE------             90.4993
```

FIGURE 9.9. Path Performance for Node SPCK

```
            CFD  0.1   0.2   0.3   0.4   0.5   0.6   0.7   0.8   0.9   1.0
6.3127  I----I----I----I----I----I----I----I----I----I----I MIN
        I                                                        0.000
6.3127  I
        I*                                                       0.020
6.4336  I
        I****                                                    0.078
6.5546  I
        I*********                                               0.194
6.6755  I
        I**************                                          0.304
6.7965  I
        I*******************                                     0.428
6.9175  I
        I**************************                              0.546
7.0384  I
        I*******************************                         0.636
7.1594  I
        I*************************************                    0.696
7.2803  I
        I********************************************            0.740
7.4013  I
        I***********************************************         0.778
7.5223  I
        I*************************************************       0.810
7.6432  I
        I***************************************************     0.842
7.7642  I
        I******************************************************  0.874
7.8851  I
        I****************************************************** *0.898
8.0061  I
        I*******************************************************0.924
8.1271  I
        I********************************************************0.948
8.2480  I
        I*********************************************************0.960
8.3690  I
        I**********************************************************0.976
8.4899  I
        I***********************************************************0.992
8.6109  I
        I************************************************************0.994
8.7319  I
        I*************************************************************0.998
8.8528  I
        I**************************************************************1.000
8.9738  I
        I**************************************************************1.000
8.9738  I----I----I----I----I----I----I----I----I----I----I MAX


        NO OBS------------      500   STD ERROR-         0.5490
        COEF OF VARIATION-     0.08   MEAN------         7.1403
        KURTOSIS (BETA 2)-     3.22   MEDIAN----         7.0022
        PEARSONIAN SKEW---     0.47   MODE------         6.8812
```

FIGURE 9.10.   Path Performance for Node ACCK

```
           CFD  0.1   0.2   0.3   0.4   0.5   0.6   0.7   0.8   0.9   1.0
33.9284  I----I----I----I----I----I----I----I----I----I----I MIN
         I                                                      0.000
33.9284  I
         I****                                                  0.080
34.9337  I
         I*******                                               0.134
35.9390  I
         I**********                                            0.208
36.9443  I
         I*************                                         0.284
37.9496  I
         I*******************                                   0.356
38.9549  I
         I*********************                                 0.402
39.9602  I
         I**********************                                0.442
40.9655  I
         I*************************                             0.490
41.9708  I
         I****************************                          0.526
42.9761  I
         I*******************************                       0.562
43.9814  I
         I***********************************                   0.594
44.9867  I
         I**************************************                0.636
45.9920  I
         I****************************************              0.678
46.9973  I
         I*******************************************           0.722
48.0026  I
         I**********************************************        0.766
49.0079  I
         I***********************************************       0.794
50.0132  I
         I**************************************************    0.822
51.0185  I
         I****************************************************  0.842
52.0238  I
         I******************************************************  0.880
53.0291  I
         I********************************************************  0.932
54.0344  I
         I**********************************************************  0.968
55.0397  I
         I*************************************************************1.000
56.0450  I
         I*************************************************************1.000
56.0450  I----I----I----I----I----I----I----I----I----I----I MAX


           NO OBS------------     500   STD ERROR-         6.6167
           COEF OF VARIATION-     0.15  MEAN------        43.3691
           KURTOSIS (BETA 2)-     1.83  MEDIAN----        42.3459
           PEARSONIAN SKEW---     1.31  MODE------        34.6871
```

FIGURE 9.11.   Path Performance for Node ECCK

```
        CFD   0.1   0.2   0.3   0.4   0.5   0.6   0.7   0.8   0.9   1.0
0.6459  I----I----I----I----I----I----I----I----I----I----I MIN
        I                                                         0.000
0.6459  I
        I                                                         0.006
0.6553  I
        I*                                                        0.012
0.6648  I
        I*                                                        0.018
0.6743  I
        I*                                                        0.028
0.6837  I
        I***                                                      0.058
0.6932  I
        I*****                                                    0.096
0.7027  I
        I******                                                   0.122
0.7121  I
        I********                                                 0.162
0.7216  I
        I**********                                               0.204
0.7310  I
        I**************                                           0.276
0.7405  I
        I*****************                                        0.332
0.7500  I
        I********************                                     0.386
0.7594  I
        I***********************                                  0.416
0.7689  I
        I**************************                               0.470
0.7783  I
        I*********************************                        0.558
0.7878  I
        I*****************************************                0.630
0.7973  I
        I**************************************************       0.718
0.8067  I
        I*****************************************************    0.822
0.8162  I
        I*****************************************************************  0.908
0.8257  I
        I***********************************************************************  0.962
0.8351  I
        I****************************************************************************0.992
0.8446  I
        I*****************************************************************************1.000
0.8540  I
        I*****************************************************************************1.000
0.8540  I----I----I----I----I----I----I----I----I----I----I MAX

        NO OBS------------     500   STD ERROR-          0.0455
        COEF OF VARIATION-    0.06   MEAN------          0.7721
        KURTOSIS (BETA 2)-    2.32   MEDIAN----          0.7825
        PEARSONIAN SKEW---    0.86   MODE------          0.8112
```

FIGURE 9.12.   Path Performance for Node STCK

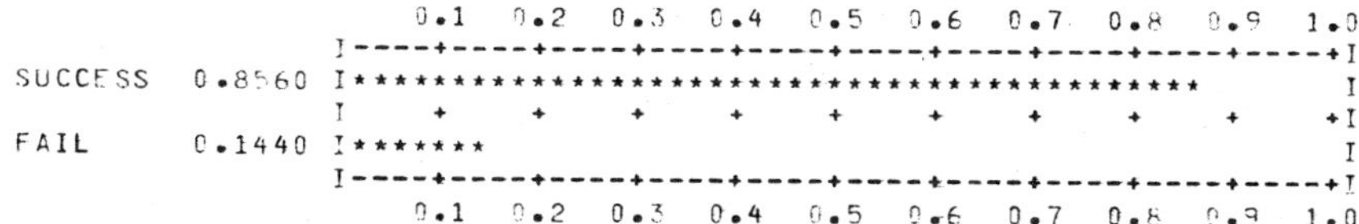

FIGURE 9.13.   Optimum Terminal Node Index, 500 Iterations

Figure 9.7, depicting the overall cost for node SYSTEM, shows that the expected cost should be approximately $26.2 million, well below the $30 million limitation. Further analysis shows approximately a 98.6% likelihood that the car will cost less than $30 million, or only a 1.4% chance that the cost will exceed available funds.

Figure 9.8 shows that the car could be expected to weigh about 1,750 pounds, with a standard error of approximately 105 pounds. This projection is slightly less than the desired 1,800 pounds of total weight.

Figure 9.9 shows an expected top speed of slightly over 86 miles per hour, well in excess of the desired and required 75 miles per hour. In fact, the simulation results show no outcomes less than approximately 77 miles per hour.

Acceleration, shown in figure 9.10, also betters the desired 10 seconds and the required 13 seconds in all cases; the figure shows an expected acceleration of 7 seconds and a maximum of roughly 9 seconds.

The economy figures, illustrated in figure 9.11, show an expected fuel economy of slightly over 43 miles per gallon. The distribution of results shows that about 99% of the outcomes bettered the required 35 miles per gallon, and approximately 60% bettered the desired level of 40 miles per gallon.

Concerning the car's stability index, plotted on figure 9.12, the expected or mean value is shown to be 0.77, which exceeds both the required 0.65 and the desired 0.75 levels. Closer examination shows that less than 1% of the outcomes failed to meet the required 0.65 index level, while fully 33.2% failed to meet the desired 0.75 level.

Figure 9.13 shows the total percentage of success and fail outcomes (with failure owing to the car's not meeting the required level in any one of the six criteria: time, cost, or the performance criteria—speed, acceleration, economy, and stability). Results show that a successful outcome occurred in 85.6% of the iterations, and a failure occurred in the remaining 14.4%.

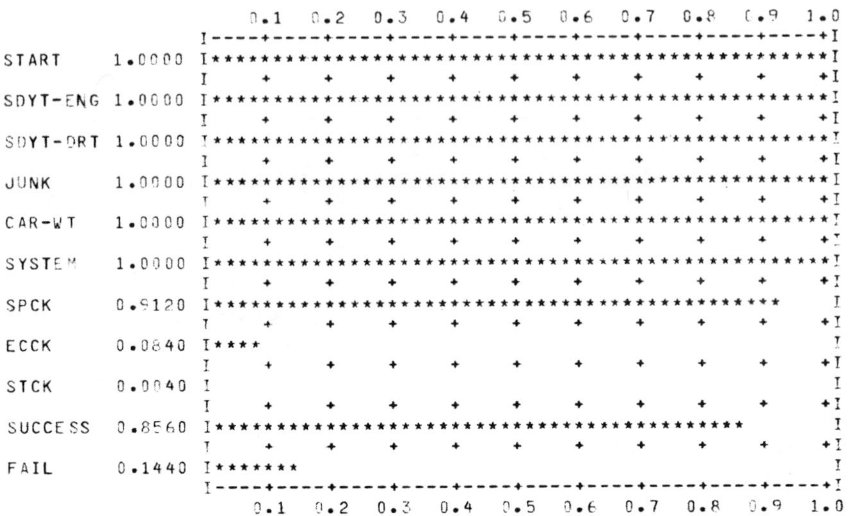

FIGURE 9.14.   Nodes Critical/Optimum Path Index, 500 Paths

RESULTS AND CONCLUSIONS

The results show that the Z-car project has approximately an 86% chance of success and a 14% risk of failure. Further analysis would indicate which of the time, cost, or performance factors resulted in the 72 failures out of the 500 iterations. From our analysis to this point, it is likely that the 20-month requirement yielded the majority of the failures and that the budget caused a smaller number of failures. The performance criteria of economy and stability each resulted in failure in 1% or fewer of the cases—not a very significant figure.

Figure 9.14 plots the percentage of the time the various nodes of the network were on the critical/optimum path (in this case, the time-related critical path). This type of output is also available for each of the arcs, or activities, of the network.

In short, we can see that various degrees of achievement of the specified time, cost, and performance parameters are important information and, in fact, serve as the definition of project success and failure. No longer does the decisionmaker have to assume that, for example, achievement of performance factors are given and that the decision relates to their time and/or cost of achievement. Performance factors are fully as variable as are time and cost, and the decision consists of trading off levels of all three parameters to arrive at an optimal result.

10 MULTIPLE PERFORMANCE ATTRIBUTES WITH CONSTRAINTS

VERT-3 provides the decisionmaker with a powerful tool for evaluating alternative or competing concepts, projects, or approaches. For example, projects or investments with useful life spans of varying lengths can be compared in terms of numerous measures of performance (results), cost, and time, and with full consideration of influences such as inflation and discounting. The following example represents a simplified actual project selection problem with multiple performance attributes and constraints.

PROJECT SELECTION TASK

A certain organization has three attractive investment projects that it would like to fund. Capital limitations, however, preclude the funding of all three, the company feels that it can fund only two of the projects. Information concerning the three projects has been developed by the company's marketing, finance, and engineering departments, and each project will require an initial investment that is subject to a degree of variability. In addition, each will provide a revenue stream that changes from year to year and possesses variability within any given year. The same is true with regard to operating

157

expenses for each of the investment alternatives. Management requires that the projects must have a return (after taxes and depreciation) of at least $5 million per year with a 90% confidence level. That is, the two projects the company can afford to fund must show a 90% chance of returning $5 million in after-tax income per year; otherwise, the projects will be dropped. If more than one combination of projects meet this criterion, the pair with the highest expected after-tax income will be selected.

For purposes of simplicity, each of the projects is assumed to have a two-year useful life with straight-line depreciation over this period—that is, no inflation or discounting is considered. A simulated performance factor (after-tax income) is calculated by year for each project. Yearly results for each pair of projects are computed to see which pairs exceed $5 million for each year. The relative frequency of a pair's exceeding the $5 million cutoff for both years is the pair's chance of exceeding the cutoff, or its confidence level.

Estimates of the investment cost, revenue by year, and operating expense by year are shown in table 10.1. Investment costs for each project are triangularly distributed. Candidate number one has the highest likely cost but

Table 10.1. Project Financial Data by Year

Project Number	Year	Investment	Revenue	Expense
1	0	T(8.5, 10.5, 12.0)		
	1		U(16.0, 26.0)	T(9.0, 10.5, 12.0)
	2		U(24.0, 34.0)	T(8.0, 10.0, 13.5)
2	0	T(7.0, 8.5, 11.0)		
	1		U(17.5, 22.5)	T(6.5, 8.5, 12.5)
	2		U(26.0, 31.0)	T(10.0, 15.0, 18.0)
3	0	T(6.5, 10.0, 13.5)		
	1		U(26.5, 31.5)	T(11.0, 15.0, 17.0)
	2		U(17.5, 22.5)	T(6.0, 8.0, 10.0

NOTE: T = triangularly distributed, with the minimum, most likely, and maximum given in that order; U = uniformly distributed, with the minimum and maximum given in that order.

the smallest range of values; number 3 has the lowest likely cost but the largest variation. Revenues are uniformly distributed, with numbers one and two increasing in the second year, while number three decreases. Candidate one has a higher range of values than do the other two. Expenses are triangularly distributed, with no consistently superior alternative from year to year.

VERT-3 MODEL

The VERT network constructed for this problem is shown in figure 10.1. The node labeled "START" gives birth to the network flow, which immediately expands into the three separate flows going through arcs BP1, BP2, and BP3. These three arcs represent the investment required for building the additional plant needed to accommodate these projects. The data used by these three arcs is listed in the *Investment* column of table 10.1. Nodes

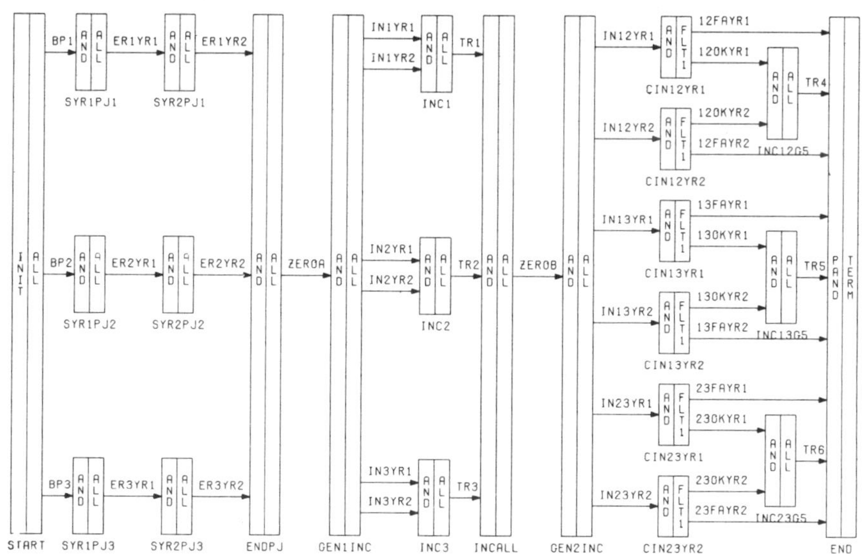

FIGURE 10.1. VERT-3 Project Selection Network

SYR1PJ1, SYR1PJ2, and SYR1PJ3 represent the combined events of the completion of the construction of the additional plant and the start of the first year of operation, represented by the arcs ER1YR1, ER2YR1, and ER3YR1. These three arcs carry the revenues (performances) and expenses (costs) data listed in the *Revenue* and *Expense* columns of table 10.1, plus an entry equal to 1, for time, which represents one year of operation.

Nodes SYR2PJ1, SYR2PJ2, and SYR2PJ3 perform a function similar to the function of nodes SYR1PJ1, SYR1PJ2, and SYR1PJ3. They separate the first year of operation from the second year of operation. Arcs ER1YR2, ER2YR2, and ER3YR2 carry the second-year expense and revenue data for the three projects in the same manner that arcs ER1YR1, ER2YR1, and ER3YR1 carried the first-year data.

The event depicting the end of the second year of operation (the life of the three projects) is represented by the node ENDPJ. This node carries the total time, expense (cost), and revenue (performance) associated with all three projects. All the remaining arcs and nodes in the network are concerned with the generation of the yearly project incomes and the checking out of all the two-at-a-time project combination incomes to see if they exceed the $5 million-per-year level.

Arc ZEROA zeroes out the time, expense (cost), and revenue (performance) values flowing to node GEN1INC. ZEROA accomplishes this task by using the transformations to reach back to node ENDPJ and multiply the time, expense, and revenue values carried on ENDPJ by minus one to generate primary time, expense, and revenue values equal in magnitude, but opposite in sign, to those carried by ENDPJ. Thus, when the time, expense, and revenue values flow from node ENDPJ on to arc ZEROA to combine with ZEROA's primary time, expense, and revenue values to generate the cumulative time, expense, and revenue values, these cumulative values all sum to 0. ZEROA's cumulative values then flow on to GEN1INC. Thus, GEN1INC will pass zero time, expense, and revenue values on to its output arcs.

Arcs IN1YR1, IN1YR2, IN2YR1, IN2YR2, IN3YR1, and IN3YR2 generate the yearly incomes of each of the three projects. The income generated on arc IN1YR1 is computed through use of mathematical relationship numbers 33 and 29 in the following two-step manner (see table 10.2): (1) The expense (cost) generated on arc ER1YR1 is added to one year of straight-line depreciation, which is computed by dividing the plant costs generated on arc BP1 by the two-year life of the project. (2) To compute after-tax income, the preceding cost is subtracted from the revenue (performance) generated on arc ER1YR1, and this sum is multiplied by ½. Income values are generated for arcs IN1YR2, IN2YR1, IN2YR2, IN3YR1, and IN3YR2 through use of the same steps and transformations. The creation of these

Table 10.2.   Initial Computer Run of Figure 10.1

```
PROBLEM IDENTIFICATION CARD OPTION-------------------------        1

TYPE OF INPUT OPTION-----------------------------------        0

TYPE OF OUTPUT OPTION----------------------------------        0

COSTING AND PRUNING OPTION-----------------------------        0

FULL PRINT TRIP OPTION---------------------------------        0

CORRELATION COMPUTATION AND PLOT OPTION--------------------        0

COST-PERFORMANCE TIME INTERVAL OPTION----------------------        0

COMPOSITE TERMINAL NODE MINIMUMS AND MAXIMUMS OPTION-------        0

TERMINAL NODE LISTING OPTION---------------------------        15

INITIAL SEED------------------------------------------        0

NUMBER OF ITERATIONS-----------------------------------        1000

YEARLY INTEREST RATE USED FOR INFLATION ADJUSTMENTS--------        0.0

YEARLY INTEREST RATE USED FOR PRESENT VALUE DISCOUNTING----        0.0

TIME FACTOR WHICH CONVERTS PROGRAM TIME TO A YEARLY BASE---        0.0

                                      TIME      COST      PERF


TERMINAL NODE SELECTION WEIGHTS       0.0       0.0       1.00

CRITICAL - OPTIMUM PATH WEIGHTS       0.0       0.0       0.0

INITIAL VALUES                        0.0       0.0       0.0


BP1       START    SYR1PJ1   1.0 BUILD PLANT FOR PROJECT 1
BP1       DCOST 1    3.0       8.5      12.0      10.5

BP2       START    SYR1PJ2   1.0 BUILD PLANT FOR PROJECT 2
BP2       DCOST 1    3.0       7.0      11.0       8.5

BP3       START    SYR1PJ3   1.0 BUILD PLANT FOR PROJECT 3
BP3       DCOST 1    3.0       6.5      13.5      10.0

ER1YR1    SYR1PJ1  SYR2PJ1   1.0 EXPENSES AND REVENUES FOR PROJECT 1, YEAR 1
ER1YR1    DTIME 1    1.0       1.0
ER1YR1    DPERF 1    2.0      16.0      26.0
ER1YR1    DCOST 1    3.0       9.0      14.0      12.0

ER2YR1    SYR1PJ2  SYR2PJ2   1.0 EXPENSES AND REVENUES FOR PROJECT 2, YEAR 1
ER2YR1    DTIME 1    1.0       1.0
ER2YR1    DPERF 1    2.0      17.5      22.5
ER2YR1    DCOST 1    3.0       6.5      12.5       8.5

ER3YR1    SYR1PJ3  SYR2PJ3   1.0 EXPENSES AND REVENUES FOR PROJECT 3, YEAR 1
ER3YR1    DTIME 1    1.0       1.0
ER3YR1    DPERF 1    2.0      26.5      31.5
ER3YR1    DCOST 1    3.0      11.0      17.0      15.0

ER1YR2    SYR2PJ1  ENDPJ     1.0 EXPENSES AND REVENUES FOR PROJECT 1, YEAR 2
ER1YR2    DTIME 1    1.0       1.0
ER1YR2    DPERF 1    2.0      24.0      34.0
ER1YR2    DCOST 1    3.0       8.0      13.5      11.0

ER2YR2    SYR2PJ2  ENDPJ     1.0 EXPENSES AND REVENUES FOR PROJECT 2, YEAR 2
ER2YR2    DTIME 1    1.0       1.0
ER2YR2    DPERF 1    2.0      26.0      31.0
ER2YR2    DCOST 1    3.0      10.0      18.0      15.0
```

(cont.)

Table 10.2.   (Cont.)

```
ER3YR2   SYR2PJ3 ENDPJ      1.0 EXPENSES AND REVENUES FOR PROJECT 3, YEAR 2
ER3YR2   DTIME 1    1.0         1.0
ER3YR2   DPERF 1    2.0        17.5         22.5
ER3YR2   DCOST 1    3.0         6.0         10.0          8.0

ZERDA    ENDPJ   GEN1INC  1.0 ZEROS OUT TIME, COST AND PERFORMANCE
ZERDA    RTIME 1 1STENDPJ    K-1.0     K 1.0
ZERDA    RPERF 1 1SPENDPJ    K-1.0     K 1.0
ZERDA    RCOST 1 1SCENDPJ    K-1.0     K 1.0

IN1YR1   GEN1INC INC1       1.0 INCOME FOR PROJECT 1, YEAR 1
IN1YR1   RPERF 1330CER1YR1  CBP1        K 2.0        29SPER1YR1     1.0      K 0.5

IN2YR1   GEN1INC INC2       1.0 INCOME FOR PROJECT 2, YEAR 1
IN2YR1   RPERF 1330CER2YR1  CBP2        K 2.0        29SPER2YR1     1.0      K 0.5

IN3YR1   GEN1INC INC3       1.0 INCOME FOR PROJECT 3, YEAR 1
IN3YR1   RPERF 1330CER3YR1  CBP3        K 2.0        29SPER3YR1     1.0      K 0.5

IN1YR2   GEN1INC INC1       1.0 INCOME FOR PROJECT 1, YEAR 2
IN1YR2   RPERF 1330CER1YR2  CBP1        K 2.0        29SPER1YR2     1.0      K 0.5

IN2YR2   GEN1INC INC2       1.0 INCOME FOR PROJECT 2, YEAR 2
IN2YR2   RPERF 1330CER2YR2  CBP2        K 2.0        29SPER2YR2     1.0      K 0.5

IN3YR2   GEN1INC INC3       1.0 INCOME FOR PROJECT 3, YEAR 2
IN3YR2   RPERF 1330CER3YR2  CBP3        K 2.0        29SPER3YR2     1.0      K 0.5

TR1      INC1    INCALL     1.0 TRANSPORTATION ARC

TR2      INC2    INCALL     1.0 TRANSPORTATION ARC

TR3      INC3    INCALL     1.0 TRANSPORTATION ARC

ZERDB    INCALL  GEN2INC  1.0 ZEROS OUT PERFORMANCE
ZERDB    RPERF 1 1SPINCALL   K-1.0     K 1.0

IN12YR1 GEN2INC CIN12YR1 1.0 PROJECTS 1&2 IN YEAR 1
IN12YR1 RPERF 1 5SPIN1YR1  PIN2YR1   K 0.0

IN12YR2 GEN2INC CIN12YR2 1.0 PROJECTS 1&2 IN YEAR 2
IN12YR2 RPERF 1 5SPIN1YR2  PIN2YR2   K 0.0

IN13YR1 GEN2INC CIN13YR1 1.0 PROJECTS 1&3 IN YEAR 1
IN13YR1 RPERF 1 5SPIN1YR1  PIN3YR1   K 0.0

IN13YR2 GEN2INC CIN13YR2 1.0 PROJECTS 1&3 IN YEAR 2
IN13YR2 RPERF 1 5SPIN1YR2  PIN3YR2   K 0.0

IN23YR1 GEN2INC CIN23YR1 1.0 PROJECTS 2&3 IN YEAR 1
IN23YR1 RPERF 1 5SPIN2YR1  PIN3YR1   K 0.0

IN23YR2 GEN2INC CIN23YR2 1.0 PROJECTS 2&3 IN YEAR 2
IN23YR2 RPERF 1 5SPIN2YR2  PIN3YR2   K 0.0

120KYR1 CIN12YR1INC12G5 1.0 INCOME OF 1 + 2, HIGH ENOUGH FOR YEAR 1
120KYR1 FILT1 1                                          5.0        900000.

130KYR1 CIN13YR1INC13G5 1.0 INCOME OF 1 + 3, HIGH ENOUGH FOR YEAR 1
130KYR1 FILT1 1                                          5.0        900000.

230KYR1 CIN23YR1INC23G5 1.0 INCOME OF 2 + 3, HIGH ENOUGH FOR YEAR 1
230KYR1 FILT1 1                                          5.0        900000.

120KYR2 CIN12YR2INC12G5 1.0 INCOME OF 1 + 2, HIGH ENOUGH FOR YEAR 2
120KYR2 FILT1 1                                          5.0        900000.

130KYR2 CIN13YR2INC13G5 1.0 INCOME OF 1 + 3, HIGH ENOUGH FOR YEAR 2
```

Table 10.2. (Cont.)

```
13OKYR2 FILT1 1                                                5.0      900000.

23OKYR2 CIN23YR2INC23G5  1.0 INCOME OF 2 + 3, HIGH ENOUGH FOR YEAR 2
23OKYR2 FILT1 1                                                5.0      900000.

12FAYR1 CIN12YR1END      1.0 INCOME OF 1 + 2, TOO LOW FOR YEAR 1

13FAYR1 CIN13YR1END      1.0 INCOME OF 1 + 3, TOO LOW FOR YEAR 1

23FAYR1 CIN23YR1END      1.0 INCOME OF 2 + 3, TOO LOW FOR YEAR 1

12FAYR2 CIN12YR2END      1.0 INCOME OF 1 + 2, TOO LOW FOR YEAR 2

13FAYR2 CIN13YR2END      1.0 INCOME OF 1 + 3, TOO LOW FOR YEAR 2

23FAYR2 CIN23YR2END      1.0 INCOME OF 2 + 3, TOO LOW FOR YEAR 2

TR4     INC12G5 END      1.0 TRANSPORTATION ARC

TR5     INC13G5 END      1.0 TRANSPORTATION ARC

TR6     INC23G5 END      1.0 TRANSPORTATION ARC

ENDARC

START   1  2                     START THE NETWORK

SYR1PJ1 2  2                     START YEAR 1 OF PROJECT 1

SYR1PJ2 2  2                     START YEAR 1 OF PROJECT 2

SYR1PJ3 2  2                     START YEAR 1 OF PROJECT 3

SYR2PJ1 2  2                     START YEAR 2 OF PROJECT 1

SYR2PJ2 2  2                     START YEAR 2 OF PROJECT 2

SYR2PJ3 2  2                     START YEAR 2 OF PROJECT 3

ENDPJ   2  2                     INITIAL COMPUTATION DONE

GEN1INC 2  2                     PREPARE TO GENERATE INCOMES BY YEAR BY PROJECT

INC1    2  2 4 1                 NET INCOME FOR PROJECT 1

INC2    2  2 4 2                 NET INCOME FOR PROJECT 2

INC3    2  2 4 3                 NET INCOME FOR PROJECT 3

INCALL  2  2                     NET INCOME FOR ALL THREE PROJECTS COMBINED

GEN2INC 2  2                     PREPARE TO GENERATE TWO PROJECT COMBINATION INCOMES

CIN12YR12  4                     CHECK TO SEE IF INCOME OF PJ 1 + 2 BIG ENOUGH, YR 1

CIN13YR12  4                     CHECK TO SEE IF INCOME OF PJ 1 + 3 BIG ENOUGH, YR 1

CIN23YR12  4                     CHECK TO SEE IF INCOME OF PJ 2 + 3 BIG ENOUGH, YR 1

CIN12YR22  4                     CHECK TO SEE IF INCOME OF PJ 1 + 2 BIG ENOUGH, YR 2

CIN13YR22  4                     CHECK TO SEE IF INCOME OF PJ 1 + 3 BIG ENOUGH, YR 2

CIN23YR22  4                     CHECK TO SEE IF INCOME OF PJ 2 + 3 BIG ENOUGH, YR 2

INC12G5 2  2 4 4                 NET PRESENT VALUE OF PROJECTS 1&2

INC13G5 2  2 4 5                 NET PRESENT VALUE OF PROJECTS 1&3
```

(cont.)

Table 10.2. (Cont.)

```
INC23G5 2   2  4  6            NET PRESENT VALUE OF PROJECTS 2&3

END      3   115              END OF THE NETWORK

ENDNODE


PATH PERFORMANCE
                 FOR NODE  INC1
NO OBS =              1000    AVE =        8.75528
MIN =        2.68795         MAX =        14.4579


PATH PERFORMANCE
                 FOR NODE  INC2
NO OBS =              1000    AVE =        8.09703
MIN =        3.46568         MAX =        12.4814


PATH PERFORMANCE
                 FOR NODE  INC3
NO OBS =              1000    AVE =        8.35223
MIN =        4.28817         MAX =        13.1982


PATH PERFORMANCE
                 FOR NODE  INC12G5
NO OBS =               559    AVE =        18.2200
MIN =        12.1374         MAX =        24.4327


PATH PERFORMANCE
                 FOR NODE  INC13G5
NO OBS =               838    AVE =        17.6214
MIN =        10.9011         MAX =        25.2049


PATH PERFORMANCE
                 FOR NODE  INC23G5
NO OBS =               979    AVE =        16.5418
MIN =        10.5807         MAX =        22.6500

        LAST RANDOM NUMBER SEED =    651015587
```

incomes represents the introduction of another dimension of performance in addition to the already currently active dimension of revenue.

The AND logic of node INC1 combines the two years of income for project number one. Thus, INC1 will carry the total income stream generated during the life of project number one. Similarly, INC2 and INC3 will carry the total income stream for projects two and three, respectively.

Arcs TR1, TR2, and TR3 are transport arcs. They merely carry the project incomes from nodes INC1, INC2, and INC3 to node INCALL. Therefore, node INCALL carries the combined project-life incomes of all three projects.

Arc ZEROB zeroes out the combined income flowing on to it from node INCALL. This task is accomplished in nearly the same manner in which ZEROA zeroed out the flow coming to it. The only difference between these processes is that ZEROA zeroed out the time, cost, and performance flow, while ZEROB zeroed out only the performance flow (income). Hence, node GEN2INC will have zero time, cost, and performance values flowing to it and will likewise pass zero time, cost, and performance values on to its output arcs.

The combined first-year incomes for projects one and two are generated on arc IN12YR1. This task is accomplished through use of the transformations to reach back to arcs IN1YR1 and IN2YR1 to sum these two first-year incomes together. The combined income of projects one and two for the second year is similarly generated. The yearly income values for the remaining project combinations are similarly generated on arcs IN13YR1, IN13YR2, IN23YR1, and IN23YR2. These arcs pass the income values they have generated directly on to their lone output nodes. Thus, nodes CIN12YR1, CIN13YR1, CIN23YR1, CIN12YR2, CIN13YR2, and CIN23YR2 carry the combined incomes for projects one and two, one and three, and two and three for both of the two years of operation.

The FILTER 1 output logic of node CIN12YR1 compares the combined first-year incomes of projects one and two with management's yearly income constraint of $5 million. If this combined income is equal to or exceeds $5 million, the network flow is routed to arc 12OKYR1. Otherwise, the flow is sent out to arc 12FAYR1. The remaining combined-income nodes perform the same function as was just described for node CIN12YR1.

The failure flows resulting from the FILTER 1 logic of nodes CIN12YR1, CIN12YR2, CIN13YR1, CIN13YR2, CIN23YR1, and CIN23YR2 flow directly to the lone terminal node END. However, the success flows travel first to intermediate nodes INC12G5, INC13G5, or INC23G5, whose AND input logic combines the incomes for both years of operation for each of the project combinations with an income for both years of operation that exceeds $5 million. If the income for only one of these two years exceeds the $5 million level, the AND input logic of nodes INC12G5, INC13G5, or INC23G5 will not allow network flow to progress beyond this point. Thus, the node will remain inactive for that iteration. Hence, by observing the number of times nodes INC12G5, INC13G5, and INC23G5 are active over a large number of iterations, one can determine what the chances are that each of the project combinations will exceed the $5 million-per-year income level.

Arcs TR4, TR5, and TR6 are transport arcs that will carry the combined income flows exceeding $5 million for each of the project combinations directly on to terminal node END. Terminal node END provides an end point for the network; it does not, however, provide any useful information.

RESULTS AND CONCLUSIONS

Upon network simulation for 1,000 iterations, nodes INC12G5, INC13G5, and INC23G5 were active for 550, 838, and 979 iterations, respectively. Hence, project combination two and three was the only one that fell within the 90% confidence range of having yearly incomes in excess of $5 million. If more combinations had fallen within this income requirement, the mean income values observed on nodes INC1, INC2, and INC3 could have been manually combined to ascertain which combination yielded the highest life-time income.

The actual problem from which this preceding example was drafted included additional complications: candidate projects with varying lives subject to the additional constraints of a minimum yearly cash flow and a maximum total cash outlay. This problem in its full scope appear to typify the financial environments of many corporations (see Raiffa 1968).

VERT-3 was able to model the complete problem nearly as easily and with as much realism as its shortened version. However, when applying VERT-3 to the complete problem, analysts used an additional feature: The independent parts of the development, such as the plant construction initially required, were networked by themselves. The results of these individual plant networks were substituted into the main network through use of VERT-3's histogram input capability. This two-step procedure can greatly simplify complex, voluminous problems and thus make them easier to illustrate and convey.

$\mathbf{11}$ NEW-PRODUCT DEVELOPMENT DECISIONS

The problem discussed in this chapter illustrates the application of VERT-3 to a new-product development decision. The company in question has learned that one of its major competitors is working on a new product that, if successfully introduced, could provide the competitor with a significant advantage in the marketplace. If the competitor were to introduce the new product before this company's own version of the product was ready, damage would be done to its present reputation and image of innovation and progressive leadership in the industry. In addition, the largest share of the market would automatically flow to the firm introducing the new product first.

Management estimates that the competitor will be ready to enter production with their new product in 12 months. Therefore, the company must be able to announce its product and enter production at least 2 months earlier to gain the advantages described above. Simultaneous introduction would offer neither firm any particular advantage, but late introduction (2 months or more behind the leader) would be definitely detrimental.

As part of a crash program, four design groups will be simultaneously working in parallel efforts on four different but equally desirable designs, each of which has an engineering-assessed 85% probability of successful completion. The product design that is the first to be conceptualized and computer evaluated will be fabricated, tested, and carried into initial production. Management is very anxious to know the likelihood of getting this product into production within 10 months. Further, to aid in the determination of the unit price, management would like an estimate of the total cost through the initial production run. Time and cost estimates rendered by the appropriate departments of this company are exhibited in table 11.1.

Table 11.1. New-Product Development Data

		Completion Time & Cost			
Project	Distribution Type	Minimum	Maximum	Most Likely or Mean	Standard Deviation
Design A	Normal	2.5 months	6.7 months	4.5 months	0.7 months
	Triangular	$2.0 M	$5.0 M	$3.0 M	
Design B	Normal	2.4 months	7.2 months	4.8 months	0.8 months
	Triangular	$2.5 M	$7.0 M	$4.0 M	
Design C	Triangular	3.0 months	8.0 months	4.6 months	
	Triangular	$1.8 M	$5.0 M	$2.9 M	
Design D	Triangular	3.2 months	7.0 months	4.4 months	
	Triangular	$2.2 M	$4.9 M	$3.2 M	
Fabrication	Uniform	2.0 months	2.5 months		
	Normal	$3.0 M	$4.0 M	$3.5 M	$0.4 M
Testing	Uniform	2.0 months	3.0 months		
	Constant	$5.0 M	$5.0 M		
Production	Triangular	5.5 months	8.0 months	6.0 months	
	Constant	$80.0 M	$80.0 M + variable of $5.0M/month		

VERT-3 MODEL

The pictorial network layout of this problem as illustrated in figure 11.1 shows the four competing parallel design efforts (arcs DESIGNA, DESIGNB, DESIGNC, and DESIGND) being started off by node START and then flowing on into node DES/FAB. Since the design arcs each carry probabilities of successful completion of less than 1, there exists the possibility that any or all of them may be failures for a given simulation iteration. In that case, the flow out of node DES/FAB will travel on arc FAILDES, the design failure arc. This arc flows into the sink for the network, which is node FAIL. However, if at least one of the design arcs is successfully completed, node DES/FAB routes the network flow to the fabrication activity (arc FAB). If more than one of the design arcs is successfully completed, the OR input logic of node DES/FAB will send the flow on to its output logic for processing as soon as the first design arc is completed, since time is of the essence. Node FAB/TEST simply acts as a separator between the fabrication activity (FAB) and the test activity (TEST). After testing is completed, the FILTER 1 output logic of node TIMECHECK checks to see if the time limit of 10 months has been exceeded. If the time limit has been exceeded, the network flow is routed to arc FAILTIME, which carries the flow to the failure node (FAIL). If the time limit is not exceeded, TIMECHECK routes the flow the the production activity (PROD), which, in turn, routes the flow into the success sink for the network, which is node SUCCESS.

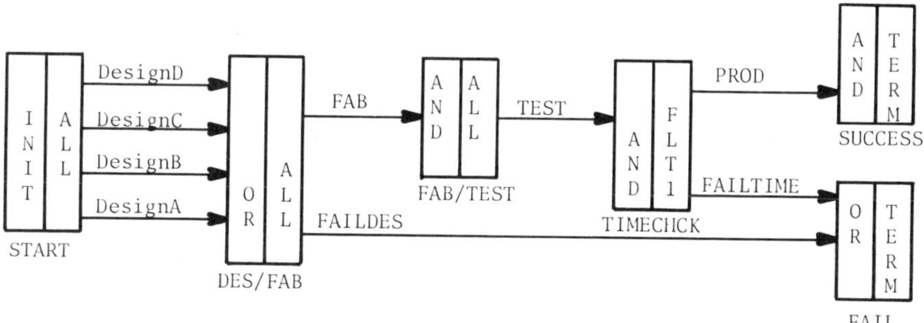

FIGURE 11.1   New-Product Development Network

RESULTS AND CONCLUSIONS

Table 11.2 and figures 11.2 and 11.3 show that the SUCCESS node was realized 982 times out of the 1,000 iterations; the FAIL node, only 18 times. Thus, there is only about a 2% chance that the new-product development effort will fail. Additionally, the arc critical path listings in figure 11.4 show that the failures that did occur all resulted from exceeding the 10-month time limit rather than from not having acceptable designs, because arc FAILDES was never on the critical path. This arc never occurred, and arcs FAB and TEST were on the critical path 100% of the time. Therefore, the only source of failure was arc FAILTIME, which accounted for all 18 failures.

Table 11.2.   Selected Output Data

```
NUMBER OF ITERATIONS-------------------------------------------   1000

YEARLY INTEREST RATE USED FOR INFLATION ADJUSTMENTS---------      0.0

YEARLY INTEREST RATE USED FOR PRESENT VALUE DISCOUNTING----      0.0

TIME FACTOR WHICH CONVERTS PROGRAM TIME TO A YEARLY BASE---      0.0

                                          TIME     COST     PERF

TERMINAL NODE SELECTION WEIGHTS           1.00     0.0      0.0

CRITICAL - OPTIMUM PATH WEIGHTS           1.00     0.0      0.0

INITIAL VALUES                            0.0      0.0      0.0

DESIGNA START     DES/FAB 0.85 FEASIBILITY DEVELOPMENT OF DESIGN A
DESIGNA DTIME 1     4.0          2.5          6.7          4.6          0.7
DESIGNA DCOST 1     3.0          2.0          5.0          3.0

DESIGNB START     DES/FAB 0.85 FEASIBILITY DEVELOPMENT OF DESIGN B
DESIGNB DTIME 1     4.0          2.4          7.2          4.8          0.8
DESIGNB DCOST 1     3.0          2.5          7.0          4.0

DESIGNC START     DES/FAB 0.85 FEASIBILITY DEVELOPMENT OF DESIGN C
DESIGNC DTIME 1     3.0          3.0          8.0          4.6
DESIGNC DCOST 1     3.0          1.8          5.0          2.9

DESIGND START     DES/FAB 0.85 FEASIBILITY DEVELOPMENT OF DESIGN D
DESIGND DTIME 1     3.0          3.2          7.0          4.4
DESIGND DCOST 1     3.0          2.2          4.9          3.2

FAB        DES/FAB FAB/TEST 1.0 FABRICATION OF THE NEW DESIGN
FAB        DTIME 1     2.0          2.0          2.5
FAB        DCOST 1     4.0          3.0          4.0          3.5          0.4

FAILDES DES/FAB FAIL      1.0 TRANSPORTS FAILURE FLOW OF THE DESIGN EFFORT

TEST       FAB/TESTTIMECHCK 1.0 FINAL PROVE OUT TEST OF NEW PRODUCT
TEST       DTIME 1     2.0          2.0          3.0
TEST       DCOST 1     1.0          5.0

PROD       TIMECHCKSUCCESS   1.0 INITIAL PRODUCTION OF THE NEW PRODUCT
PROD       DTIME 1     3.0          5.5          8.0          6.0
PROD       RCOST 123STPROD   K 5.0     K 80.0
PROD       FILT1 1     0.0          10.0
```

Table 11.2.   (Cont.)

```
FAILTIMETIMECHCKFAIL       1.0 TRANSPORTS FAILURE FLOW OF EXCEEDING THE TIME LIMIT
ENDARC

START   1  2                   START THE NETWORK

DES/FAB 4  2                   GOING FROM DESIGN TO FABRICATION

FAB/TEST2  2                   GOING FROM FABRICATION TO TESTING

TIMECHCK2  4                   CHECK TO SEE IF TOO MUCH TIME HAS BEEN EXPENDED

SUCCESS 2  1 1                 SUCCESS SINK

FAIL    4  1 1                 FAILURE SINK

ENDNODE

        NETWORK TIME FOR NODE SUCCESS
NO OBS =              982   AVE =            15.1785
MIN =         13.1478       MAX =            17.6341

        NETWORK TIME FOR NODE FAIL
NO OBS =               18   AVE =            10.2032
MIN =         10.0014       MAX =            10.7199

        NETWORK TIME FOR THE COMPOSITE TERMINAL NODE
NO OBS =             1000   AVE =            15.0889
MIN =         10.0014       MAX =            17.6341

        OVERALL COST FOR THE COMPOSITE TERMINAL NODE
NO OBS =             1000   AVE =            133.390
MIN =         20.3318       MAX =            144.129
```

Further examination of figure 11.4 shows that arc DESIGNA was on the critical path 33.5% of the time—an indication that design A was the first one completed in 33.5% of the runs. Design B was first 27.6% of the time; design D, 21.5%; and design C, 17.4%. These figures represent for each design the likelihood of being the one that will be selected for subsequent fabrication, testing, and (if the time limit has not been exceeded) production.

From table 11.2 we can see that the average overall cost for the composite terminal node is $133,390. However, the total cost ranges from a minimum of $20,332 to a maximum of $144,129. What is the source of this large variation? In table 11.1 the cost of production is $80,000 plus $5,000 per month. Therefore, those iterations in which production began incurred a fixed cost of $80,000, whereas iterations that either failed or barely finished the test phase were spared this additional expense. In reality, management would probably want to know the total of the design, fabrication, and testing costs rather than adding in some amount of production costs. These figures could be easily obtained if desired.

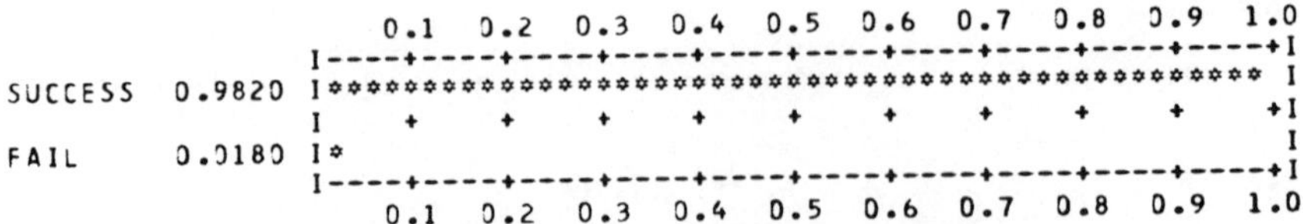

FIGURE 11.2.   Optimum Terminal Node Index, 1,000 Iterations

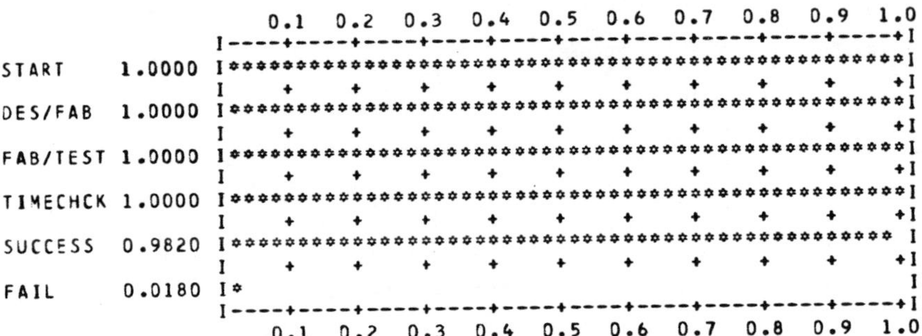

FIGURE 11.3.   Nodes Critical/Optimum Path Index, 1,000 Paths

FIGURE 11.4.   Arcs Critical/Optimum Path Index, 1,000 Paths

The oversimplified new-product development example includes only a few of VERT-3's features and options, but does illustrate the type of planning analysis that can be performed with the technique. In strategic planning, information concerning results the manager can expect is critical, as is the range and likelihood of possible outcomes. VERT-3 can assist the manager by providing this information.

12 DECISION TREE APPLICATIONS

Decision trees represent an attempt to include realism in decision situations by adding consideration of several possible chance events. Expected-value calculations are then used to determine which of a set of alternate decisions is most desirable. Let us examine an often-used decision tree example to illustrate some of the weaknesses present in the decision tree approach, as well as some of the advantages stochastic techniques such as VERT-3 can offer.

DECISION TREE PROBLEM

The decision tree problem shown in figure 12.1 is similar to one described by Magee (1964*a*, 1964*b*) and by Weston and Brigham (1969). Management is faced with the problem of deciding whether to build a large plant costing $500,000 or a small plant costing $200,000. After the plant is built, one of three chance events will occur: There exists a 50% chance of high demand, a 30% chance of medium demand, and a 20% chance of low demand. (Returns under these various outcomes are shown on the figure.) Further, if the small plant is built and a high demand is experienced during the first two

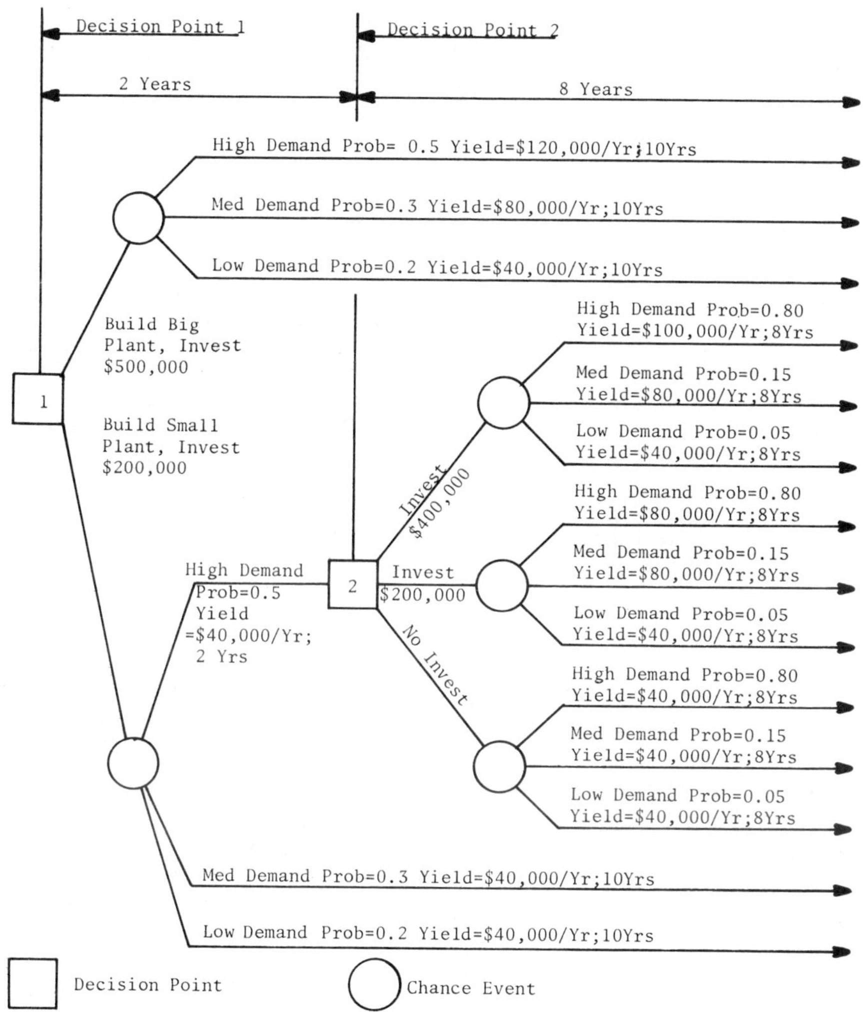

FIGURE 12.1.   Decision Tree Layout of Investment Choices

years of operation, management will consider increasing the plant size. The three options are: a $400,000 addition, a $200,000 addition, or no addition. After this point, one of three chance events will occur for the remaining eight years (with returns shown on the figure): There exists an 80% chance of high demand, a 15% chance of medium demand, and a 5% chance of low demand.

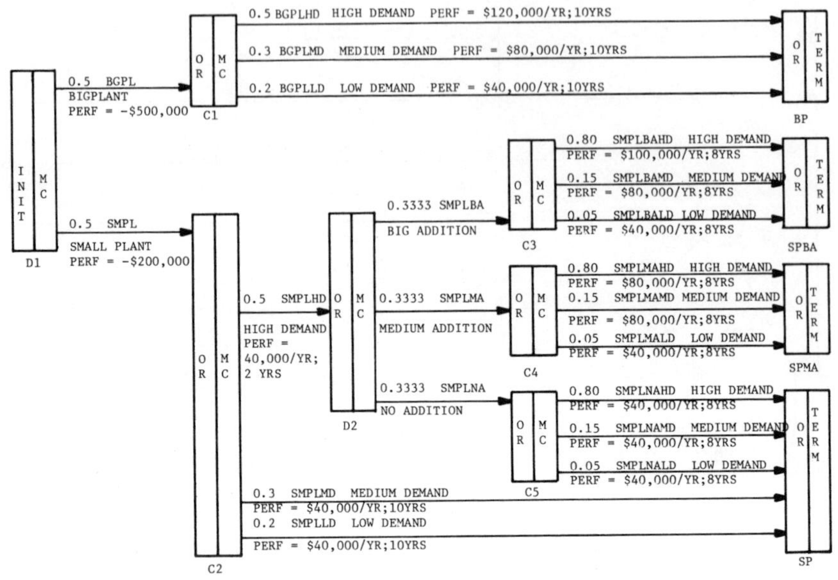

FIGURE 12.2.   VERT Network of the Decision Tree

To determine the best course of action, the analyst can readily apply the "rollback" method as described by Magee (1964*a,* 1964*b*). However, the analyst can also apply VERT-3 by letting both the chance and the decision events be represented by nodes having **OR** input logic and **MONTE CARLO** output logic (see figure 12.2). The exceptions are the node starting the network (D1), which has **INITIAL** input logic, and the nodes finishing the network (BP, SPBA, SPMA, and SP), which have **TERMINAL** output logic. The chance nodes (C1–C5) use the probability weights assigned to the various alternative demand levels to initiate their output arcs. Equally likely probability weights are used to initiate the output arcs of the decision nodes (D1 and D2). A further expedient that the analyst may take when using VERT-3 on decision tree problems similar to this one is to enter both the cost and the yield data as performance-type data by prefixing the cost data with a minus (−) sign and the performance data with no sign or with a plus (+) sign.

In each simulation iteration the network flow starts out in the initial node, D1, and as a result of the **MONTE CARLO** output logic carried by most of the nodes in the network, the flows travel stochastically through the network to one of four possible terminal nodes. These nodes, which are labeled BP (big plant), SPBA (small plant–big addition), SPMA (small plant–medium addition), and SP (small plant), collect all the possible plant configuration outcomes. When the network is simulated a large number of times (1,000), the statistics accumulated on the terminal nodes closely approximate the results obtained through the "rollback" method.

VERT-3 MODIFICATIONS

A major weakness of the decision tree approach is that only a few possible outcomes can be considered. For instance, the previous example included only three sales levels as possibilities, with a single profit figure for each.

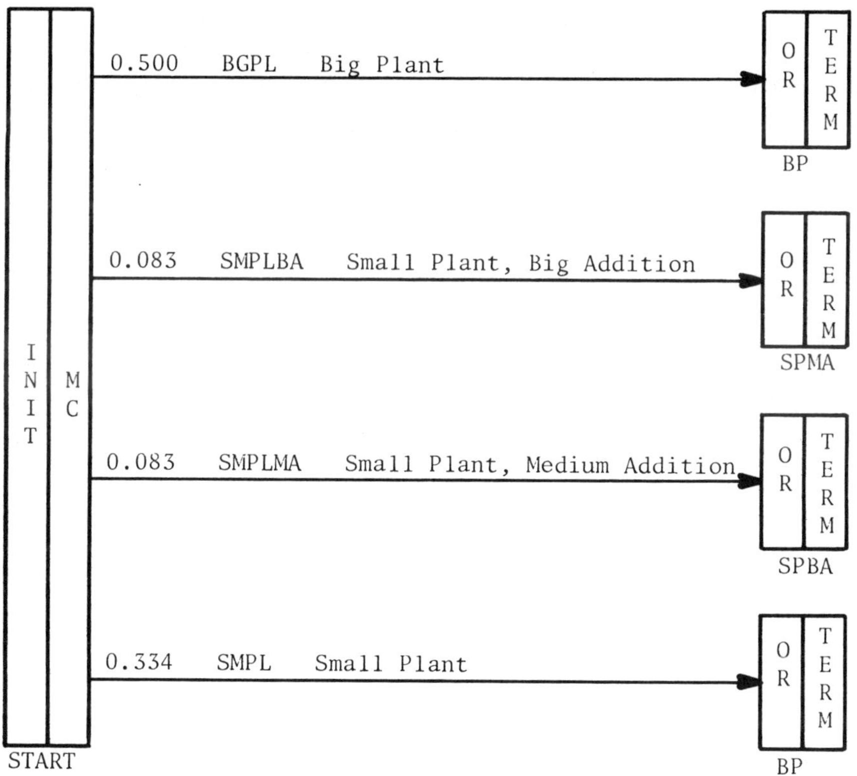

FIGURE 12.3.   Compressed VERT Decision Tree Network

Also, plant investments were considered to be point estimates as well. Such a hypothesis is highly unrealistic; there is likely to be a range of plant costs and a continuous range (rather than only three levels) of sales and profits. In addition, sales and profit levels could vary from year to year in accordance with general economic and/or industry cyclicality. VERT-3 can incorporate a continuous distribution of possible outcomes and of economic and cyclical variation, as well as the complexities due to inflation and to varying interest rates.

In addition to being able to express costs, sales, and profits as random variables with a continuous distribution, VERT-3 does not consider construction time as a deterministic variable, since it is unlikely that construction times will be equal for plants of different sizes and plant additions. Also, VERT-3 will provide a distribution of likely outcomes for each alternative rather than just a point-expected value. The manager thus has a much better "picture" upon which to make the decision.

While the inclusion of the factors we have discussed is vitally important in conducting a realistic decision analysis, perhaps the most useful contribution VERT-3 can make to the solution of decision tree problems lies in its ability to squeeze out branch repetition. For example, by employing VERT-3's histogram input feature to carry the yield data as a function of the probabilistic demand level, the analyst could compress the preceding problem into the four-arc and five-node network exhibited in figure 12.3. Decision trees involving a large number of factors typically develop into massive jungles of repetition, but VERT-3's wide array of operands generally enables the modeler to maintain a close, one-for-one parallel between the real world and the model representation and thus to avoid redundancies and to model with a wider problem scope.

13 ANALYSIS OF ALTERNATIVE ENERGY SOURCES

As we have mentioned several times, VERT-3 and other stochastic networking techniques are especially powerful in the evaluation of alternative concepts and approaches. The example presented in this chapter is a hypothetical situation that illustrates more of VERT-3's features than any of the previous examples; however, one should keep in mind that this example may still be simpler than an actual application. For instance, the relatively straightforward presentation of the data for this problem tends to underemphasize the data collection task typically associated with problems modeled by VERT-3; data collection and elicitation usually prove to be the most difficult and time-consuming parts of the analysis. The numerous time, cost, and performance values associated with the various activities in a typical network usually are not known precisely; however, they can be represented by a probability distribution. Procedures useful for eliciting such data have been suggested by Dalkey (1969), Northrop (1970), and Raiffa (1968).

PROBLEM DESCRIPTION

The following problem is based on Moeller and Digman (1981*a*). The Federal Power Commission has retained a consulting firm to study the development of methods for meeting our nation's future electric power needs.

Three alternative methods of power generation are under consideration: nuclear fusion, nuclear fission, and coal gasification. The consulting firm's main task is to estimate the probability of successfully developing at least one of these three methods. In addition, the commission would like the following estimates: (1) the overall time required and the cost incurred for completion of the entire project, and (2) the amount of money needed for each 5-year period over the project's entire 20-year budgeting horizon. These budget figures should be set at a level where a 75% chance exists that sufficient funds will be available over the entire 20 years, and a 90% chance exists that enough money will be available for the first 5-year period. Additionally, the confidence level associated with having enough money available for the remaining four budgeting periods should be uniformly proportionally less than for the first period.

The commission has imposed time and cost limitations of 7 years and $70 million on the research and development (R&D) phase of each method. Failure of any of the three development efforts to complete the R&D phase, or failure to stay within the time and cost constraints, will result in the effort's failure. After successful completion of the R&D phase, commission engineers require that pilot generating stations be built and run to prove and compare the efforts. Required are one station for coal gas, one (selected from two different designs) for fission, and four (all based on the same design) for fusion. (The commission engineers feel that the number of pilot stations requested for testing is commensurate with the development risks inherent in each effort. For example, the fusion development requires the creation of special new alloys that will be able to withstand the high temperatures associated with the fusion process; the fission process has radioactive leakage problems that both, or at least one, of the two different pilot plants should be able to circumvent. In the event that the coal gas or the fission pilot stations fail, the development efforts will be abandoned. The fusion development effort will be considered a failure if more than one of the four stations fail.

Environmentalists have set a design goal for the temperature of the cooling water discharge at 10°F. above the ambient temperature of the receiving lake or stream. Temperature increases greater than 20° are considered unacceptable, while temperature increases under the 10° tolerance mark should merit a bonus. Additionally, power stations must have a high reliability—that is, they must operate without major breakdowns, which cause blackouts. Commission engineers have set a reliability design goal of 90% uptime.

Because funds are limited, only two of the three concepts can be carried to the final phase of development, the shock test phase. The concept that exhibits the lowest performance will be eliminated before this phase, since

Table 13.1. Power-Generating Projects: Research and Development Data

Power-Generating Method	Distribution Type	Time in Years				Cost in Millions (T = time; C = cost)	Probability of Success
		Minimum	Maximum	Mean	Standard Deviation		
Fusion	Normal	4.0	7.5	5.75	1.25	$C = 100 \, (\mathrm{LOG}_{10}(T))$	0.65
Fission	Normal	3.0	7.0	5.0	0.55	$C = T^2 - T + 50$	0.85
Coal Gas	Normal	3.0	6.5	4.75	0.25	$C = 100 + 20 \, (\mathrm{SIN}(T))$	0.95

Table 13.2. Power-Generating Projects: Pilot Station Data

Power-Generating Method	Distribution Type	Minimum Time & Cost	Maximum Time & Cost	Mean or Most Likely Time & Cost	Standard Deviation	Probability of Success
Fusion	Uniform Gamma	8 years $40 million	12 years $60 million	$52 million	$5 million	0.84
Fission	Triangular Lognormal	5 years $40 million	10 years $50 million	8 years $44 million	$3 million	0.92
Coal Gas	Triangular Normal	10 years $20 million	12 years $40 million	11.4 years $27 million	$4 million	0.97

Table 13.3.   Power-Generating Projects: Station Performance Data

Power-Generating Method	Distribution Type	Cooling Water Temperature above Ambient; Station Reliability			No. Exponential Deviates or Standard Deviation
		Minimum	Maximum	Mean	
Fusion	Erlang	5°F.	15°F.	11°F.	8°F.
	Normal	0.70	0.99	0.84	0.01
Fission	Erlang	3°F.	14°F.	9°F.	6°F.
	Normal	0.80	0.99	0.88	0.02
Coal Gas	Erlang	2°F.	12°F.	7°F.	7°F.
	Histogram: Data per figure 13.1				

NOTE: Engineers assess relative importance of above factors as follows: Cooling water = 0.2; station reliability = 0.8.

that concept would be least likely to pass the shock test. There are 72%, 88%, and 93% probabilities, respectively, that the fusion, fission, and coal gas processes will pass the shock test.

Upon completion of the shock tests, the commission will select the winning candidate from the two remaining concepts. The commission members' first preference is the fusion process, followed by the fission process and the coal gas process; this ranking is based on abundance of supplies of

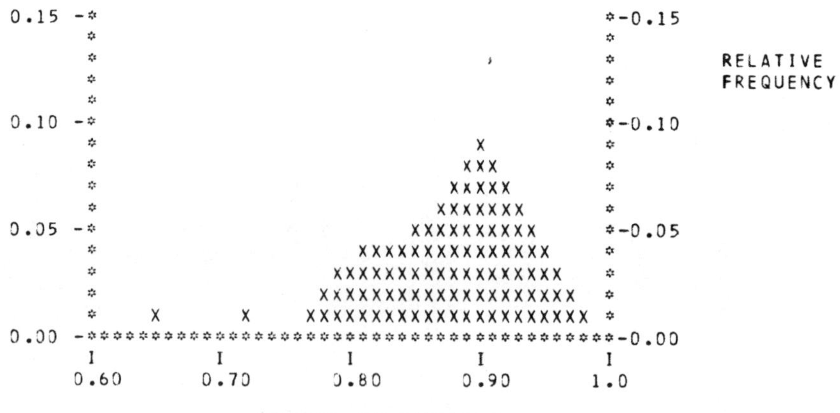

FIGURE 13.1.   Coal Gas Reliability Histogram

the raw materials used by each process. Additional data, supplied by either the commission engineers or the consulting firm, are given in tables 13.1 through 13.3 and in figure 13.1.

NETWORK ANALYSIS

The following narrative is designed to help guide the reader through the pictorial network layout (figure 13.2) and computer solution of the problem described above. Node START initiates three parallel independent network flows, which represent the R&D effort being expended on the three electrical power–generating methods under study. The flow through arc RDFU (fusion) causes that arc to generate a success or fail status via the usual MONTE CARLO procedure, based on the 0.65 probability of successful completion given in table 13.1. This arc also generates the time consumed and the cost incurred (also specified in table 13.1). Likewise, the flows through arcs RDFI and RDCO create the generation of success or fail status and time and cost values for the fission and the coal gas methods, respectively. In a real development project, these three arcs would represent a number of activities and decisions and would likely be expanded into individual subnetworks to attain an adequate level of analytical resolution.

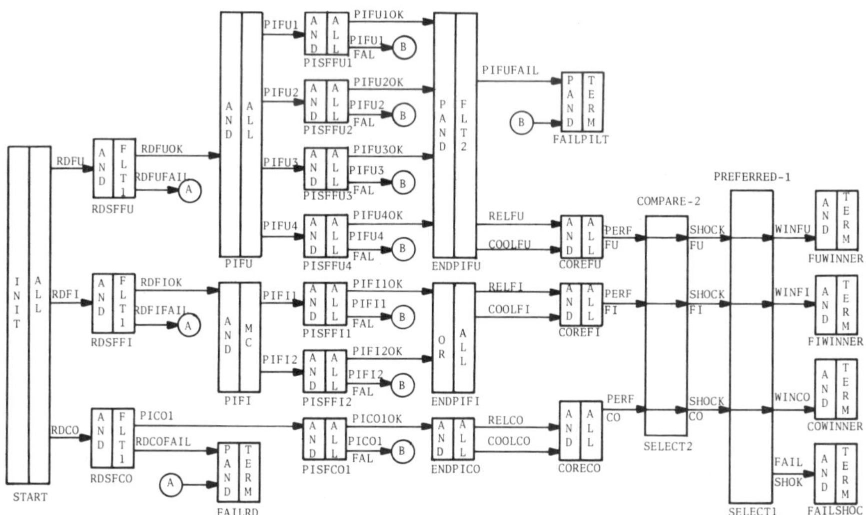

FIGURE 13.2. Future Electric Power–Generating Methods Network

These arcs would then be used to input the results attained in the lower-level subnetworks to the higher-level network.

Node RDSFFU acts as a success/fail determinator of the fusion research and development effort being modeled in arc RDFU. If this effort is successfully completed (that is, if arc RDFU has a success status) within 7 years and $70 million (these two constraints are being carried by arc RDFUOK), node RDSFFU will route the network flow to arc RDFUOK, the success path. This path leads to the pilot plant testing of the successful fusion R&D effort. If the R&D effort is not successful, or if it exceeds the time and cost limitations, the effort is considered a failure; node RDSFFU will then route the fusion network flow to arc RDFUFAIL, the failure path, which terminates further fusion work. Nodes RDSFFI and RDSFCO perform similar functions for the fission and coal gas network flows.

The ALL output logic of node PIFU acts as a network flow expansion device. The four arcs going out of node PIFU represent the four pilot generating stations required to prove out the fusion process. The MONTE CARLO output logic of node PIFI acts as a selector for determining which of the equally likely pilot generating station designs will be used to prove out the fission process.

Arcs PIFU1, PIFU2, PIFU3, and PIFU4 each carry the fusion plant performance data listed in table 13.3. Arcs PIFI1 and PIFI2 each carry the fission data, while PICO1 carries the coal gas data listed in the table. These arcs represent the activities of constructing and running seven pilot generating stations—four based on fusion, two on fission, and one on coal gas. These arcs generate the time consumed, the cost incurred, and the degree of probability of successfully constructing and running each pilot station.

Nodes PISFFU1, PISFFU2, PISFFU3, PISFFU4, PISFFI1, PISFFI2, and PISFCO1 individually route their input flows to either a success path or a fail path, depending on whether their lone input arc was a success or a failure. Every time an input arc has a probability of successful completion of less than 1.0, there is a chance that it will fail. In the event that arc PIFU1 fails, for example, node PISFFU1 will then direct the network flow to arc PIFU1FAL. If PIFU1 is successfully completed, the network flow will then be directed to arc PIFU1OK.

Arcs PIFU1OK, PIFU2OK, PIFU3OK, and PIFU4OK, which carry the successful fusion pilot station flows, converge into node ENDPIFU. Thus, the four parallel flows that went through the four pilot station developments now converge back into one flow through node ENDPIFU (providing none of these four fusion pilot station flows were routed to the failure sink by one of the failure arcs). Arcs PIFU1OK, PIFU2OK, PIFU3OK, and PIFU4OK are "transportation arcs," which carry the network flow from their input to their output nodes without contributing any time, cost, or performance value to the flow. Arcs PIFI1OK, PIFI2OK, and PICO1OK perform analogous functions for the fission and coal gas structures. The node ENDPIFU directs the network flow either to arcs RELFU and COOLFU or to arc PIFUFAIL, depending on the number of successfully completed arcs coming into this node (no more than one fusion pilot failure will be tolerated).

Thus, if three or more fusion pilot stations were running successfully (three or more successfully completed input arcs), the output flow would be passed on to arcs RELFU and COOLFU. The OR input logic of node ENDPIFI requires that one fission pilot station must run successfully before the fission flow will be allowed to flow to arcs RELFI and COOLFI. Node ENDPICO will not be realized unless the coal gas pilot station has been run successfully. If the coal gas flow enters node ENDPICO, this node will pass the flow on to arcs RELCO and COOLCO. Arcs RELFU, RELFI, and RELCO generate station reliability data for the successful fusion, fission, and coal gas processes, respectively. (The data for RELFU and RELFI come from table 13.3, while the data for RELCO comes from figure 13.1.) Arcs COOLFU, COOLFI, and COOLCO generate the cooling water temperature differential above the ambient temperature for the fusion, fission and coal gas processes, respectively. (The data for these arcs come from table 13.3.)

Node COREFU accumulates the fusion performance values generated by arcs RELFU and COOLFU, in addition to carrying the current time and cost values accumulated throughout the network. Nodes COREFI and CORECO accomplish similar tasks for the fission and coal gas alternatives.

Arc PERFFU converts the performance values generated on arcs RELFU and COOLFU into a single weighted value by using VERT's transformations. (These normalization bases come from the problem description, while the preference weights are listed at the bottom of table 13.3.) Arcs PERFFI and PERFCO accomplish similar tasks for the fission and coal gas flows, as PERFFU does for the fusion network flow.

Node SELECT2, with its COMPARE unit logic, narrows the network flow down to two parallel flows if all three generating processes are still being considered. Since the selection weight for node SELECT2 is 1.0 for performance and 0.0 for time and cost, the criterion for flow elimination consists of dropping the flow carrying the lowest performance. If only one or two flows are left coming into SELECT2, this node acts as a transitory device by passing the input arcs' flow directly to their corresponding output arc.

Arcs SHOCKFU, SHOCKFI, and SHOCKCO model the pilot station shock testing for each of the respective generating processes. These arcs simply add a constant time and cost to each of the network flows and have a probability of success of 72%, 88%, and 93%, respectively.

The node SELECT1, with its PREFERRED unit logic, selects the winning generating method from the two methods potentially remaining. SELECT1's PREFERRED logic will give preference to arc WINFU over arc WINFI and to arc WINFI over arc WINCO. This preference is in accord with the commission's observance of the abundance of supplies of raw materials for each of the various electric power–generating methods. Arcs WINFU, WINFI, and WINCO are transportation arcs that constitute the last leg of a successful project completion to terminal nodes FUWINNER, FIWINNER, and COWINNER, respectively. In the event that any of the shock tests fail, SELECT1 will send its flow out arc FAILSHOK on to terminal node FAILSHOC, the last of the three possible failure sinks in which the network flow can terminate.

Terminal node FAILSHOC is a final repository for the network flow. Unlike the other two failure terminal nodes (FAILRD and FAILPILT), FAILSHOC has a higher-priority class designation number. FAILPILT will be selected as the winning terminal node only if there are no flows going into terminal nodes FUWINNER, FIWINNER, COWINNER, or FAIL-SHOC. FAILRD's class designation number is lower than all the other terminal nodes and will be chosen as the winning terminal node only in the event that all the other terminal nodes do not have flows coming into them. This example problem illustrates that the class designation ability of VERT-3 permits one to ignore failure flows not critical enough to terminate a problem. However, if too many individual failures do occur, the problem will be terminated at the last possible project failure point that can occur. The PAND input logic of FAILRD and FAILPILT also aids the network to be run out as far as possible before failure occurs.

RESULTS AND CONCLUSIONS

The computer input for the power-generating problem is shown in the addendum to this chapter. This input yields the outputs shown in figures 13.3 through 13.13 and in table 13.4. The optimum terminal node index bar chart (figure 13.3) indicates that the probability that one of the three generating methods under study will be successfully developed is 94.8% (FUWINNER + FIWINNER + COWINNER = 7.1% + 33.6% + 54.1%). There exist approximately a 1% chance of failure in the pilot plant

Table 13.4   Cost Confidence Balance among Selected Time Periods,
26 Iterations

CFD NO	TIME INTERVAL COVERED		CONFIDENCES	COST INTERPOLATED FOR THE
	START	STOP	COMPUTED	CONFIDENCES COMPUTED
1.	0.0 –	5.00	0.89	201.693024
2.	5.00 –	10.00	0.75	61.7453156
3.	10.00 –	15.00	0.60	22.8625183
4.	15.00 –	20.00	0.45	13.8157101
		SUM OF ABOVE COSTS		300.116211
5.	0.0 –	20.00	0.75	300.456543

test phase (FAILPILT), a 4% chance of failure in the shock test phase
(FAILSHOC), and virtually no chance of failure in the R&D phase
(FAILRD).

The cumulative frequency distribution (CFD) of the network time for the
composite terminal node (see figure 13.4) indicates that the project will terminate somewhere within the time span of 11.08 and 17.53 years. A clustering of times between 15 and 17 years accounts for approximately 94% of the
observations.

The CFD of the overall cost for the composite terminal node (figure
13.5) indicates that the project will cost somewhere between $223.2 and
$509.2 million. Two definite areas of concentration exist in this distribu-
tion—that is, approximately 85% of the observations lie within the first
area of concentration between $223.2 and $327.2 million; the remaining
15% of the observations fall between $418.2 and $509.2 million. The path
cost CFDs for terminal nodes FUWINNER, FIWINNER, and CO-
WINNER (figures 13.6 through 13.8) indicate that fusion development
costs considerably more than fission or coal gas and has a much wider variance; these facts account for the shape of the total project cost CFD.

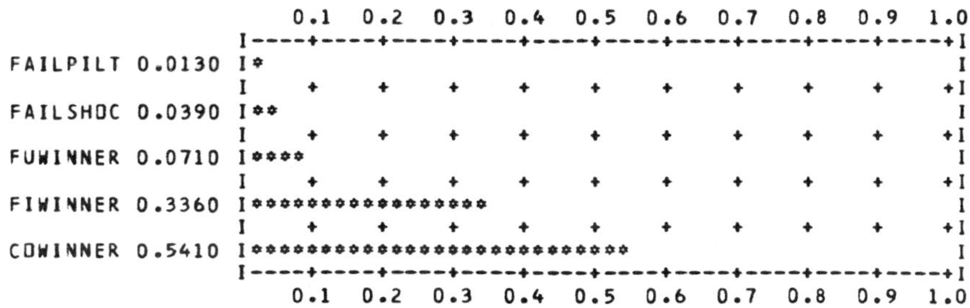

FIGURE 13.3.   Optimum Terminal Index Chart

```
        CFD  0.1  0.2  0.3  0.4  0.5  0.6  0.7  0.8  0.9  1.0
11.08   I----I----I----I----I----I----I----I----I----I----I MIN
        I                                                     0.005
12.00   I
        I                                                     0.005
12.23   I
        I                                                     0.006
12.45   I
        I                                                     0.007
12.68   I
        I                                                     0.007
12.91   I
        I                                                     0.007
13.14   I
        I                                                     0.007
13.36   I
        I                                                     0.007
13.59   I
        I                                                     0.007
13.82   I
        I                                                     0.007
14.05   I
        I                                                     0.008
14.27   I
        I                                                     0.008
14.50   I
        I                                                     0.009
14.73   I
        I*                                                    0.020
14.95   I
        I**                                                   0.044
15.18   I
        I*****                                                0.096
15.41   I
        I********                                             0.165
15.64   I
        I***************                                      0.271
15.86   I
        I**********************                               0.432
16.09   I
        I*******************************                      0.627
16.32   I
        I*****************************************            0.787
16.55   I
        I*************************************************    0.910
16.77   I
        I****************************************************** 0.981
17.00   I
        I******************************************************1.000
17.53   I----I----I----I----I----I----I----I----I----I----I MAX
        NO OBS------------     1000   STD ERROR-       .6070
        COEF OF VARIATION-     0.04   MEAN------       16.10
        KURTOSIS (BETA 2)-     19.66  MEDIAN----       16.16
        PEARSONIAN SKEW---     16.20  MODE------       .1620
```

FIGURE 13.4.   Network Time for the Composite Terminal Node

One of the project guidelines was that the overall budget requested be at
the 75% confidence level, while the budget for the first 5-year period be set
at the 90% confidence level. Inspection of figure 13.5 shows a 76.2% probability that the cost will be $301.2 million or less. In order to determine the

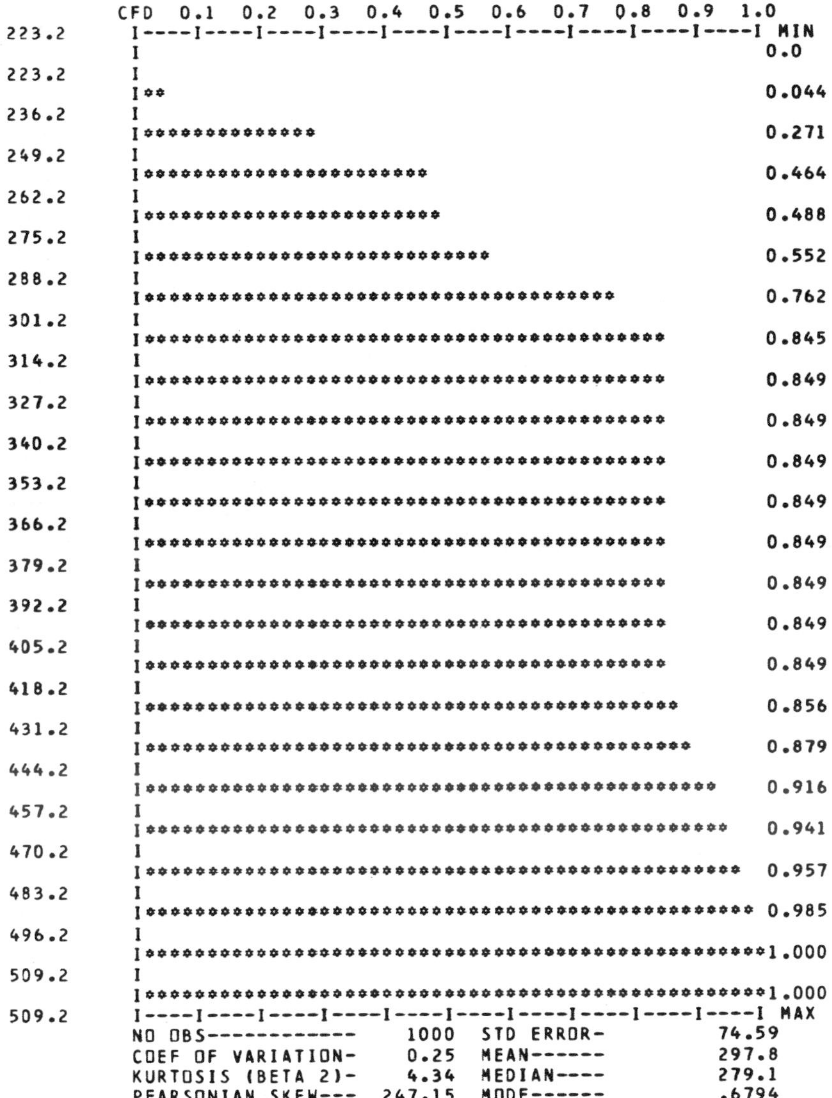

FIGURE 13.5.   Overall Cost for the Composite Terminal Node

exact cost figures corresponding to the specified confidence levels, the program was given initial arbitrary confidence-level values of 0.90, 0.80, 0.70, and 0.60 for time periods one through four and a fixed 0.75 confidence value for the entire period. After simulating the network and then generating a

```
         CFD  0.1  0.2  0.3  0.4  0.5  0.6  0.7  0.8  0.9  1.0
209.4    I----I----I----I----I----I----I----I----I----I----I MIN
         I                                                     0.0
209.4    I
         I*                                                    0.014
213.5    I
         I*                                                    0.014
217.6    I
         I*                                                    0.028
221.7    I
         I****                                                 0.070
225.8    I
         I********                                             0.169
229.9    I
         I**************                                       0.282
234.0    I
         I******************                                   0.352
238.1    I
         I********************                                 0.408
242.2    I
         I**********************                               0.423
246.2    I
         I**********************                               0.423
250.3    I
         I**********************                               0.423
254.4    I
         I**********************                               0.423
258.5    I
         I***********************                              0.451
262.6    I
         I*************************                            0.493
266.7    I
         I**************************                           0.521
270.8    I
         I*****************************                        0.592
274.9    I
         I********************************                     0.634
279.0    I
         I*****************************************            0.761
283.1    I
         I*********************************************        0.845
287.2    I
         I***************************************************  0.930
291.3    I
         I****************************************************** 0.944
295.4    I
         I*******************************************************1.000
299.5    I
         I*******************************************************1.000
299.5    I----I----I----I----I----I----I----I----I----I----I MAX
         NO OBS-------------      71   STD ERROR-         26.31
         COEF OF VARIATION-     0.10   MEAN------         259.6
         KURTOSIS (BETA 2)-     1.45   MEDIAN----         267.6
         PEARSONIAN SKEW---   281.72   MODE------         .8403
```

FIGURE 13.6.   Path Cost for Node FUWINNER

CFD for each of the four budgeting periods (shown in figures 13.9 through 13.12) and for the total project (figure 13.13), VERT-3 was able to select budget values that met the project guidelines. Table 13.4 shows the total cost corresponding to the 75% confidence level ($300.5 million), as well as

```
       CFD  0.1  0.2  0.3  0.4  0.5  0.6  0.7  0.8  0.9  1.0
109.5   I----I----I----I----I----I----I----I----I----I----I MIN
        I                                                     0.0
109.5   I
        I                                                     0.003
110.4   I
        I                                                     0.006
111.3   I
        I                                                     0.006
112.2   I
        I                                                     0.009
113.2   I
        I*                                                    0.015
114.1   I
        I***                                                  0.060
115.0   I
        I****                                                 0.080
115.9   I
        I*******                                              0.131
116.8   I
        I**********                                           0.190
117.7   I
        I***************                                      0.292
118.6   I
        I********************                                 0.381
119.5   I
        I************************                             0.464
120.4   I
        I******************************                       0.586
121.3   I
        I*************************************                0.693
122.2   I
        I**********************************************       0.756
123.1   I
        I***************************************************  0.824
124.0   I
        I****************************************************** 0.878
124.9   I
        I*******************************************************  0.932
125.8   I
        I******************************************************** 0.970
126.7   I
        I********************************************************* 0.982
127.6   I
        I********************************************************** 0.988
128.5   I
        I********************************************************1.000
129.5   I
        I********************************************************1.000
129.5   I----I----I----I----I----I----I----I----I----I----I MAX
        NO OBS------------      336   STD ERROR-       3.435
        COEF OF VARIATION-      0.03  MEAN------       120.6
        KURTOSIS (BETA 2)-      2.82  MEDIAN----       120.5
        PEARSONIAN SKEW---    121.05  MODE------       .1334
```

FIGURE 13.7.   Path Cost for Node FIWINNER

the cost corresponding to the approximately 90% confidence level for the
first 5-year time period ($201.7 million). Also, it was possible to schedule
activities within budget periods two through four so that the confidence level declined uniformly from period to period, as was desired. This solution

```
          CFD  0.1  0.2  0.3  0.4  0.5  0.6  0.7  0.8  0.9  1.0
 90.26    I----I----I----I----I----I----I----I----I----I----I MIN
          I                                                     0.0
 90.26    I
          I*                                                    0.017
 91.31    I
          I****                                                 0.072
 92.36    I
          I******                                               0.124
 93.40    I
          I*********                                            0.183
 94.45    I
          I**************                                       0.274
 95.49    I
          I*****************                                    0.344
 96.54    I
          I************************                             0.460
 97.58    I
          I******************************                       0.579
 98.63    I
          I***********************************                  0.671
 99.67    I
          I*****************************************            0.784
 100.7    I
          I*******************************************          0.861
 101.8    I
          I***********************************************      0.902
 102.8    I
          I**************************************************   0.933
 103.9    I
          I***************************************************  0.967
 104.9    I
          I**************************************************** 0.982
 105.9    I
          I***************************************************** 0.989
 107.0    I
          I*****************************************************0.993
 108.0    I
          I*****************************************************0.994
 109.1    I
          I*****************************************************0.996
 110.1    I
          I*****************************************************0.998
 111.2    I
          I*****************************************************0.998
 112.2    I
          I*****************************************************1.000
 113.3    I
          I*****************************************************1.000
 113.3    I----I----I----I----I----I----I----I----I----I----I MAX
          NO OBS------------    541  STD ERROR-       3.717
          COEF OF VARIATION-    0.04  MEAN------      97.97
          KURTOSIS (BETA 2)-    3.22  MEDIAN----      97.85
          PEARSONIAN SKEW---   97.65  MODE------      .8636E-01
```

FIGURE 13.8.   Path Cost for Node COWINNER

```
        CFD  0.1  0.2  0.3  0.4  0.5  0.6  0.7  0.8  0.9  1.0
180.1    I----I----I----I----I----I----I----I----I----I----I MIN
         I                                                      0.005
185.0    I
         I                                                      0.009
186.2    I
         I*                                                     0.021
187.4    I
         I***                                                   0.052
188.5    I
         I*****                                                 0.106
189.7    I
         I*********                                             0.175
190.9    I
         I************                                          0.250
192.1    I
         I*******************                                   0.354
193.3    I
         I***********************                               0.450
194.5    I
         I*******************************                       0.557
195.6    I
         I*************************************                 0.654
196.8    I
         I*******************************************           0.742
198.0    I
         I**************************************************    0.818
199.2    I
         I*****************************************************   0.858
200.4    I
         I******************************************************  0.885
201.5    I
         I*******************************************************  0.905
202.7    I
         I********************************************************  0.921
203.9    I
         I*********************************************************  0.936
205.1    I
         I**********************************************************  0.949
206.3    I
         I***********************************************************  0.967
207.5    I
         I************************************************************ 0.976
208.6    I
         I************************************************************* 0.985
209.8    I
         I*************************************************************0.994
211.0    I
         I**************************************************************1.000
212.8    I----I----I----I----I----I----I----I----I----I----I MAX
         NO OBS------------     1000   STD ERROR-      5.116
         COEF OF VARIATION-     0.03   MEAN------      195.5
         KURTOSIS (BETA 2)-     3.72   MEDIAN----      195.0
         PEARSONIAN SKEW---   195.07   MODE------      .8996E-01
```

FIGURE 13.9.   Budget Period One: Positive Cost Incurred, Years 0.00–5.00

```
              CFD  0.1  0.2  0.3  0.4  0.5  0.6  0.7  0.8  0.9  1.0
  9.680       I----I----I----I----I----I----I----I----I----I----I MIN
              I                                                      0.0
  9.680       I
              I***                                                   0.057
 17.26        I
              I********                                              0.153
 24.84        I
              I***************                                       0.296
 32.42        I
              I*********************                                 0.426
 39.99        I
              I**************************                            0.554
 47.57        I
              I*******************************                       0.663
 55.15        I
              I************************************                  0.763
 62.73        I
              I****************************************              0.830
 70.31        I
              I*******************************************           0.848
 77.89        I
              I*********************************************         0.849
 85.47        I
              I*********************************************         0.849
 93.05        I
              I**********************************************        0.850
100.6         I
              I***********************************************       0.859
108.2         I
              I************************************************      0.874
115.8         I
              I**************************************************    0.905
123.4         I
              I***************************************************   0.936
130.9         I
              I****************************************************  0.949
138.5         I
              I*****************************************************  0.964
146.1         I
              I****************************************************** 0.989
153.7         I
              I******************************************************0.998
161.3         I
              I******************************************************0.999
168.8         I
              I******************************************************1.000
176.4         I
              I******************************************************1.000
176.4         I----I----I----I----I----I----I----I----I----I----I MAX
              NO OBS------------    1000  STD ERROR-        35.91
              COEF OF VARIATION-    0.66  MEAN------        54.26
              KURTOSIS (BETA 2)-    4.15  MEDIAN----        44.32
              PEARSONIAN SKEW---   30.77  MODE------        .6540
```

FIGURE 13.10.   Budget Period Two: Positive Cost Incurred, Years 5.00–10.00

```
       CFD  0.1   0.2   0.3   0.4   0.5   0.6   0.7   0.8   0.9   1.0
 8.503   I----I----I----I----I----I----I----I----I----I----I MIN
         I                                                       0.0
 8.503   I
         I*********************                                   0.422
14.07    I
         I*******************************                         0.528
19.63    I
         I***********************************                     0.652
25.20    I
         I**********************************************          0.794
30.76    I
         I************************************************        0.843
36.32    I
         I*************************************************       0.847
41.89    I
         I**************************************************      0.849
47.45    I
         I**************************************************      0.849
53.02    I
         I**************************************************      0.849
58.58    I
         I**************************************************      0.849
64.15    I
         I**************************************************      0.849
69.71    I
         I**************************************************      0.849
75.28    I
         I**************************************************      0.849
80.84    I
         I***************************************************     0.853
86.41    I
         I****************************************************    0.861
91.97    I
         I*****************************************************   0.883
97.53    I
         I*******************************************************  0.919
103.1    I
         I******************************************************** 0.953
108.7    I
         I*********************************************************0.980
114.2    I
         I*********************************************************0.990
119.8    I
         I*********************************************************0.995
125.4    I
         I*********************************************************1.000
130.9    I
         I*********************************************************1.000
130.9    I----I----I----I----I----I----I----I----I----I----I MAX
         NO OBS------------    1000    STD ERROR-        31.91
         COEF OF VARIATION-    1.03    MEAN------        31.05
         KURTOSIS (BETA 2)-    4.69    MEDIAN----        17.37
         PEARSONIAN SKEW---   11.68    MODE------         .6068
```

FIGURE 13.11.   Budget Period Three: Positive Cost Incurred, Years 10.00–15.00

```
        CFD  0.1   0.2   0.3   0.4   0.5   0.6   0.7   0.8   0.9   1.0
.5393   I----I----I----I----I----I----I----I----I----I----I MIN
        I                                                      0.0
.5393   I
        I*                                                     0.013
2.639   I
        I*                                                     0.022
4.738   I
        I**                                                    0.031
6.837   I
        I**                                                    0.034
8.937   I
        I*****                                                 0.094
11.04   I
        I*******************                                   0.408
13.14   I
        I***************************                           0.537
15.23   I
        I***************************                           0.543
17.33   I
        I***************************                           0.548
19.43   I
        I********************************                      0.631
21.53   I
        I*********************************************         0.854
23.63   I
        I************************************************      0.919
25.73   I
        I*************************************************     0.943
27.83   I
        I**************************************************    0.964
29.93   I
        I**************************************************    0.977
32.03   I
        I***************************************************   0.984
34.13   I
        I***************************************************0.992
36.23   I
        I****************************************************0.997
38.33   I
        I****************************************************0.999
40.43   I
        I****************************************************0.999
42.53   I
        I****************************************************0.999
44.63   I
        I*****************************************************1.000
46.73   I
        I*****************************************************1.000
46.73   I----I----I----I----I----I----I----I----I----I----I MAX
        NO OBS------------     987   STD ERROR-        6.771
        COEF OF VARIATION-     0.39  MEAN------        17.34
        KURTOSIS (BETA 2)-     3.10  MEDIAN----        13.84
        PEARSONIAN SKEW---    12.25  MODE------         .7518
```

FIGURE 13.12.   Budget Period Four: Positive Cost Incurred, Years 15.00–20.00

```
              CFD  0.1   0.2   0.3   0.4   0.5   0.6   0.7   0.8   0.9   1.0
 223.2        I----I----I----I----I----I----I----I----I----I----I MIN
              I                                                        0.0
 223.2        I
              I**                                                      0.044
 236.2        I
              I*************                                           0.271
 249.2        I
              I*********************                                   0.464
 262.2        I
              I**********************                                  0.486
 275.2        I
              I*************************                               0.548
 288.2        I
              I*********************************                       0.762
 301.2        I
              I************************************                    0.845
 314.2        I
              I*************************************                   0.849
 327.2        I
              I*************************************                   0.849
 340.2        I
              I*************************************                   0.849
 353.2        I
              I*************************************                   0.849
 366.2        I
              I*************************************                   0.849
 379.2        I
              I*************************************                   0.849
 392.2        I
              I*************************************                   0.849
 405.2        I
              I*************************************                   0.849
 418.2        I
              I*************************************                   0.856
 431.2        I
              I**************************************                  0.878
 444.2        I
              I***************************************                 0.915
 457.2        I
              I****************************************                0.941
 470.2        I
              I*****************************************               0.956
 483.2        I
              I*******************************************             0.985
 496.2        I
              I*********************************************1.000
 509.2        I
              I*********************************************1.000
 509.2        I----I----I----I----I----I----I----I----I----I----I MAX
              NO OBS------------    1000    STD ERROR-       74.69
              COEF OF VARIATION-    0.25    MEAN------       298.0
              KURTOSIS (BETA 2)-    4.33    MEDIAN----       280.0
              PEARSONIAN SKEW---  247.15    MODE------        .6802
```

FIGURE 13.13. Total Budget Period: Positive Cost Incurred, Years 0.00–20.00

consisted of budgeting $201.7 million, $61.7 million, $22.9 million, and $13.8 million for periods one through four to yield a total of $300.1 million for the entire development period.

The pilot plant example illustrates more of VERT-3's features and capabilities than did the preceding examples. Of particular value to management is the ability to develop cost figures by budget period. Cost limits can also be set by budget period, and the technique will adjust activities (arcs) to stay within the budget limitations, as was done in this example.

ADDENDUM: THE COMPUTER INPUT

```
PROBLEM IDENTIFICATION CARD OPTION---------------------------           1

TYPE OF INPUT OPTION-------------------------------------------           0

TYPE OF OUTPUT OPTION------------------------------------------           4

COSTING AND PRUNING OPTION-------------------------------------           1

FULL PRINT TRIP OPTION-----------------------------------------           0

CORRELATION COMPUTATION AND PLOT OPTION-----------------------           0

COST-PERFORMANCE TIME INTERVAL OPTION-------------------------           1

COMPOSITE TERMINAL NODE MINIMUMS AND MAXIMUMS OPTION-------           1

TERMINAL NODE LISTING OPTION----------------------------------           6

INITIAL SEED--------------------------------------------------      435459

NUMBER OF ITERATIONS------------------------------------------        1000

YEARLY INTEREST RATE USED FOR INFLATION ADJUSTMENTS--------         0.0

YEARLY INTEREST RATE USED FOR PRESENT VALUE DISCOUNTING----         0.0

TIME FACTOR WHICH CONVERTS PROGRAM TIME TO A YEARLY BASE---         0.0

                                    TIME      COST      PERF

TERMINAL NODE SELECTION WEIGHTS     1.00      0.0       0.0

CRITICAL - OPTIMUM PATH WEIGHTS     0.0       0.0       0.0

INITIAL VALUES                      0.0       0.0       0.0

COST-PERFORMANCE TIME INTERVAL DATA

      0.0        5.0      185.0      211.0                    0.90      0.0005
      5.0       10.0                                          0.80      0.002
     10.0       15.0                                          0.70      0.004
     15.0       20.0                                          0.60      0.006
      0.0       20.0                                          0.75
ENDCTPR

MINIMUMS AND MAXIMUMS FOR THE COMPOSITE TERMINAL NODE

     12.0       17.0
```

```
RDFU      START    RDSFFU  0.65 FUSION RESEARCH AND DEVELOPMENT
RDFU      DTIME 1    4.0          4.0          7.5          5.75         1.25
RDFU      RCOST 115SK   100.0 TRDFJ    K    1.0

RDFI      START    RDSFFI  0.85 FISSION RESEARCH AND DEVELOPMENT
RDFI      DTIME 1    4.0          3.0          7.0          5.0          0.55
RDFI      RCOST 1 10TRDFI     TRDFI    K    1.0   6S    1.0   K   50.0   TRDFI

RDCO      START    RDSFCO  0.97 COAL GAS RESEARCH AND DEVELOPMENT
RDCO      DTIME 1    4.0          3.0          6.5          4.75         0.25
RDCO      RCOST 1 1SK    80.0 K  1.0    K    1.0   16SK  20.0   TRDCO     K  1.0

RDFJOK   RDSFFU   PIFU    1.0  FUSION R&D OKAY, TRANSPORT TO PILOT TESTS
RDFUOK   FILT1 1   0.0          7.0          0.0          70.0

RDFIOK   RDSFFI   PIFI    1.0  FISSION R&D OKAY, TRANSPORT TO PILOT TEST
RDFIOK   FILT1 1   0.0          7.0          0.0          70.0

RDFJFAILRDSFFU  FAILRD  1.0  FUSION R&D FAILED, GO TO R&D FAILURE SINK

RDFIFAILRDSFFI  FAILRD  1.0  FISSION R&D FAILED, GO TO R&D FAILURE SINK

RDCOFAILRDSFCO  FAILRD  1.0  COAL GAS R&D FAILED, GO TO R&D FAILURE SINK

PIFU1     PIFU     PISFFU1 0.84 DEVELOPMENT OF FUSION STATION #1
PIFU1     DTIME 1    2.0          8.0         12.0
PIFU1     DCOST 1    6.0         40.0         60.0         52.0          5.0

PIFU2     PIFU     PISFFU2 0.84 DEVELOPMENT OF FUSION STATION #2
PIFU2     DTIME 1    2.0          8.0         12.0
PIFU2     DCOST 1    6.0         40.0         60.0         52.0          5.0

PIFU3     PIFU     PISFFU3 0.84 DEVELOPMENT OF FUSION STATION #3
PIFU3     DTIME 1    2.0          8.0         12.0
PIFU3     DCOST 1    6.0         40.0         60.0         52.0          5.0

PIFU4     PIFU     PISFFU4 0.84 DEVELOPMENT OF FUSION STATION #4
PIFU4     DTIME 1    2.0          8.0         12.0
PIFU4     DCOST 1    6.0         40.0         60.0         52.0          5.0

PIFI1     PIFI     PISFFI1 0.92 DEVELOPMENT OF FISSION STATION #1
PIFI1     DTIME 1    3.0          5.0         10.0          8.0
PIFI1     DCOST 1    5.0         40.0         50.0         44.0          3.0
PIFI1     M     1    0.5

PIFI2     PIFI     PISFFI2 0.92 DEVELOPMENT OF FISSION STATION #2
PIFI2     DTIME 1    3.0          5.0         10.0          8.0
PIFI2     DCOST 1    5.0         40.0         50.0         44.0          3.0
PIFI2     M     1    0.5

PICO1     RDSFCO   PISFCO1 0.97 DEVELOPMENT OF COAL GAS PILOT STATION #1
PICO1     FILT1 1    0.0          7.0          0.0         70.0
PICO1     DTIME 1    3.0         10.0         12.0         11.4
PICO1     DCOST 1    4.0         20.0         40.0         27.0          4.0

PIFU1OK PISFFU1 ENDPIFU 1.0  FUSION PILOT STATION #1 SUCCESSFUL

PIFU2OK PISFFU2 ENDPIFU 1.0  FUSION PILOT STATION #2 SUCCESSFUL

PIFU3OK PISFFU3 ENDPIFU 1.0  FUSION PILOT STATION #3 SUCCESSFUL

PIFU4OK PISFFU4 ENDPIFU 1.0  FUSION PILOT STATION #4 SUCCESSFUL

PIFI1OK PISFFI1 ENDPIFI 1.0  FISSION PILOT STATION #1 SUCCESSFUL

PIFI2OK PISFFI2 ENDPIFI 1.0  FISSION PILOT STATION #2 SUCCESSFUL

PICO1OK PISFCO1 ENDPICO 1.0  COAL GAS PILOT STATION #1 SUCCESSFUL

PIFU1FALPISFFU1 FAILPILT1.0  FUSION PILOT STATION #1 FAILED
```

(cont.)

```
PIFU2FALPISFFU2 FAILPILT1.0  FUSION PILOT STATION #2 FAILED

PIFU3FALPISFFU3 FAILPILT1.0  FUSION PILOT STATION #3 FAILED

PIFU4FALPISFFU4 FAILPILT1.0  FUSION PILOT STATION #4 FAILED

PIFI1FALPISFFI1 FAILPILT1.0  FISSION PILOT STATION #1 FAILED

PIFI2FALPISFFI2 FAILPILT1.0  FISSION PILOT STATION #2 FAILED

PICO1FALPISFCO1 FAILPILT1.0  COAL GAS PILOT STATION #1 FAILED

PIFUFAILENDPIFU FAILPILT1.0  TOO MANY FUSION PILOT STATIONS FAILED

RELFU   ENDPIFU COREFU   1.0  STATION RELIABILITY-FUSION
RELFU   FILT2 1   3.0        4.0
RELFU   DPERF 1   4.0        0.70       0.99       0.85       0.01

COOLFU  ENDPIFU COREFU   1.0  COOLING WATER TEMPERATURE-FUSION
COOLFU  FILT2 1   3.0        4.0
COOLFU  DPERF 1   8.0        5.0        15.0       11.0       8.0

RELFI   ENDPIFI COREFI   1.0  STATION RELIABILITY-FISSION
RELFI   DPERF 1   4.0        0.80       0.99       0.88       0.02

COOLFI  ENDPIFI COREFI   1.0  COOLING WATER TEMPERATURE-FISSION
COOLFI  DPERF 1   8.0        3.0        14.0       9.0        6.0

RELCO   ENDPICO CORECO   1.0  STATION RELIABILITY-COAL GAS
RELCO   HPERF 1   0.629      0.01       0.631      0.0        0.679      0.01
RELCO   HPERF 2   0.681      0.0        0.745      0.01       0.755      0.01
RELCO   HPERF 3   0.765      0.01       0.775      0.01       0.785      0.01
RELCO   HPERF 4   0.795      0.02       0.805      0.02       0.815      0.03
RELCO   HPERF 5   0.825      0.03       0.835      0.04       0.845      0.04
RELCO   HPERF 6   0.855      0.05       0.865      0.06       0.875      0.07
RELCO   HPERF 7   0.885      0.08       0.895      0.09       0.905      0.08
RELCO   HPERF 8   0.915      0.07       0.925      0.06       0.935      0.05
RELCO   HPERF 9   0.945      0.04       0.955      0.04       0.965      0.03
RELCO   HPERF10   0.975      0.02       0.985      0.01       0.995

COOLCO  ENDPICO CORECO   1.0  COOLING WATER TEMPERATURE-COAL GAS
COOLCO  DPERF 1   8.0        2.0        12.0       7.0        7.0

PERFFU  COREFU  SELECT2 1.0  AGGREGATE THE PERFORMANCE FOR FUSION
PERFFU  RPERF 1 1SPCOREFU  K -1.0   K   1.0    2SK  10.0   K  0.2    PCOOLFU
PERFFU  RPERF 2 2SPRELFU   K  0.8   K   0.90

PERFFI  COREFI  SELECT2 1.0  AGGREGATE THE PERFORMANCE FOR FISSION
PERFFI  RPERF 1 1SPCOREFI  K -1.0   K   1.0    2SK  10.0   K  0.2    PCOOLFI
PERFFI  RPERF 2 2SPRELFI   K  0.8   K   0.90

PERFCO  CORECO  SELECT2 1.0  AGGREGATE THE PERFORMANCE FOR COAL GAS
PERFCO  RPERF 1 1SPCORECO  K -1.0   K   1.0    2SK  10.0   K  0.2    PCOOLCO
PERFCO  RPERF 2 2SPRELCO   K  0.8   K   0.90

SHOCKFU SELECT2 SELECT1 0.72 SHOCK TEST FOR FUSION
SHOCKFU RTIME 1 1SK  0.2   K   1.0    K   1.0
SHOCKFU DCOST 1    1.0        10.0

SHOCKFI SELECT2 SELECT1 0.88 SHOCK TEST FOR FISSION
SHOCKFI DTIME 1    1.0        0.2
SHOCKFI DCOST 1    1.0        10.0

SHOCKCO SELECT2 SELECT1 0.93 SHOCK TEST FOR COAL GAS
SHOCKCO DTIME 1    1.0        0.2
SHOCKCO DCOST 1    1.0        10.0

WINFU   SELECT1 FUWINNER1.0  FUSION IS THE WINNER

WINFI   SELECT1 FIWINNER1.0  FISSION IS THE WINNER

WINCO   SELECT1 COWINNER1.0  COAL GAS IS THE WINNER
```

```
FAILSHCKSELECT1 FAILSHOC1.0  TOTAL FAILURE OF THE SHOCK TEST

ENDARC

START    1   2                    STARTING POINT FOR THE NETWORK

RDSFFU   2   4                    R&D SUCCESS-FAIL DETERMINATOR-FUSION

RDSFFI   2   4                    R&D SUCCESS-FAIL DETERMINATOR-FISSION

RDSFCO   2   4                    R&D SUCCESS-FAIL DETERMINATOR-COAL GAS
PIFU     2   2                    INITIATION PILOT STATION-CONSTRUCTION-FUSION

PIFI     2   3                    INITIATION PILOT STATION-CONSTRUCTION-FISSION

FAILRD   3 3115                   FAILURE OF R&D EFFORT

PISFFU1 2   2                     PILOT SUCCESS-FAIL DETERMINATOR STATION 1-FUSION

PISFFU2 2   2                     PILOT SUCCESS-FAIL DETERMINATOR STATION 2-FUSION

PISFFU3 2   2                     PILOT SUCCESS-FAIL DETERMINATOR STATION 3-FUSION

PISFFU4 2   2                     PILOT SUCCESS-FAIL DETERMINATOR STATION 4-FUSION

PISFFI1 2   2                     PILOT SUCCESS-FAIL DETERMINATOR STATION 1-FISSION

PISFFI2 2   2                     PILOT SUCCESS-FAIL DETERMINATOR STATION 2-FISSION

PISFCO1 2   2                     PILOT SUCCESS-FAIL DETERMINATOR STATION 1-COAL GAS

ENDPIFU 3   5                     END OF PILOT STATION EXERCISE-FUSION

ENDPIFI 4   2                     END OF PILOT STATION EXERCISE-FISSION

ENDPICO 2   2                     END OF PILOT STATION EXERCISE-COAL GAS

FAILPILT3 2115                    FAILURE OF PILOT STATIONS TESTS

COREFU   2   2                    COOL + RELIABILITY-RAW GENERATES-FUSION

COREFI   2   2                    COOL + RELIABILITY-RAW GENERATES-FISSION

CORECO   2   2                    COOL + RELIABILITY-RAW GENERATES-COAL GAS

SELECT2 5  -2            1.0 SELECT 2 METHODS FOR SHOCK TESTING

SELECT1 6  -1                SELECT THE MOST DESIRED METHOD

FAILSHOC2 1115                    FAILURE OF THE SHOCK TEST

FUWINNER2  1 2                    SUCCESSFULL PROJECT COMPLETION VIA FUSION

FIWINNER2  1 2                    SUCCESSFULL PROJECT COMPLETION VIA FISSION
FIWINNERHIST  15.00    17.00              275.00  500.00   0.90    1.35

COWINNER2  1 2                    SUCCESSFULL PROJECT COMPLETION VIA COAL GAS

ENDNODE

W A R N I N G  NO.  6144

W A R N I N G  NO.  6156

     OPTIMUM TERMINAL NODE INDEX - NO. ITERATIONS = 1000
```

14 ANALYSIS OF STRATEGIC DECISIONS: *Mergers and Acquisitions*

Stochastic network techniques—VERT, in particular—are "naturals" for analysis of strategic decisions. We have seen that VERT-3 can analyze alternative courses of action in advance by incorporating various relationships and probabilistic data and can generate stochastic output information in terms of performance, costs, time, and risks. While each of the problems examined thus far—development of the Z-car, the plant investment decision, the new-product development problem, decision trees, and the alternative energy sources analysis—could be viewed from a strategic perspective, certain decisions are perhaps more "strategic" in nature. Such decisions would pertain to mergers and acquisitions, diversification, business unit analysis, industry analyses, product life-cycle considerations, scenario analysis, and other situations.

As we stated earlier, strategic decision making requires managers to deal with four key parameters when they evaluate alternative courses of action:

1. *Results* (including return on investment, sales, profits, and other measures of performance relating to available alternatives);
2. *Costs* (resources required, both in total and by time period);

202

3. *Time* (the schedule for developing, implementing,and producing the alternatives);
4. *Risk* (the likelihood of successful and unsuccessful outcomes for the time, cost, and results parameters).

Managers charged with strategic decision making realize that significant interactions exist among these four parameters; they are not independent. Also, incomplete or inadequate information about the alternatives may be the rule rather than the exception and may make the strategic planner's task still more difficult.

Perhaps one of the more clearly "strategic" decisions involves a company's merger and/or acquisition activities. This chapter describes a hypothetical merger/acquisition situation with alternatives that represent varying degrees of synergy, reward, stability, and risk.

THE MERGER/ACQUISITION PROBLEM

MDCO is a well-established, profitable, conservatively growing, and healthy company. Its management is considering growth via the merger/acquisition route and wants to evaluate which (if either) of two companies it should acquire. One company, XRX, is a relatively small, rapidly growing company with a somewhat variable growth rate. The other company, ABC, is larger and more stable than XRX, although equally profitable.

MDCO's objective is to maximize its stock market quotation over the next 10 years; it hopes for certain synergies with the acquisition candidates. Probabilistic estimates have been developed for the economy and for financial markets for the next 10 years, as well as for the various merger candidates: MDCO alone/XRX and ABC alone, MDCO and XRX merged, and MDCO and ABC merged. These data are shown in table 14.1. MDCO currently has 5 times the sales of XRX and 2 1/2 times those of ABC. MDCO can be expected to grow at 10% per year, with a plus or minus 2% variation. XRX will grow at 20%, with an 8% expected volatility, while ABC will grow only at 5% with a ½% volatility expected. Acquisition of XRX will raise MDCOs growth rate to 13% and its variability to 4%; acquisition of ABC, however, will lower MDCOs growth rate to 9% and keep the 2% variation. Acquisition of either company will improve MDCOs profit margin, but the MDCO/XRX marriage will somewhat increase profit variation. MDCO stock currently sells for 11.5 times its earnings; XRX stock, for 12 times its earnings; ABC's PE ratio is 9.5. Management feels that XRX

Table 14.1. Merger/Acquisition Problem: Probabilistic Data Estimates

Item	MDCO Alone	XRX Alone	MDCO + XRX	ABC Alone	MDCO + ABC
Current sales	$250 million	$50 million		$100 million	
Growth rate	$10\% \pm 2\%$/yr.	$20\% \pm 8\%$	$13\% \pm 4\%$	$5\% \pm \frac{1}{2}\%$	$9\% \pm 2\%$
Profit	$12\% \pm 3\%$ of sales	$20\% \pm 5\%$ of sales	$15\% \pm 4\%$ of sales	$20\% \pm 3\%$ of sales	$15\% \pm 3\%$ of sales
Tax rate	40%	35%	40%	45%	42%
Current earnings	$18 million	$6.5 million		$11 million	
Shares	3 million	1 million	4.3–5 million	2 million	3.75–4.1 million
EPS	$6.00	$6.50		$5.50	
Dividends	35% of EPS	None	20% of EPS	50% of EPS	40% of EPS
Stock price	$69.00	$78.00		$52.25	
Consistency	0.80	0.60	0.69	0.90	0.78
Cost of acquisition			1.3–2 million shares		0.75–1.1 million shares

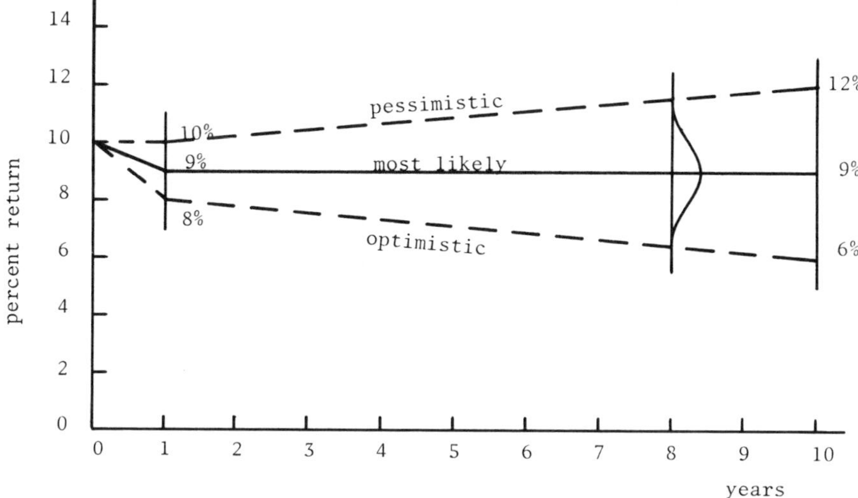

FIGURE 14.1.   Estimated AAA Bond Rates

could be acquired for 1.3 to 2.0 million shares of MDCO, while ABC would
be available for the bargain price of 0.75 to 1.1 million shares of MDCO.

MDCO's financial advisers have developed a mathematical model of its
per share stock price, which is to be used as the criterion function in analysis
of the three alternative decisions. This model calculates stock price as follows:

$$\text{Price} = EPS(10 + 10GC - 10B) + \frac{D}{B},$$

where EPS = earnings per share, G = growth rate of sales, C = growth
consistency, B = AAA bond rate of return, and D = dividends per share.

Growth rate is a stochastic variable that varies around a projected base
figure. Profit is assumed to be a certain percentage of sales, again including
a stochastic component. Shares of stock are known for MDCO, XRX, and
ABC but can vary for the merger alternatives since a stock trade is assumed
as the price of the merger or acquisition. Various tax rates are assumed in
order to allow determination of after-tax earnings. Earnings per share can
be calculated from the values of after-tax earnings divided by the possible
shares of stock resulting from the terms of the acquisition. Dividends per
share are determined from assumed pay-out rates. Bond yields are currently
10% and are likely to vary in the future according to the distribution shown
in figure 14.1. Other data for the three companies and the three alternatives
(don't merge, acquire XRX, acquire ABC) are also included in table 14.1.

VERT-3 ANALYSIS AND RESULTS

The pictorial network layout of the problem is illustrated in figure 14.2, which shows the three alternatives (arcs SHARE P1, SHARE P2, and SHARE P3) emanating from node START and then flowing into nodes SPALT 1, SPALT 2, and SPALT 3. The arcs SHARE P1, SHARE P2, and SHARE P3 generate the stock price per share for each of the three decision alternatives by incorporating all of the probabilistic data included in table 14.1. Nodes SPALT 1, SPALT 2, and SPALT 3 accumulate this information and prepare the distributions of likely outcomes shown in figures 14.3, 14.4, and 14.5. Arcs T1, T2, and T3 are merely transport arcs that carry the results to the terminal node, END.

Simulating the problem 250 times yielded the results shown in the figures. Figure 14.3 depicts the distribution of stock prices resulting from the nonmerger alternative. Examination indicates a mean per share price of $220.50, with a standard error of $71.69 and a range of $74.11 to $506.90. The acquisition of XRX (shown in figure 14.4) yields a mean per share price of $260.00, with a standard error of $120.50 and a range of $84.45 to $665.90. Acquisition of ABC (shown in figure 14.5) results in a mean per share price of $260.60, with a standard error of $66.90 and a range of $114.50 to $440.70.

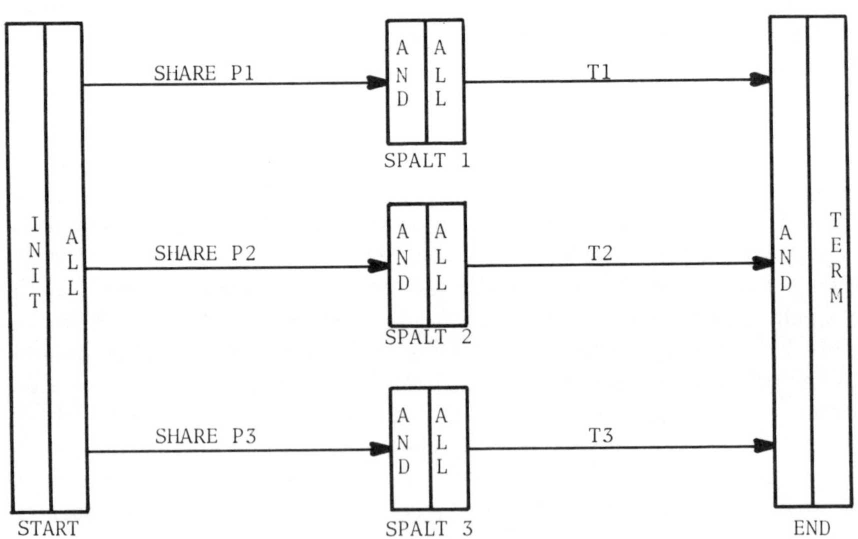

FIGURE 14.2.   VERT Acquisition/Merger Network

```
PATH PERFORMANCE FOR MODE SPALT1


                  CFD  0.1   0.2   0.3   0.4   0.5   0.6   0.7   0.8   0.9   1.0
   53.49          I----I----I----I----I----I----I----I----I----I----I MIN
                  I                                                       0.000
   53.49          I
                  I                                                       0.004
   74.11          I
                  I*                                                      0.024
   94.72          I
                  I***                                                    0.056
   115.3          I
                  I***                                                    0.068
   135.9          I
                  I*******                                                0.156
   156.6          I
                  I**************                                         0.272
   177.2          I
                  I*********************                                  0.420
   197.8          I
                  I**************************                             0.536
   218.4          I
                  I*************************************                  0.652
   239.0          I
                  I*********************************************          0.764
   259.6          I
                  I*************************************************      0.860
   280.2          I
                  I****************************************************   0.896
   300.8          I
                  I*******************************************************0.916
   321.4          I
                  I*******************************************************0.936
   342.1          I
                  I*******************************************************0.964
   362.7          I
                  I*******************************************************0.980
   383.3          I
                  I*******************************************************0.984
   403.9          I
                  I*******************************************************0.984
   424.5          I
                  I*******************************************************0.984
   445.1          I
                  I*******************************************************0.984
   465.7          I
                  I*******************************************************0.992
   486.3          I
                  I*******************************************************1.000
   506.9          I
                  I*******************************************************1.000
   506.9          I----I----I----I----I----I----I----I----I----I----I MAX


         NO OBS-------------     250   STD ERROR-          71.69
         COEF OF VARIATION-     0.33   MEAN------          220.5
         KURTOSIS (BETA 2)-     5.19   MEDIAN----          212.2
         PEARSONIAN SKEW---     0.46   MODE------          187.5
```

FIGURE 14.3.   Alternative One: No Merger

```
PATH PERFORMANCE FOR NODE SPALT2

                  CFD  0.1   0.2   0.3   0.4   0.5   0.6   0.7   0.8   0.9   1.0
     56.76        I----I----I----I----I----I----I----I----I----I----I----I MIN
                  I                                                            0.000
     56.76        I
                  I*                                                           0.024
     84.45        I
                  I****                                                        0.080
     112.1        I
                  I*******                                                     0.144
     139.8        I
                  I***********                                                 0.244
     167.5        I
                  I****************                                            0.349
     195.2        I
                  I**********************                                      0.469
     222.9        I
                  I***************************                                 0.536
     250.6        I
                  I********************************                            0.620
     278.3        I
                  I***********************************                         0.692
     305.9        I
                  I**************************************                      0.760
     333.6        I
                  I*****************************************                   0.812
     361.3        I
                  I********************************************                0.848
     389.0        I
                  I***********************************************             0.884
     416.7        I
                  I**************************************************          0.900
     444.4        I
                  I****************************************************        0.920
     472.1        I
                  I******************************************************      0.964
     499.7        I
                  I********************************************************    0.976
     527.4        I
                  I**********************************************************  0.984
     555.1        I
                  I*********************************************************** 0.988
     582.8        I
                  I************************************************************0.992
     610.5        I
                  I************************************************************0.996
     638.2        I
                  I************************************************************1.000
     665.9        I
                  I************************************************************1.000
     665.9        I----I----I----I----I----I----I----I----I----I----I----I MAX

             NO OBS------------     250   STD ERROR-          120.5
             COEF OF VARIATION-     0.46  MEAN------          260.0
             KURTOSIS (BETA 2)-     3.23  MEDIAN----          240.6
             PEARSONIAN SKEW---     0.48  MODE------          201.7
```

FIGURE 14.4.   Alternative Two: Acquisition of XRX

```
PATH PERFORMANCE FOR NODE SPALT3

                CFD   0.1   0.2   0.3   0.4   0.5   0.6   0.7   0.8   0.9   1.0
    98.97       I----I----I----I----I----I----I----I----I----I----I MIN
                I                                                       0.000
    98.97       I
                I                                                       0.008
   114.5        I
                I*                                                      0.012
   130.0        I
                I*                                                      0.024
   145.6        I
                I**                                                     0.040
   161.1        I
                I*****                                                  0.100
   176.6        I
                I*********                                              0.172
   192.2        I
                I***********                                            0.220
   207.7        I
                I****************                                       0.328
   223.3        I
                I*********************                                  0.408
   238.8        I
                I**************************                             0.508
   254.3        I
                I*********************************                      0.596
   269.9        I
                I***************************************                0.664
   285.4        I
                I******************************************             0.720
   300.9        I
                I*********************************************          0.776
   316.5        I
                I***********************************************        0.832
   332.0        I
                I**************************************************     0.880
   347.5        I
                I****************************************************   0.916
   363.1        I
                I******************************************************** 0.952
   378.6        I
                I********************************************************* 0.976
   394.1        I
                I********************************************************** 0.984
   409.7        I
                I*********************************************************** 0.988
   425.2        I
                I**********************************************************1.000
   440.7        I
                I***********************************************************1.000
   440.7        I----I----I----I----I----I----I----I----I----I----I MAX

                NO OBS------------      250    STD ERROR-        66.90
                COEF OF VARIATION-      0.26   MEAN------        260.6
                KURTOSIS (BETA 2)-      2.58   MEDIAN----        252.4
                PEARSONIAN SKEW---      0.63   MODE------        218.3
```

FIGURE 14.5.   Alternative Three: Acquisition of ABC

While the results do not give the decisionmaker an absolute indication of the "best" alternative, they do provide a "picture" of the likely results. A comparison of alternative one (no acquisition) and alternative three (acquisition of ABC) indicates that alternative three could be expected to yield a higher mean stock price and less variation. Most decisionmakers would probably select alternative three over alternative one on this basis. Comparison of alternative two (acquisition of XRX) and alternative three shows that both have essentially the same mean stock price, but alternative two possesses appreciably high variation. Whether the decisionmaker would be attracted to, or repelled by, this increased risk is an individual matter and probably depends upon other factors.

CONCLUSIONS

Stochastic networking techniques have the ability to deal systematically with the variation inherent in uncertain, probabilistic situations. These techniques offer great potential to the decisionmaker faced with uncertain situations, particularly of a strategic nature. Because the techniques provide a method for incorporating much of the possible variation and for illustrating the full range of possible and likely outcomes, the decisionmaker is on firmer ground when he or she makes critical decisions in the area of strategic planning. The oversimplified merger/acquisition example presented in this chapter illustrates the application of an advanced stochastic networking technique—VERT-3—to one type of strategic decision. Applications have also been developed in other areas, including new-product development, R&D project selection, project management, and evaluation of marketing strategies.

IV NETWORK MODELING IN PERSPECTIVE

As we pointed out in the introduction to part I, the focus of this book is on network analysis techniques for use by managers. Our specific focus has been on an advanced stochastic networking technique—the Venture Evaluation and Review Technique, third generation (VERT-3)—that can be used by public- and private-sector managers and analysts in a variety of strategic planning projects and situations.

Part I presented a background or environment for analysis, part II investigated the VERT-3 concept in detail, and part III applied VERT-3 to a variety of application areas. Part IV presents conclusions and predictions about future directions for VERT.

15 CONCLUSIONS AND FUTURE DIRECTIONS

We have been concerned to this point with the analytical functions of managers, the concepts involved in management of new projects or ventures, the major types of network models for management use, and the Venture Evaluation and Review Technique. We have described the general concepts, network construction and logic, input requirements, output reports, and computer mechanics of VERT-3. We have shown simplified applications of VERT-3 in a host of settings—project management, plant investments, new-product development, decision trees, evaluation of alternative concepts, and merger/acquisition analysis.

Perhaps at this point it would be helpful to take a step or two back for a brief view of the entire forest. First we will recap VERT-3's main values and limitations and then examine what is currently being done in regard to new developments and applications. Lastly, we will present our view of the future, clouded as it may be.

VALUES OF VERT-3

The main value of VERT-3 is that it enables the manager to deal with the three parameters of prime interest—results (or performance), cost, and time—and the risks inherent in each. While other stochastic networking techniques exist, VERT-3 is, to our knowledge, the only one that can treat

213

fully and equally the three key parameters and their interactions. None is accorded a secondary or tag along status. Further, none need be a discrete, all-or-none condition; VERT allows analysis of degrees of completion, levels of performance, ranges of results, levels of costs (by budget period, if desired), and time requirements.

In large part, the ability of a technique to represent the real world in terms of a network lies in the flexibility of its nodes. Basic VERT made a quantum jump in this area with the introduction of six new types of node logics. But perhaps the most significant innovation was the introduction of its mathematical relationships. VERT-3 has the capability of establishing a mathemetical relationship between any given arc's time and/or cost and/or performance and any other arc's and/or node's time and/or cost and/or performance. Thus, any two points within the network can be tied together by a mathematical relationship selected by the user from the array of mathematical relationships available in VERT-3. Additionally, these same mathematical relationships can be used to establish relationships among the time, cost, and performance variables of a given arc. VERT-3 is designed to be very open-ended and comprehensive when used to establish relationships among network parameters.

Favorable and unfavorable outcomes can be analyzed in terms of their likelihood or risk. The network can be branched into alternative scenarios, which can then be studied in more detail from a Bayesian, or "what if" perspective. This capability is particularly valuable for contingency planning and for scenarios with a low probability of occurrence (where the results may be buried in the average or total range of outcomes). For example, a nuclear generating station accident or a plane crash may be a "one-in-a-million" event. With simulation models the impact of such an event, significant as it might be, would probably be buried because of the event's low probability of occurrence. However, in some cases it might be helpful to look at such occurrences or scenarios as if they had happened.

VERT-3 enables the manager to deal with several types of decisions —strategic, project, and concept evaluaton. It is relatively easy to see the contribution VERT-3 can make in project and concept (or alternative) evaluation environments. More difficult to see, however, may be the contribution VERT-3 can make to the analysis of strategic alternatives for an organization. Strategic decision making is typically top-level management and high-level corporate planning staff territory; it is also the area in which VERT-3 can potentially make its greatest contribution. VERT-3 can help to "structure the unstructured," pursue the outcomes of "logical incrementalism," and reduce the amount of "muddling" behavior and "seat-of-the-pants" decisions. In short, VERT-3 provides a *realistic* quantitative tool that can have its greatest input at higher levels of the organization.

LIMITATIONS OF VERT-3

No technique is perfect; nothing can be all things for all people. In fact, great caution must be taken not to oversell a system like VERT. One need not look very far to find instances of overkill—for example, computers, operations research, PERT, MBO, and so on. In each case, overly zealous "missionaries" oversold the item or concept with inflated claims about its power, results, or capabilities. When users found these claims to be untrue, disappointment and disillusionment resulted, *even though* use of the system may have brought about an improvement over previous conditions. Inflated expectations result in disappointment and can actually retard the use of potentially valuable concepts.

What are some of VERT-3's shortcomings or limitations? Perhaps the main shortcoming flows from one of VERT-3's advantages: A great deal of flexibility requires a wide variety of options and possible input preparations. While neither version of VERT is overly difficult to use, the ability to use VERT effectively takes some time because of all the options and features with which one needs to become familiar. The user must develop a degree of familiarity, a way of thinking, or a "mental set" to use VERT efficiently; occasional users may have to "retrain" themselves.

Operationally, perhaps the most persistent problem involves collecting the data required to describe the time, cost, and results variables. To take full advantage of VERT, one needs to employ realistic data and distributions; in other cases, perhaps, a simpler technique would suffice. Running simple, "ball-park" data through a more sophisticated simulation technique does not necessarily improve the validity of the output (even though the output may appear more valid) or of the decisions made as a result of the output.

A second operational difficulty involves obtaining accurate estimates of time, cost and results from knowledgeable sources. The age-old bugaboo, overoptimistic estimating, can negate the results of the best analytical technique. Perhaps this type of estimating is less a limitation of VERT than it is a human bias, but it affects VERT's results nonetheless.

NEW DEVELOPMENTS

Several developments in the area of stochastic networking systems warrant mention. Pritsker and Associates continues to refine and improve its GERT family of techniques. One of the latest developments has been the introduction of Q-GERT, which augments GERT with queuing and decision capabilities. Briefly, Q-GERT was developed for modeling in graphic form and for studying "the procedural aspects of manufacturing, defense, and service systems" (Pritsker 1977). While Q-GERT includes many of the features of previous GERT-related systems, some features have been deleted and others added. Halting of activites, counter types, and activity costs have been deleted, while user-written program inserts, user statistics collection, and subnetwork generation and editing have been added.

Most recently, Pritsker and Associates has added a new technique, Simulation Language for Alternative Modeling (SLAM), to its arsenal. Like its predecessors, SLAM possesses a networking capability and permits significantly more flexibility than other techniques by allowing the user to do part of the coding (see Pritsker and Pegden 1979). The result should be a system more directly geared to the user's needs; however, preparation of part of the coding requires a fairly high degree of skill on the user's part, as well as an additional investment of resources.

Like other systems, VERT continues to evolve. While the authors feel that VERT and VERT-2 already offer significantly more capability to the user than any other technique, changes continue to be made in the system. The latest, proprietary version of VERT is called VERT-3—essentially, a completely new program with many new features that required a complete recoding of the software. The authors sincerely believe that VERT-3 can model, without any loss of realism, any problem that does not involve mathematics beyond the mathematical relationships given in chapter 6. The mathematical relationship capability of VERT-3 has grown to the point where flows can be isolated, started, stopped, or altered in nearly any conceivable way. For example, separate flows can be established and maintained for each of the pertinent performance characteristics desired if the analyst does not want to join these characteristics together in one single value. Cost can be given a similar treatment. However, going beyond three dimensions generally proves difficult for the manager.

More nodes have been added to VERT-3, and many of the nodes in the initial versions have been altered or condensed to facilitate modeling and the achievement of realism. To facilitate budgeting, VERT-3 accumulates cost into a histogram form between user-selected time intervals such as years or months. VERT-3 now is able to print out slack times in histogram form. It can also highlight the optimal path in addition to the critical path and thus help the manager determine where best to expend resources. VERT-3 carries two types of cost, which are known as path cost and overall cost. Their names are indicative of the type of information they produce. If a mathematical relationship happened to create a dependency situation on an arc where, for example, cost was dependent on time and time on performance, the user was required by the initial version of VERT to specify the order for processing its time, cost, and performance inputs in order to keep from violating this dependency. VERT-3 now atuomatically determines these dependencies. To facilitate data handling, VERT-3 has an automated storage and retrieval capability. The initial version of VERT frequently ran out of core when loading large problems; a core-saving data storage pointer system now exists in VERT-3. VERT-3 has been structured to be both very usable and sufficiently powerful to handle any problem not requiring high-level math.

NEW APPLICATIONS

VERT continues to expand its base of applications, not only to a larger *quantity* of applications, but—more significantly—to new *types* of applications. VERT's initial applications involved the analysis of competing concepts for weapon system development, including tanks, helicopters, fighter planes, artillery, self-propelled howitzers, electronic sensors, air defense systems, and others. More recently, applications have broadened to include many other types of planning, including flood control programs, pollution abatement methods, earthquake analysis, railyard switching operations, fault tree analysis, war gaming, and production line balancing. Applications currently underway or planned include analysis of mergers and acquisitions, diversification alernatives, business expansion, new-product development, engineering concept evaluation, contingency planning, and the impact of changing environmental conditions. At the corporate level, one planned application involves the interrelationship of strategic business units (SBUs) and a firm's strategic posture. Included is the funding of potential high-growth units ("stars") out of the profits of other units ("cash cows") and the elimination of low producers ("dogs"). Planners intend the analysis to consider the likely growth/decline rates of the various units, to investigate the impact of proposed actions on cash flow, timing, and other factors, and to yield the optimal "strategic posture" or portfolio of business units for the firm.

THE FUTURE OF VERT

Although VERT is quite promising and has experienced modest growth, its use is not as widespread as might be expected. One reason is external to the technique; many managers are not familiar with quantitative risk assessment techniques and have not had experience in using the outputs of a risk analysis. While managers think in terms of ranges of outcomes, they are used to seeing (and they perhaps demand) point estimates. Obviously, probability distributions depict risks more accurately than do point estimates; however, many managers appear to resist probability distributions because of their unfamiliarity. Increased familiarity will ease some of this reluctance, and growing numbers of better-trained managers and analysts will accelerate the process.

As the use of VERT increases, so will the number and type of applications. Users are invited to feed back their experiences, which can be comminicated to other current and potential users. The technique has great potential; the prime beneficiaries of that potential have been, and will continue to be, the users.

APPENDIX:
VERT-3 Computer Program

```
      IMPLICIT INTEGER*4(I-N)
     OCOMMON EMARK(11),SETUP,FINT,DINT,FACTOR,STIME,SCOST,SPERF,CTIME,
     1CCOST,CPERF,BTIME,BCOST,BPERF,TIMESM,UNFORM,ITRIP1(22),ITRIP2(22),
     2KNT(6),INPT,IOUT,IPNH,IWF1,IWF2,IWF3,IWF4,IBLK,ITRACE,ICUT,ISEED,
     3IRROR,ISAVE,NODESM,MAXTAG,ICOUNT,JCOUNT,NTRIP,ICORR,LCOMP,NWARN,
     4NRK,NRKS,SLAK,MODE,INFC,INFP,DISC,DISP,CRIT,KGEN,MED,SINGLE,DOUBLE
      LOGICAL SLAK,MODE,INFC,INFP,DISC,DISP,CRIT,KGEN,MED,SINGLE,DOUBLE
      ISEED = 0
C
C     MAIN IS THE QUARTERBACK
C
 1111 CALL LOADF
      IF(IRROR.GT.0) GO TO 1111
      CALL LOADT
      CALL LOADA
      IF(IRROR.GT.0) GO TO 1111
      CALL LOADNR
      IF(IRROR.GT.0) GO TO 1111
      CALL DONET
      IF(IRROR.GT.0) GO TO 1111
      CALL OUTFLO
      IF(ICORR.GT.0) CALL CORR
      GO TO 1111
      END
      SUBROUTINE ERROR(I,J,K,L)
      IMPLICIT INTEGER*4(I-N)
     OCOMMON EMARK(11),SETUP,FINT,DINT,FACTOR,STIME,SCOST,SPERF,CTIME,
     1CCOST,CPERF,BTIME,BCOST,BPERF,TIMESM,UNFORM,ITRIP1(22),ITRIP2(22),
     2KNT(6),INPT,IOUT,IPNH,IWF1,IWF2,IWF3,IWF4,IBLK,ITRACE,ICUT,ISEED,
     3IRROR,ISAVE,NODESM,MAXTAG,ICOUNT,JCOUNT,NTRIP,ICORR,LCOMP,NWARN,
     4NRK,NRKS,SLAK,MODE,INFC,INFP,DISC,DISP,CRIT,KGEN,MED,SINGLE,DOUBLE
      LOGICAL SLAK,MODE,INFC,INFP,DISC,DISP,CRIT,KGEN,MED,SINGLE,DOUBLE
      INTEGER*2 I
C
C     THIS SUBROUTINE LISTS ERROR AND WARNING MESSAGES
C
      IF(L.EQ.NWARN) GO TO 1144
      IRROR = IRROR + 1
      WRITE (IOUT,1122) IRROR, I, J, K
 1122 FORMAT (1H0, I4, 17H.  E R R O R   NO., I6, 2X, 2A4)
      IF(L.NE.IBLK) WRITE (IOUT,1133) L
 1133 FORMAT (1H+, 37X, I6)
      RETURN
 1144 WRITE (IOUT,1155) I, J, K
 1155 FORMAT (1H0, 7X, 18HW A R N I N G   NO., I6, 2X, 2A4)
      RETURN
      END
      SUBROUTINE SEEK (J,ENDX)
      IMPLICIT INTEGER*4(I-N)
     OCOMMON EMARK(11),SETUP,FINT,DINT,FACTOR,STIME,SCOST,SPERF,CTIME,
     1CCOST,CPERF,BTIME,BCOST,BPERF,TIMESM,UNFORM,ITRIP1(22),ITRIP2(22),
     2KNT(6),INPT,IOUT,IPNH,IWF1,IWF2,IWF3,IWF4,IBLK,ITRACE,ICUT,ISEED,
     3IRROR,ISAVE,NODESM,MAXTAG,ICOUNT,JCOUNT,NTRIP,ICORR,LCOMP,NWARN,
     4NRK,NRKS,SLAK,MODE,INFC,INFP,DISC,DISP,CRIT,KGEN,MED,SINGLE,DOUBLE
     OCOMMON/ARCS/ASTORE( 2800),UTIMEA( 350),TIMEA( 350),UCOSTA( 350),
     1COSTA( 350),UPERFA( 350),PERFA( 350),WORK( 350),ISTATE( 350),
     2NODEI( 350),NODEO( 350),ICRITA( 350),KEEPC( 350),KEEPP( 350),
     3IARC1( 350),IARC2( 350),IPOINT( 350),JPOINT( 350),ISLAK( 350),
     4KARC,LARC,MARC,NARC,ITALC,ITALP,ISTAR
      LOGICAL SLAK,MODE,INFC,INFP,DISC,DISP,CRIT,KGEN,MED,SINGLE,DOUBLE

      THIS SUBROUTINE SPOTS INFORMATION IN VARIABLE ARC STORAGE (ASTORE)
```

```fortran
C
      M = J + 2
      ITALC = J + 1
      ITALP = 0
      DO 1188 N=M,KARC
      IF(ASTORE(N) - ENDX)1177,1166,1188
 1166 ITALP = N - ITALC - 1
 1177 RETURN
 1188 IF(ASTORE(N).LT.EMARK(11)) ITALC = N
      RETURN
      END
      SUBROUTINE LOADF
      IMPLICIT INTEGER*4(I-N)
 0COMMON EMARK(11),SETUP,FINT,DINT,FACTOR,STIME,SCOST,SPERF,CTIME,
 1CCOST,CPERF,BTIME,BCOST,BPERF,TIMESM,UNFORM,ITRIP1(22),ITRIP2(22),
 2KNT(6),INPT,IOUT,IPNH,IWF1,IWF2,IWF3,IWF4,IBLK,ITRACE,ICUT,ISEED,
 3IRROR,ISAVE,NODESM,MAXTAG,ICOUNT,JCOUNT,NTRIP,ICORR,LCOMP,NWARN,
 4NRK,NRKS,SLAK,MODE,INFC,INFP,DISC,DISP,CRIT,KGEN,MED,SINGLE,DOUBLE
 0COMMON/TRIALS/STORET( 1000,4),TERM(10,8),KPOINT(10),NODET( 1000),
 1MTERM,NTERM,MITER,ITER
 0COMMON/ARCS/ASTORE( 2800),UTIMEA( 350),TIMEA( 350),UCOSTA( 350),
 1COSTA( 350),UPERFA( 350),PERFA( 350),WORK( 350),ISTATE( 350),
 2NODEI( 350),NODEO( 350),ICRITA( 350),KEEPC( 350),KEEPP( 350),
 3IARC1( 350),IARC2( 350),IPOINT( 350),JPOINT( 350),ISLAK( 350),
 4KARC,LARC,MARC,NARC,ITALC,ITALP,ISTAR
 0COMMON/NODES/TIMEN( 200),COSTN( 200),PERFN( 200),NSTORE( 5400),
 1NODE1( 200),NODE2( 200),LOGI( 200),LOGO( 200),NSTATE( 200),
 2NARCI( 200),NARCO( 200),ISTAT( 200),INSTAT( 200),ICRITN( 200),
 3NPOINT( 200),NSLAK( 200),JUMP( 200),KNODE,LNODE,MNODE,NNODE,MTAG,
 4NTAG
 0COMMON/INTERN/XMIN(20,4),XMAX(20,4),HMIN(20,4),HMAX(20,4),
 1HAVE(20,4),IOBS(20),MHIST,NHIST
 0COMMON/SLACK/RMIN(20),RMAX(20),SMIN(20),SMAX(20),SAVE(20),
 1JOBS(20),MSLACK,NSLACK
 0COMMON/CPGAP/T1(10),T2(10),CONFL(10),CONFS(10),CAVE(10),CSMIN(10),
 1CSMAX(10),CHMIN(10),CHMAX(10),PAVE(10),PSMIN(10),PSMAX(10),
 2PHMIN(10),PHMAX(10),KCOBS(10),KPOBS(10),MCPGAP,NCPGAP,ICPGAP
      DIMENSION IDA(17), IDB(17), IDX(14), NUMB(4)
      LOGICAL SLAK,MODE,INFC,INFP,DISC,DISP,CRIT,KGEN,MED,SINGLE,DOUBLE
      INTEGER*2LOW,IUP
      DATA IDA,IDB,IDX/4HM      ,4HMTIM,4HMCOS,4HMPER,4HFILT,4HFILT,4HFILT,
 14HSLAK,4HDTIM,4HHTIM,4HRTIM,4HDCOS,4HHCOS,4HRCOS,4HDPER,4HHPER,4HR
 2PER,1H ,1HE,1HT,1HF,1H1,1H2,1H3,1H ,1HE,1HE,1HE,1HT,1HT,1HT,1HF,1H
 3F,1HF,4HENDA,4HENDN,4H----,4HHIST,4HSUBT,4HSLAK,4H*#.$,1HK,1HE,
 41HN,1HD,1HC,1HI,1HB/
      DATA NUMB/1H1,1H2,1H3,1H4/
C
C     IF THE PROGRAM HAS NOT BEEN INITIALIZED, DO SO AT THIS TIME.
C        SETUP IS USED FOR INITIALIZING.  THE LARGEST FLOAT POINT NUMBER
C        THE COMPUTER BEING USED CAN HANDLE SHOULD BE ASSIGNED TO SETUP.
C        ZERO THE ERROR COUNTER AND ASSIGN BLANK TO IBLK.
C
      IRROR = 0
      M = 0
      N = ISEED
      INFC = .TRUE.
      INFP = .TRUE.
      DISC = .TRUE.
      DISP = .TRUE.
      MED  = .TRUE.
      IF(ISEED.GT.0) GO TO 1411
      M = 1
```

```
      SETUP = 1.0E38
      EMARK(1)  = -999991.0
      EMARK(2)  = -999992.0
      EMARK(3)  = -999993.0
      EMARK(4)  = -999994.0
      EMARK(5)  = -999995.0
      EMARK(6)  = -999996.0
      EMARK(7)  = -999997.0
      EMARK(8)  = -999998.0
      EMARK(9)  = -999999.0
      EMARK(10)= -999990.0
      EMARK(11)= -999900.0
      IBLK = IDB(1)
      NWARN = IDX(7)
C
C     I/O SET UP - INPT = READER, IOUT = PRINTER, IPNH = PUNCH, IWF1 =
C        MASTER FILE, IWF2 = INTERMEDIATE FILE, IWF3 = TRANSPORT FILE
C        NO. 1, IWF4 = TRANSPORT FILE NO. 2. (TO AVOID COMMENT DELETIONS
C        THESE ASSIGNMENTS ARE DISTRIBUTED AMONG THE FOLLOWING COMMENTS)
C
      INPT = 5
C
C     NOTE---THE DIMENSIONS ASSIGNED TO THE COMMON AND CERTAIN DIMENSION
C            ARRAYS AND THE VALUES ASSIGNED TO CHECK VARIABLES MITER,
C            MTERM, LARC, MARC, MNODE, LNODE, MTAG AND MHIST MUST BE
C            KEPT CONSISTANT WITH THE FOLLOWING MNEMONIC DIMENSIONING
C            SCHEME.  AN AUXILIARY PROGRAM CALLED DIMEN DOES THIS TASK.
C
      IOUT = 6
C
C     COMMON/TRIALS/STORET(MITER,4),TERM(MTERM,8),KPOINT(MTERM),
C     1NODET(MITER)---ETC. AS ABOVE
C
C      IN SUBROUTINE OUTFLO, MEDIAN AND MS
C      COMMON/MEDAN/Z(MITER),---ETC. AS SHOWN IN THOSE SUBROUTINES
C
      IPNH = 7
C
C     COMMON/ARCS/ASTORE(LARC),UTIMEA(MARC),TIMEA(MARC),UCOSTA(MARC),
C     1COSTA(MARC),UPERFA(MARC),PERFA(MARC),WORK(MARC),ISTATE(MARC),
C     2NODEI(MARC),NODEO(MARC),ICRITA(MARC),KEEPC(MARC),KEEPP(MARC),
C     3IARC1(MARC),IARC2(MARC),IPOINT(MARC),JPOINT(MARC),ISLAK(MARC),---
C     4ETC. AS ABOVE
C
      IWF1 = 8
C
C     COMMON/NODES/TIMEN(MNODE),COSTN(MNODE),PERFN(MNODE),NSTORE(LNODE+
C     1MTAG),NODE1(MNODE),NODE2(MNODE),LOGI(MNODE),LOGO(MNODE),
C     2NSTATE(MNODE),NARCI(MNODE),NARCO(MNODE),INSTAT(MNODE),
C     3ISTAT(MNODE),ICRITN(MNODE),NPOINT(MNODE),NSLAK(MNODE),
C     4JUMP(MNODE),---ETC. AS ABOVE
C
C      IN SUBROUTINE DONET
C      DIMENSION SLACKA(MARC),SLACKN(MNODE),IPATHA(MARC),IPATHN(MNODE)
C
      IWF2 = 9
C
C     COMMON/INTERN/XMIN(MHIST,4),XMAX(MHIST,4),HMIN(MHIST,4),
C     1HMAX(MHIST,4),HAVE(MHIST,4),IOBS(MHIST)---ETC. AS ABOVE
C
C     COMMON/SLACK/RMIN(MSLACK),RMAX(MSLACK),SMIN(MSLACK),SMAX(MSLACK),
C     1SAVE(MSLACK),JOBS(MSLACK)---ETC. AS ABOVE
```

```
C
C         0COMMON/CPGAP/T1(MCPGAP),T2(MCPGAP),CONFL(MCPGAP),CONFS(MCPGAP),
C         1CAVE(MCPGAP),CSMIN(MCPGAP),CSMAX(MCPGAP),CHMIN(MCPGAP),
C         2CHMAX(MCPGAP),PAVE(MCPGAP),PSMIN(MCPGAP),PSMAX(MCPGAP),
C         3PHMIN(MCPGAP),PHMAX(MCPGAP),KCOBS(MCPGAP),KPOBS(MCPGAP),---ETC. AS
C         4ABOVE
C
          IWF3 = 10
C
C         IN SUBROUTINE OUTFLO
C         0DIMENSION HSTO(MHIST,4), HTOS(MHIST,4), IHIST(MHIST,4,28),
C         1SSTO(MSLACK), STOS(MSLACK), JHIST(MSLACK,28)---ETC.
C         SHOWN IN SUBROUTINE OUTFLO
C
          IWF4 = 11
C
C         MITER = MAX. NO. OF ITERATIONS
C         MARC  = MAX. NO. OF ARCS
C         LARC  = MAX. SPACE IN GENERAL ARC STORAGE ARRAY ASTORE
C         MNODE = MAX. NO. OF NODES
C         LNODE = MAX. SPACE IN GENERAL NODE STORAGE ARRAY NSTORE
C         MTAG  = MAX. SPACE IN THE BACK END OF NSTORE ALLOCATED TO THE STO-
C                 RAGE OF THE SINGLE COUNTING COST AND PERFORMANCE VECTORS
C         MHIST = MAX. NO. OF INTERNAL NODE HISTOGRAMS ALLOWED
C         MTERM = MAX. NO. OF TERMINAL NODE HISTOGRAM CARDS
C         MSLACK= MAX. NO. OF SLACK HISTOGRAMS ALLOWED
C         MCPGAP= MAX. NO. OF COST-PERFORMANCE INTERVAL HISTOGRAMS ALLOWED
C
          MITER = 1000
          MARC  =  350
          LARC  = 2800
          MNODE =  200
          LNODE = 1800
          MTAG  = 3600
          MHIST =   20
          MTERM =   10
          MSLACK=   20
          MCPGAP=   10
C
C         THE CHECK VARIABLES MUST HAVE THE FOLLOWING MINIMUM SIZES TO EN-
C             ABLE USING THE ABOVE ARRAYS IN AUXILIARY CAPACITIES.   IRROR
C             COUNTS THE NUMBER OF ERRORS ENCOUNTERED.
C
          IF(MITER.LT.150)CALL ERROR (1300,IBLK,IBLK,IBLK)
          IF(MARC.LT.100) CALL ERROR (1311,IBLK,IBLK,IBLK)
          IF(LARC.LT.200) CALL ERROR (1322,IBLK,IBLK,IBLK)
          IF(MNODE.LT.60) CALL ERROR (1333,IBLK,IBLK,IBLK)
          IF(LNODE.LT.200)CALL ERROR (1344,IBLK,IBLK,IBLK)
          IF(MTAG.LT.200) CALL ERROR (1355,IBLK,IBLK,IBLK)
          IF(MHIST.LT.1)  CALL ERROR (1366,IBLK,IBLK,IBLK)
          IF(MTERM.LT.1)  CALL ERROR (1377,IBLK,IBLK,IBLK)
          IF(MSLACK.LT.1) CALL ERROR (1388,IBLK,IBLK,IBLK)
          IF(MCPGAP.LT.1) CALL ERROR (1399,IBLK,IBLK,IBLK)
          IF(MCPGAP.GT.MARC/2) CALL ERROR (1400,IBLK,IBLK,IBLK)
          IF(IRROR.GT.0) GO TO 2940
C
C         THIS SUBROUTINE READS IN AND LOADS DATA ON THE FILES, LOADS ARC
C             AND NODE NAMES IN CORE STORAGE, CHECKS FOR ALPHA DATA IN
C             NUMERIC FIELDS AND ZEROS OUT STORAGE ARRAYS FOR LOADA
C
     1411 0READ (INPT,1422,END=2944) I, ISTAR, ITRACE, ICUT, NTRIP, ICORR,
          1ICPGAP, K, LCOMP, L8080, ISEED, ITER, FINT, DINT, FACTOR, STIME,
```

```
     2SCOST, SPERF, CTIME, CCOST, CPERF, BTIME, BCOST, BPERF
1422 FORMAT (8I1, I2, I1, I8, I5, 2F4.2, 7F3.2, 3F9.0)
     DO 1433 L=1,20
1433 NODET(L) = IBLK
     IF(I.GE.0.AND.I.LE.1) GO TO 1450
     CALL ERROR (1444,IBLK,IBLK,IBLK)
     GO TO 2940
1450 IF(I.EQ.1) READ (INPT,1455,END=2922) (NODET(L), L=1,20)
1455 FORMAT (20A4)
     0WRITE (IOUT,1466) (NODET(L), L=1,20), I, ISTAR, ITRACE, ICUT,
     1NTRIP, ICORR, ICPGAP, K, LCOMP, L8080, ISEED, ITER
14660FORMAT (1H1, 20A4//35H PROBLEM IDENTIFICATION CARD OPTION,25(1H-),
     1I11//21H TYPE OF INPUT OPTION, 39(1H-), I11//22H TYPE OF OUTPUT OP
     2TION, 38(1H-), I11//27H COSTING AND PRUNING OPTION, 33(1H-), I11//
     323H FULL PRINT TRIP OPTION, 37(1H-), I11//40H CORRELATION COMPUTAT
     4ION AND PLOT OPTION, 20(1H-), I11//38H COST-PERFORMANCE TIME INTER
     5VAL OPTION, 22(1H-), I11//53H COMPOSITE TERMINAL NODE MINIMUMS AND
     6 MAXIMUMS OPTION, 7(1H-), I11//29H TERMINAL NODE LISTING OPTION,
     731(1H-), I11//26H 80-80 LIST OF INPUT CARDS, 34(1H-), I11//13H INI
     8TIAL SEED, 47(1H-), I11//21H NUMBER OF ITERATIONS, 39(1H-), I11//)
     0WRITE (IOUT,1477) FINT, DINT, FACTOR, STIME, SCOST, SPERF, CTIME,
     1CCOST, CPERF, BTIME, BCOST, BPERF
14770FORMAT (52H YEARLY INTEREST RATE USED FOR INFLATION ADJUSTMENTS,
     18(1H-), F11.2//56H YEARLY INTEREST RATE USED FOR PRESENT VALUE DIS
     2COUNTING, 4(1H-), F11.2//57H TIME FACTOR WHICH CONVERTS PROGRAM TI
     3ME TO A YEARLY BASE, 3(1H-), F11.2//1H , 36X, 4HTIME, 5X, 4HCOST,
     45X, 4HPERF//32H TERMINAL NODE SELECTION WEIGHTS, 3F9.2//32H CRITIC
     5AL - OPTIMUM PATH WEIGHTS, 3F9.2//15H INITIAL VALUES, 17X,3F9.2/)
C
C    DETERMINE IF SINGLE OR DOUBLE (MULTIPLE) COUNTING OF ACTIVITIES
C        IS WANTED.
C
     SINGLE = .FALSE.
     DOUBLE = .FALSE.
     IF(ICUT.GT.0) GO TO 1480
     DOUBLE = .TRUE.
     ICUT = 2
     GO TO 1485
C
3911 KARC = KARC + 2
     IF(KARC.GT.LARC) GO TO 3177
     ASTORE(KARC-1) = IADJ
     ASTORE(KARC) = EMARK(10) - TIMESM
     LOGI(N) = 1
C
C    THIS SECTION STORES HISTOGRAMS
C
3922 IF(NSTATE(I-1).LE.0) GO TO 4033
     KARC = KARC + 1
     IF(KARC.GT.LARC) GO TO 3177
     ASTORE(KARC) = 15.0
     CALL SEREAD (NSTATE(I-1),IE,JE,IADJ)
     IF(NODESM.GT.0) GO TO 3933
     KARC = KARC - 1
     GO TO 4033
3933 IF(KARC.GT.LARC) GO TO 3177
     J = (NODESM/2)*2
     IF(NODESM.NE.J) GO TO 3955
     CALL ERROR (3944,IE,JE,IBLK)
     GO TO 4022
3955 IF(NODESM.GE.5) GO TO 3977
     CALL ERROR (3966,IE,JE,IBLK)
```

```
      GO TO 4022
C
C     DO PROBABILITY ENTRIES SUM TO 1 AND ARE LEFT HAND CELL BOUNDARIES
C        SMALLER THAN OR EQUAL TO THE RIGHT HAND CELL BOUNDARIES
C
 3977 UNFORM = 0.0
      DO 3999 J=3,NODESM,2
      IF(TIMEA(J-1).LT.0.0) CALL ERROR (3988,IE,JE,IBLK)
      UNFORM = UNFORM + TIMEA(J-1)
 3999 IF(TIMEA(J-2).GT.TIMEA(J)) CALL ERROR (4000,IE,JE,IBLK)
      IF(UNFORM.GT.0.99.AND.UNFORM.LT.1.01) GO TO 4022
      CALL ERROR (4011,IE,JE,IBLK)
C
C     ENTER THE END OF SEGMENT INDICATOR FOR HISTOGRAM INPUT
C
 4022 KARC = KARC + 2
      IF(KARC.GT.LARC) GO TO 3177
      ASTORE(KARC-1) = IADJ
      ASTORE(KARC) = EMARK(10) - TIMESM
      LOGI(N) = 1
C
C     THIS SECTION STORES MATHEMATICAL RELATIONSHIPS IN ASTORE
C
 4033 IF(NSTATE(I).LE.0) GO TO 4388
      MAXTAG = 0
      M = NSTATE(I)
      DO 4066 J=1,M
      MAXTAG = MAXTAG + 2
      IF(MAXTAG.LE.MARC) GO TO 4044
      MAXTAG = 2
 4044 NODESM = MAXTAG - 1
      OREAD (IWF4,4055) K, (UTIMEA(L), TIMEA(L), UCOSTA(L), ICRITA(L),
     1ICRITN(L), COSTA(L), ISTATE(L), JPOINT(L), UPERFA(L), KEEPC(L),
     2KEEPP(L), TIMEN(L), COSTN(L), PERFN(L), L=NODESM,MAXTAG), IADJ
 4055 FORMAT (I2, 2(F2.0, 2A1, 2A4, A1, 2A4, A1, 2A4, 3F8.0), I1)
 4066 IF(K.NE.J) CALL ERROR (4077,IE,JE,IBLK)
C
C     DECODE SUM-OMIT-BACKUP INDICATOR, WORK CARRIES THE BACKUP STATUS
C
      DO 4133 J=1,MAXTAG
      OIF(UTIMEA(J).EQ.0.0.AND.ICRITA(J).EQ.IBLK.AND.ICRITN(J).EQ.IBLK.
     1AND.ISTATE(J).EQ.IBLK.AND.JPOINT(J).EQ.IBLK.AND.KEEPC(J).EQ.IBLK.
     2AND.KEEPP(J).EQ.IBLK) GO TO 4133
      WORK(J) = 0.0
      IF(TIMEA(J).NE.BACKUP) GO TO 4100
      IF(J.GT.1) GO TO 4099
      CALL ERROR (4088,IE,JE,IBLK)
      GO TO 4133
C
C     INDICATE TO THE PREVIOUS TRANSFORMATION THAT IT HAS A BACK UP,
C        THEN PASS THE SUM-OMIT INDICATOR OF THE PREVIOUS TRANSFORMATION
C        FORWARD TO THE PRESENT BACKUP TRANSFORMATION
C
 4099 WORK(J-1) = 1.0
      IF(TIMEA(J-1).LT.0.0) TIMEA(J) =-2.0
      IF(TIMEA(J-1).GT.0.0) TIMEA(J) = 2.0
      GO TO 4133
 4100 IF(TIMEA(J).NE.SUM) GO TO 4111
      TIMEA(J) = 1.0
      GO TO 4133
 4111 IF(TIMEA(J).NE.OMIT) CALL ERROR (4122,IE,JE,IBLK)
      TIMEA(J) =-1.0
```

```
 4133 CONTINUE
C
C        LOAD THE FUNCTIONAL RELATIONSHIPS DATA INTO VARIABLE ARC STORAGE
C
         DO 4377 J=1,MAXTAG
        UIF(UTIMEA(J).EQ.0.0.AND.ICRITA(J).EQ.IBLK.AND.ICRITN(J).EQ.IBLK.
       1AND.ISTATE(J).EQ.IBLK.AND.JPOINT(J).EQ.IBLK.AND.KEEPC(J).EQ.IBLK.
       2AND.KEEPP(J).EQ.IBLK) GO TO 4377
         KARC = KARC + 9
         IF(KARC.GT.LARC) GO TO 3177
C
C        CHECK AND ENTER IN STORAGE THE TRANSFORMATION NUMBER, STORE THE
C            SUM-OMIT AND BACKUP INDICATORS, THEN DECODE THE RETRIVAL CODE
C            (T=TIME, C=COST, P=PERFORMANCE OF THE NODE OR ARC FOLLOWING
C            K=CONSTANT AND BLANK=PREVIOUS TRANSFORMATION.)
C
         X = UTIMEA(J)
         IF(X.LE.0.0.AND.X.GT.100.0) CALL ERROR (4155,IE,JE,IBLK)
 4166 ASTORE(KARC-8) = X
         ASTORE(KARC-7) = TIMEA(J)
         ASTORE(KARC-6) = WORK(J)
         DO 4366 M5=1,3
         GO TO (4177,4188,4199),M5
 4177 X = UCOSTA(J)
         M1= ICRITA(J)
         M2= ICRITN(J)
         Y = TIMEN(J)
         GO TO 4200
 4188 X = COSTA(J)
         M1= ISTATE(J)
         M2= JPOINT(J)
         Y = COSTN(J)
         GO TO 4200
 4199 X = UPERFA(J)
         M1= KEEPC(J)
         M2= KEEPP(J)
         Y = PERFN(J)
 4200 DO 4211 M4=1,5
         IF(X.NE.TYPE(M4)) GO TO 4211
         ASTORE(KARC-7+2*M5) = M4
         L = M4
         GO TO 4233
 4211 CONTINUE
         CALL ERROR (4222,IE,JE,IBLK)
         GO TO 4366
C
C        DECODE THE ARC OR NODE NAME FROM WHICH DATA WILL BE RETRIVED
C
 4233 K = KARC-6+2*M5
         JCOUNT = 0
         IF(L - 4)4244,4333,4311
 4244 DO 4288 M4=1,NARC
         IF(M1.NE.IARC1(M4).OR.M2.NE.IARC2(M4)) GO TO 4288
         IF(M4.NE.ISAVE) GO TO 4277
         JCOUNT = 1
C
C        CHECK TO SEE IF THIS VARIABLE IS DEPENDENT UPON ITSELF
C
         IF(N.NE.L) GO TO 4266
         CALL ERROR (4255,IE,JE,IBLK)
         GO TO 4366
 4266 M = I - 3 + L
```

```
       ISTAT(M) = 1
  4277 ASTORE(K) = M4
       GC TO 4366
  4288 CONTINUE
C
C      IS THIS MATHEMATICALLY RELATED VARIABLE A NODE
C
       DO 4299 M4=1,NNODE
       IF(M1.NE.NODE1(M4).OR.M2.NE.NODE2(M4)) GO TO 4299
       ASTORE(K) = -M4
       GO TO 4366
  4299 CONTINUE
       CALL ERROR (4300,IE,JE,IBLK)
       GO TO 4366
C
C      CHECK AND STCRE THE TRANSFORMATION REFERENCE NUMBER OR CHECK AND
C         STORE THE CONSTANT
C
  4311 L = Y + 0.001
       IF(L.LT.0.OR.L.GE.J) CALL ERROR (4322,IE,JE,IBLK)
       GO TO 4355
  4333 IF(Y.LT.EMARK(11)) CALL ERROR (4344,IE,JE,IBLK)
  4355 ASTORE(K) = Y
  4366 CONTINUE
  4377 CONTINUE
C
C      LOAD THE END OF SEGMENT INDICATOR FOR MATHEMATICAL RELATIONSHIPS
C
       KARC = KARC + 2
       IF(KARC.GT.LARC) GC TO 3177
       ASTORE(KARC-1) = IADJ
       ASTORE(KARC) = EMARK(1) - TIMESM
       IF(JCOUNT.EQ.1) GO TO 4388
       LOGI(N) = 1
  4388 CONTINUE
C
C      SUM, CHECK AND LOAD THE DEPENDENCY INDICATORS
C
       NSTATE(1) = ISTAT(1) + ISTAT(2) + ISTAT(3)
       NSTATE(2) = ISTAT(4) + ISTAT(5) + ISTAT(6)
       NSTATE(3) = ISTAT(7) + ISTAT(8) + ISTAT(9)
       DO 4455 I=1,3
C
C      ANY DEPENDENCIES FOR THIS VARIABLE, IF NOT, WAS A NON DEPENDENT
C         INPUT MADE FOR THIS VARIABLE, IF SO NOTE THIS VIA ASSIGNING
C         A 1 TO THE FINAL DEPENDENCY VECTOR
C
       IF(NSTATE(I).GT.0) GC TO 4399
       IF(LOGI(I).NE.0) LOGC(I) = 1
       GO TO 4455
C
C      FIND THE INDEPENDENT, IS IT DEPENDENT UPON ANOTHER VARIABLE
C
  4399 N = I*3
       M = N - 2
       K = 0
       DO 4444 J=M,N
       K = K + 1
       IF(ISTAT(J).EQ.0) GO TO 4444
       IF(NSTATE(K).GT.0) GO TO 4411
C
C      THIS INDEPENDENT IS NCT DEPENDENT UPON ANOTHER VARIABLE, CHECK TO
```

```
C          SEE IF A NON DEPENDENT INPUT WAS MADE FOR THE VARIABLE, NOTE
C          THIS SECOND LEVEL DEPENDENCY VIA ASSIGNING A 2 TO THE FINAL
C          DEPENDENCY VECTOR
C
           IF(LOGI(K).EQ.0) CALL ERROR (4400,IE,JE,IBLK)
           LOGO(I) = 2
           GO TO 4444
C
C      THIS INDEPENDENT IS DEPENDENT UPON ANOTHER VARIABLE, FIND WHICH
C          VARIABLE IT IS DEPENDENT ON AND THEN CHECK TO SEE IF IT IS
C          DEPENDENT UPON THE VARIABLE BEING EXAMINED, LASTLY NOTE THIS
C          THIRD LEVEL DEPENDENCY VIA ASSIGNING A 3 TO THE FINAL
C          DEPENDENCY VECTOR
C
  4411 L = K*3 - 3
       DO 4433 NODESM=1,3
       L = L + 1
       IF(ISTAT(L).EQ.0) GO TO 4433
       IF(NODESM.EQ.I) CALL ERROR (4422,IE,JE,IBLK)
       LOGO(I) = 3
  4433 CONTINUE
  4444 CONTINUE
  4455 CONTINUE
       M = IPOINT(ISAVE) + 1
       ASTORE(M) = 100*LOGO(1) + 10*LOGO(2) + LOGO(3)
C
C      ENTER THE END OF RECORD INDICATOR
C
  4466 KARC = KARC + 1
       IF(KARC.GT.LARC) GO TO 3177
       ASTORE(KARC) = EMARK(9)
       GO TO 3011
       END
       SUBROUTINE SEREAD (NCARDS,IE,JE,IADJ)
       IMPLICIT INTEGER*4(I-N)
      0COMMON EMARK(11),SETUP,FINT,DINT,FACTOR,STIME,SCOST,SPERF,CTIME,
      1CCOST,CPERF,BTIME,BCOST,BPERF,TIMESM,UNFORM,ITRIP1(22),ITRIP2(22),
      2KNT(6),INPT,IOUT,IPNH,IWF1,IWF2,IWF3,IWF4,IBLK,ITRACE,ICUT,ISEED,
      3IRROR,ISAVE,NODESM,MAXTAG,ICOUNT,JCOUNT,NTRIP,ICORR,LCOMP,NWARN,
      4NRK,NRKS,SLAK,MODE,INFC,INFP,DISC,DISP,CRIT,KGEN,MED,SINGLE,DOUBLE
      0COMMON/ARCS/ASTORE( 2800),UTIMEA( 350),TIMEA( 350),UCOSTA( 350),
      1COSTA( 350),UPERFA( 350),PERFA( 350),WORK( 350),ISTATE( 350),
      2NODEI( 350),NODEO( 350),ICRITA( 350),KEEPC( 350),KEEPP( 350),
      3IARC1( 350),IARC2( 350),IPOINT( 350),JPOINT( 350),ISLAK( 350),
      4KARC,LARC,MARC,NARC,ITALC,ITALP,ISTAR
       LOGICAL SLAK,MODE,INFC,INFP,DISC,DISP,CRIT,KGEN,MED,SINGLE,DOUBLE
C
C      THIS SUBROUTINE READS THE MONTE CARLO, FILT1, FILT2, DISTRIBUTION
C          AND HISTOGRAM SATELLITE ARC CARDS
C
       MAXTAG = 0
       DO 4499 I=1,NCARDS
       MAXTAG = MAXTAG + 6
       IF(MAXTAG.LE.MARC) GO TO 4477
       MAXTAG = 6
  4477 NODESM = MAXTAG - 5
       READ (IWF4,4488) J, (TIMEA(K), K=NODESM,MAXTAG), IADJ, K
  4488 FORMAT (I2, 6F10.0, I1, I5)
  4499 IF(J.NE.I) CALL ERROR (4500,IE,JE,IBLK)
C
C      DETERMINE HOW MANY ELEMENTS THERE ARE AND STORE THESE ELEMENTS.
C
```

```
      NODESM = MAXTAG - 6 + (K+9)/10
      IF(NODESM.LE.0) RETURN
      IF(ICOUNT.EQ.-9999) RETURN
      DO 4522 I=1,NODESM
      KARC = KARC + 1
      IF(KARC.GT.LARC) RETURN
      IF(TIMEA(I).LT.EMARK(11)) CALL ERROR (4511,IE,JE,IBLK)
 4522 ASTORE(KARC) = TIMEA(I)
      RETURN
      END
      SUBROUTINE LOADNR
      IMPLICIT INTEGER*4(I-N)
      COMMON EMARK(11),SETUP,FINT,DINT,FACTOR,STIME,SCOST,SPERF,CTIME,
     1CCOST,CPERF,BTIME,BCOST,BPERF,TIMESM,UNFORM,ITRIP1(22),ITRIP2(22),
     2KNT(6),INPT,IOUT,IPNH,IWF1,IWF2,IWF3,IWF4,IBLK,ITRACE,ICUT,ISEED,
     3IRROR,ISAVE,NODESM,MAXTAG,ICOUNT,JCOUNT,NTRIP,ICORR,LCOMP,NWARN,
     4NRK,NRKS,SLAK,MODE,INFC,INFP,DISC,DISP,CRIT,KGEN,MED,SINGLE,DOUBLE
      COMMON/TRIALS/STORET( 1000,4),TERM(10,8),KPOINT(10),NODET( 1000),
     1MTERM,NTERM,MITER,ITER
      COMMON/ARCS/ASTORE( 2800),UTIMEA( 350),TIMEA( 350),UCOSTA( 350),
     1COSTA( 350),UPERFA( 350),PERFA( 350),WORK( 350),ISTATE( 350),
     2NODEI( 350),NODEO( 350),ICRITA( 350),KEEPC( 350),KEEPP( 350),
     3IARC1( 350),IARC2( 350),IPOINT( 350),JPOINT( 350),ISLAK( 350),
     4KARC,LARC,MARC,NARC,ITALC,ITALP,ISTAR
      COMMON/NODES/TIMEN( 200),COSTN( 200),PERFN( 200),NSTORE( 5400),
     1NODE1( 200),NODE2( 200),LOGI( 200),LOGO( 200),NSTATE( 200),
     2NARCI( 200),NARCO( 200),ISTAT( 200),INSTAT( 200),ICRITN( 200),
     3NPOINT( 200),NSLAK( 200),JUMP( 200),KNODE,LNODE,MNODE,NNODE,MTAG,
     4NTAG
      COMMON/INTERN/XMIN(20,4),XMAX(20,4),HMIN(20,4),HMAX(20,4),
     1HAVE(20,4),IOBS(20),MHIST,NHIST
      COMMON/SLACK/RMIN(20),RMAX(20),SMIN(20),SMAX(20),SAVE(20),
     1JOBS(20),MSLACK,NSLACK
      COMMON/CPGAP/T1(10),T2(10),CONFL(10),CONFS(10),CAVE(10),CSMIN(10),
     1CSMAX(10),CHMIN(10),CHMAX(10),PAVE(10),PSMIN(10),PSMAX(10),
     2PHMIN(10),PHMAX(10),KCOBS(10),KPOBS(10),MCPGAP,NCPGAP,ICPGAP
      LOGICAL SLAK,MODE,INFC,INFP,DISC,DISP,CRIT,KGEN,MED,SINGLE,DOUBLE
      DATA IS/1HS/
C
C
C     THIS SUBROUTINE LOADS THE NODE DATA, CHECKS THE PROBLEM OVER FOR
C        ERRORS AND PRINTS OUT THE STORAGE AREAS IF REQUESTED.
C
      ISAVE = 0
      MODE = .TRUE.
 4533 ISAVE = ISAVE + 1
      IF(ISAVE.GT.NNODE) GO TO 4933
      READ (IWF4,4544) LOGI(ISAVE), LOGO(ISAVE), INSTAT(ISAVE),
     1ISTAT(ISAVE), (TIMEA(I), I=1,3), M, N1, N2, N3
 4544 FORMAT (I1, I3, 2I2, 3F4.3, A1, 3I4)
      LOGII = LOGI(ISAVE)
      LOGOO = LOGO(ISAVE)
      IF(LOGII.GT.4) GO TO 4555
      IF(LOGOO.GE.11) LOGOO = 1
 4555 I = ISTAT(ISAVE)
      IF(I.LE.0) GO TO 4566
      MODE = .FALSE.
      IF(I.GT.NHIST) NHIST = I
 4566 NPOINT(ISAVE) = KNODE + 1
      IE = NODE1(ISAVE)
      JE = NODE2(ISAVE)
      IF(INSTAT(ISAVE).EQ.16) INSTAT(ISAVE) = 17
      IF(INSTAT(ISAVE).EQ.0) INSTAT(ISAVE) = 16
```

```
      UIF(INSTAT(ISAVE).LT.1.OR.INSTAT(ISAVE).GT.17)
     1CALL ERROR (4577,IE,JE,IBLK)
       IF(CTIME.NE.1.0) GO TO 4622
       IF(M.NE.IS) GO TO 4622
       SLAK = .FALSE.
       NSLACK = NSLACK + 1
       IF(NSLACK.LE.MSLACK) GO TO 4600
       IF(ITRACE.EQ.3) CALL ERROR (4588,IE,JE,NWARN)
       IF(ITRACE.NE.3) CALL ERROR (4599,IE,JE,IBLK)
       NSLACK = 1
 4600 NSLAK(ISAVE) = NSLACK
       IF(LOGII.LE.4.AND.LOGCO.EQ.1) CALL ERROR (4611,IE,JE,IBLK)
C
C      IF PRESENT, READ IN A HISTOGRAM MIN-MAX CARD
C
 4622 IF(N1.EQ.0) GO TO 4722
       IF(LOGII.LE.4.AND.LOGOO.EQ.1) GO TO 4688
       I = ISTAT(ISAVE)
       IF(I.GE.1.AND.I.LE.MHIST) GO TO 4655
       CALL ERROR (4633,IE,JE,IBLK)
 4644 I = 1
 4655 READ (IWF4,4666) (COSTA(J), PERFA(J), J=1,4)
 4666 FORMAT (8F8.0)
       DO 4677 J=1,4
       IF(COSTA(J).EQ.0.0.AND.PERFA(J).EQ.0.0) GO TO 4677
       HMIN(I,J) = COSTA(J)
       HMAX(I,J) = PERFA(J)
 4677 CONTINUE
       GO TO 4722
 4688 NTERM = NTERM + 1
       IF(NTERM.LE.MTERM) GO TO 4700
       CALL ERROR (4699,IE,JE,IBLK)
       GO TO 4644
 4700 READ (IWF4,4666) (COSTA(J), J=1,8)
       DO 4711 J=2,8,2
       I = J - 1
       IF(COSTA(I).EQ.0.0.AND.COSTA(J).EQ.0.0) GO TO 4711
       TERM(NTERM,I) = COSTA(I)
       TERM(NTERM,J) = COSTA(J)
 4711 CONTINUE
       KPOINT(NTERM) = ISAVE
C
C      IF PRESENT, READ IN A SUBTRACT CARD AND FIND THE SUBTRACT NODE NO.
C
 4722 NODESM = 0
       IF(N2.EQ.0) GO TO 4777
       IF(LOGII.GT.4.OR.LOGCO.NE.4) CALL ERROR (4733,IE,JE,IBLK)
       READ (IWF4,4744) K, J
 4744 FORMAT (16A4)
       DO 4755 I=1,NNODE
       IF(I.EQ.ISAVE) GO TO 4755
       NODESM = I
       IF(NODE1(I).EQ.K.AND.NODE2(I).EQ.J) GO TO 4777
 4755 CONTINUE
       CALL ERROR (4766,IE,JE,IBLK)
C
C      IF PRESENT, READ IN A SLACK HISTOGRAM MIN-MAX CARD
C
 4777 IF(N3.LE.0) GO TO 4811
       IF(NSLAK(ISAVE).EQ.0) CALL ERROR (4788,IE,JE,IBLK)
       IF(N3.NE.1) CALL ERROR (4799,IE,JE,IBLK)
       DO 4800 I=1,N3
```

```
      READ (IWF4,4666) X, Y
      SMIN(NSLACK) = X
 4800 SMAX(NSLACK) = Y
C
C     STORE INPUT-OUTPUT ARC ADDRESSES
C
 4811 DO 4822 I=1,NARC
      IF(NODEO(I).NE.ISAVE) GO TO 4822
      KNODE = KNODE + 1
      IF(KNODE.GT.LNODE) GO TO 4844
      NARCI(ISAVE) = NARCI(ISAVE) + 1
      NSTORE(KNODE) = I
 4822 CONTINUE
      DO 4833 I=1,NARC
      IF(NODEI(I).NE.ISAVE) GO TO 4833
      KNODE = KNODE + 1
      IF(KNODE.GT.LNODE) GO TO 4844
      NARCO(ISAVE) = NARCO(ISAVE) + 1
      NSTORE(KNODE) = I
 4833 CONTINUE
      IF(LOGCO.NE.4) GO TO 4866
      KNODE = KNODE + 1
      IF(KNODE.GT.LNODE) GO TO 4844
      NSTORE(KNODE) = NODESM
      GO TO 4866
 4844 CALL ERROR (4855,IBLK,IBLK,IBLK)
      RETURN
C
C     CHECK THE FEASIBILITY OF THE NUMBER OF OUTPUT ARCS INITIATION RE-
C        QUESTS OR NUMBER OF SERVERS AND TAG ON THE FACTOR WEIGHTS
C
 4866 IF(LOGII.LE.4) GO TO 4533
      IF(LOGII.EQ.8) GO TO 4900
      K = IABS(LOGCO)
      IF(LOGII.NE.7) GO TO 4888
      IF(K.LE.0) CALL ERROR (4877,IE,JE,IBLK)
      GO TO 4533
 4888 IF(K.EQ.0.OR.K.GT.NARCI(ISAVE)) CALL ERROR (4899,IE,JE,IBLK)
      IF(LOGII.EQ.6) GO TO 4533
 4900 UNFORM = 0.0
      DO 4911 I=1,3
      UNFORM = UNFORM + TIMEA(I)
      KNODE = KNODE + 1
      IF(KNODE.GT.LNODE) GO TO 4844
 4911 NSTORE(KNODE) = TIMEA(I)*100.0 + 0.001
      IF(ABS(UNFORM).GT.0.99.AND.ABS(UNFORM).LT.1.01) GO TO 4533
      CALL ERROR (4922,IE,JE,IBLK)
C
C     LIST THE STORAGE ARRAYS IF REQUESTED
C
 4933 IF(ITRACE.NE.3) GO TO 5088
      WRITE (IOUT,4944)
 4944 FORMAT (1H1, 20X, 29HVARIABLE ARC STORAGE (ASTORE))
      WRITE (IOUT,4955) (I, I=1,12)
 4955 FORMAT (22H0  NC.   NAME    POINTER,10I11/ 1H ,21X,2I11,9H --- ETC.)
      DO 5022 I=1,NARC
      IF(IPOINT(I).GT.0) GO TO 4977
      WRITE (IOUT,4966) I, IARC1(I), IARC2(I)
 4966 FORMAT (1H , I5, 1X, 2A4, 1X, I6, 10G11.5)
      GO TO 5022
C
C     FIND THE END OF BLOCK MARK (EMARK(9))
```

```
C
 4977 M = IPOINT(I)
      DO 4988 L=M,KARC
      K = L
      IF(ASTORE(K).EQ.EMARK(9)) GO TO 4999
 4988 CONTINUE
 4999 N = M + 9
      IF(N.GT.K)/N = K
      WRITE (IOUT,4966) I,IARC1(I),IARC2(I),IPOINT(I),(ASTORE(L),L=M,N)
 5000 IF(N.EQ.K) GO TO 5022
      M = N + 1
      N = N + 10
      IF(N.GT.K) N = K
      WRITE (IOUT,5011) (ASTORE(L), L=M,N)
 5011 FORMAT (1H , 21X, 10G11.5)
      GO TO 5000
 5022 CONTINUE
      WRITE (IOUT,5033)
 5033 FORMAT (1H1, 20X, 44HVARIABLE NODE STORAGE (FRONT PART OF NSTORE))
      WRITE (IOUT,4955) (I, I=1,12)
      DO 5077 I=1,NNODE
      M = NPOINT(I)
      K = NARCI(I) + NARCO(I) + M - 1
      IF(LOGI(I).EQ.5.OR.LOGI(I).EQ.8) K = K + 3
      IF(LOGI(I).LT.5.AND.LOGO(I).EQ.4) K = K + 1
      N = M + 9
      IF(N.GT.K) N = K
      WRITE (IOUT,5044) I,NODE1(I),NODE2(I),NPOINT(I),(NSTORE(L),L=M,N)
 5044 FORMAT (1H , I5, 1X, 2A4, 1X, I6, 10I11)
 5055 IF(N.EQ.K) GO TO 5077
      M = N + 1
      N = N + 10
      IF(N.GT.K) N = K
      WRITE (IOUT,5066) (NSTORE(L), L=M,N)
 5066 FORMAT (1H , 21X, 10I11)
      GO TO 5055
 5077 CONTINUE
C
C     CHECK THE REST OF THE CONTROL CARD PARAMETERS
C
 5088 IF(ITRACE.LT.0.OR.ITRACE.GT.3) CALL ERROR (5099,IBLK,IBLK,IBLK)
      IF(ITRACE.EQ.3.AND.ITER.GT.100) CALL ERROR (5100,IBLK,IBLK,IBLK)
      IF(ICUT.LT.0.OR.ICUT.GT.2) CALL ERROR (5111,IBLK,IBLK,IBLK)
      IF(LCOMP.EQ.0) LCOMP = 16
      IF(LCOMP.LT.1.OR.LCOMP.GT.16) CALL ERROR (5120,IBLK,IBLK,IBLK)
      IF(ISEED.EQ.0) ISEED = 435459
      IF(ITER.LE.0.OR.ITER.GT.MITER) CALL ERROR (5122,IBLK,IBLK,MITER)
      L = 0
      M = 0
      IF(INFC.AND.INFP) GO TO 5144
      IF(FINT.LE.0.0) CALL ERROR (5133,IBLK,IBLK,IBLK)
      FINT = FINT/100.0 + 1.0
      L = 1
 5144 IF(DISC.AND.DISP) GO TO 5166
      IF(DINT.LE.0.0) CALL ERROR (5155,IBLK,IBLK,IBLK)
      DINT = DINT/100.0 + 1.0
      M = 1
 5166 IF(L+M.EQ.0) GO TO 5188
      IF(FACTOR.LE.0.0) CALL ERROR (5177,IBLK,IBLK,IBLK)
 5188 IF(STIME.LT.-1.0.OR.STIME.GT.1.0) CALL ERROR (5199,IBLK,IBLK,IBLK)
      IF(SCOST.LT.-1.0.OR.SCOST.GT.1.0) CALL ERROR (5200,IBLK,IBLK,IBLK)
      IF(SPERF.LT.-1.0.OR.SPERF.GT.1.0) CALL ERROR (5211,IBLK,IBLK,IBLK)
```

```
      UNFORM = ABS(STIME + SCOST + SPERF)
      IF(UNFORM.LT.0.99.OR.UNFORM.GT.1.01)CALLERROR(5222,IBLK,IBLK,IBLK)
      IF(CTIME.LT.-1.0.OR.CTIME.GT.1.0) CALL ERROR (5233,IBLK,IBLK,IBLK)
      IF(CCOST.LT.-1.0.OR.CCOST.GT.1.0) CALL ERROR (5244,IBLK,IBLK,IBLK)
      IF(CPERF.LT.-1.0.OR.CPERF.GT.1.0) CALL ERROR (5255,IBLK,IBLK,IBLK)
      UNFORM = ABS(CTIME + CCOST + CPERF)
      IF(UNFORM.EQ.0.0) GO TO 5288
      IF(UNFORM.LT.0.99.OR.UNFORM.GT.1.01)CALLERROR(5266,IBLK,IBLK,IBLK)
      IF(NHIST.GT.MHIST) CALL ERROR (5277,IBLK,IBLK,IBLK)
C
C     CHECK NODES FOR COMPLETENESS
C         ICOUNT - USED TO COUNT NUMBER OF INITIAL NODES
C         JCOUNT - USED TO COUNT NUMBER OF TERMINAL NODES
C         NI - NO. OF INPUT ARCS
C         NO - NO. OF OUTPUT ARCS
C         NP - POINTER INTO VARIABLE NODE STORAGE
C
 5288 ICOUNT = 0
      JCOUNT = 0
      DO 6066 I=1,NNODE
      JUMP(I) = 0
      NI = NARCI(I)
      NO = NARCO(I)
 1480 SINGLE = .TRUE.
      ICUT = ICUT - 1
C
C     DETERMINE IF THE MEDIAN IS WANTED
C
 1485 IF(ITRACE.NE.4) GO TO 1488
      ITRACE = 1
      GO TO 1499
 1488 IF(ITRACE.NE.5) GO TO 1500
      ITRACE = 2
 1499 MED = .FALSE.
C
C     LOAD THE SEED GENERATED AT THE END OF THE PREVIOUS RUN IF THE
C         VALUE ENTERED FOR THE SEED FOR THE CURRENT RUN IS ZERO.  IF
C         THIS RUN IS THE FIRST BEING ENTERED IN THE COMPUTER, DETERMINE
C         WHICH GENERATOR IS WANTED
C
 1500 IF(M.EQ.1) GO TO 1511
      IF(ISEED.NE.0) GO TO 1511
      ISEED = N
      GO TO 1522
 1511 KGEN = .FALSE.
      IF(ISEED.GE.0) GO TO 1522
      KGEN = .TRUE.
      ISEED = -ISEED
C
C     READ AND LIST INPUT OPTIONS
C
 1522 IF(NTRIP.GE.0.AND.NTRIP.LE.1) GO TO 1530
      CALL ERROR (1533,IBLK,IBLK,IBLK)
      GO TO 2940
 1530 IF(ICORR.GE.0.AND.ICORR.LE.1) GO TO 1540
      CALL ERROR (1544,IBLK,IBLK,IBLK)
      GO TO 2940
 1540 IF(ICPGAP.GE.0.AND.ICPGAP.LE.3) GO TO 1550
      CALL ERROR (1555,IBLK,IBLK,IBLK)
      GO TO 2940
 1550 IF(K.GE.0.AND.K.LE.1) GO TO 1560
      CALL ERROR (1566,IBLK,IBLK,IBLK)
```

```
            GO TO 2940
    1560 IF(LCOMP.GE.0.AND.LCCMP.LE.16) GO TO 1572
         CALL ERROR (1570,IBLK,IBLK,IBLK)
         GO TO 2940
    1572 IF(L8080.GE.0.AND.L8C80.LE.1) GO TO 1575
         CALL ERROR (1574,IBLK,IBLK,IBLK)
         GO TO 2940
    1575 IF(NTRIP.GT.0)READ(INPT,1455,END=2922)(ITRIP1(L),ITRIP2(L),L=1,10)
        UIF(ICORR.GT.0) REAC (INPT,2488,END=2922)
        1(ITRIP1(L), ITRIP2(L), L=11,22)
         IF(L8080.NE.0) GO TO 1590
         IF(NTRIP.GT.0) WRITE (IOUT,1577) (ITRIP1(L), ITRIP2(L), L=1,10)
    1577OFORMAT (41H DESIGNATED ARCS AND NODES FOR FULL PRINT//1H , 20A4//)
         IF(ICORR.GT.0) WRITE (IOUT,1588) (ITRIP1(L), ITRIP2(L), L=11,22)
    1588OFORMAT (35H CORRELATICN COMBINATIONS REQUESTED//1H , 12(2A1,1X)//)
C
C       READ THE COST-PERFORMANCE TIME INTERVAL DATA
C
    1590 NCPGAP = 0
         IF(ICPGAP.EQ.0) GO TC 1722
         M = 0
         REWIND IWF4
         IF(L8080.EQ.0) WRITE (IOUT,1599)
    1599 FORMAT (36H COST-PERFORMANCE TIME INTERVAL DATA//)
    1600 READ (INPT,2488,END=2922) (NODET(L), L=1,80)
         IF(L8080.EQ.0) WRITE (IOUT,1611) (NODET(L), L=1,80)
    1611 FORMAT (1H , 80A1)
         IF(NODET(1).NE.IDX(9)) GO TO 1622
         IF(NODET(2).NE.IDX(10))GO TO 1622
         IF(NODET(3).NE.IDX(11))GO TO 1622
         IF(NODET(4).EQ.IDX(12))GO TO 1644
    1622 NCPGAP = NCPGAP + 1
         WRITE (IWF4,2488) (NODET(L), L=1,80)
         DO 1633 J=10,80,10
         I = J - 9
    1633 CALL CHECK (M,I,J)
         GO TO 1600
C
C       IF NO ERRORS, STORE THE COST-PERFORMANCE TIME INTERVAL DATA
C
    1644 L = IRROR
         IF(NCPGAP.GT.MCPGAP) CALL ERROR (1655,IBLK,IBLK,IBLK)
         IF(M.GT.0) CALL ERROR (1656,IBLK,IBLK,IBLK)
         IF(L.NE.IRROR) GO TO 1722
         REWIND IWF4
         DO 1699 L=1,NCPGAP
         CSMIN(L) = SETUP
         CSMAX(L) =-SETUP
         CHMIN(L) = SETUP
         CHMAX(L) =-SETUP
         PSMIN(L) = SETUP
         PSMAX(L) =-SETUP
         PHMIN(L) = SETUP
         PHMAX(L) =-SETUP
         CAVE(L) = 0.0
         PAVE(L) = 0.0
         KCOBS(L) = 0
         KPOBS(L) = 0
         READ (IWF4,1677) T1(L),T2(L),(WORK(I),I=1,4),CONFL(L),CONFS(L)
    1677 FORMAT (8F10.0)
         IF(T2(L).LT.RTIME) M = M + 1
         IF(T1(L).GE.T2(L)) M = M + 1
```

```
      IF(WORK(1).EQ.0.0.AND.WORK(2).EQ.0.0) GO TO 1688
      IF(WORK(1).GE.WORK(2)) M = M + 1
      CHMIN(L) = WORK(1)
      CHMAX(L) = WORK(2)
 1688 IF(WORK(3).EQ.0.0.AND.WORK(4).EQ.0.0) GO TO 1699
      IF(WORK(3).GE.WORK(4)) M = M + 1
      PHMIN(L) = WORK(3)
      PHMAX(L) = WORK(4)
      IF(CONFL(L).EQ.0.0.AND.CONFS(L).EQ.0.0) GO TO 1699
      IF(CONFL(L).LE.0.0.OR.CONFL(L).GT.1.0.OR.CONFS(L).LT.0.0)M = M + 1
 1699 CONTINUE
      IF(M.GT.0) CALL ERROR (1700,IBLK,IBLK,IBLK)
      IF(L8080.EQ.0) WRITE (IOUT,1711)
 1711 FORMAT (1H )
C
C
C     INITIALIZE FOR STORING TERMINAL NODE HISTOGRAMS MINS AND MAXS
C
 1722 NTERM = 0
      DO 1733 I=1,MTERM
      KPOINT(I) = 0
      DO 1733 J=2,8,2
      TERM(I,J-1)= SETUP
 1733 TERM(I,J) = -SETUP
      IF(K.EQ.0) GO TO 1799
C
C     CHECK AND LOAD THE COMPOSITE TERMINAL NODE HISTOGRAM MINS AND MAXS
C
      M = 0
      REWIND IWF4
      READ (INPT,2488,END=2922) (NODET(L), L=1,80)
      IF(L8080.EQ.0) WRITE (IOUT,1744) (NODET(L), L=1,80)
 17440FORMAT (54H MINIMUMS AND MAXIMUMS FOR THE COMPOSITE TERMINAL NODE/
     1/1H , 80A1/)
      DO 1755 J=10,80,10
      I = J - 9
 1755 CALL CHECK (M,I,J)
      IF(M.EQ.0) GO TO 1777
      CALL ERROR (1766,IBLK,IBLK,IBLK)
      GO TO 1799
 1777 WRITE (IWF4,2488) (NODET(L), L=1,80)
      REWIND IWF4
      READ (IWF4,1677) (COSTA(J), J=1,8)
      DO 1788 J=2,8,2
      I = J - 1
      IF(COSTA(I).EQ.0.0.AND.COSTA(J).EQ.0.0) GO TO 1788
      TERM(1,I) = COSTA(I)
      TERM(1,J) = COSTA(J)
 1788 CONTINUE
      KPOINT(1) = -1
      NTERM = 1
C
C     CHECK TYPE OF RUN REQUESTED
C
 1799 REWIND IWF1
      REWIND IWF2
      ISAVE = 0
      LASTA = IDX(7)
      LASTB = IDX(7)
      JSTAR = 0
      IF(ISTAR.EQ.0) GO TO 1820
      IF(ISTAR.EQ.3) GO TO 1800
      IF(ISTAR.NE.4) GO TO 1804
```

```
 1800 JSTAR = ISTAR
      ISTAR = ISTAR - 2
 1804 IF(ISTAR.EQ.1.OR.ISTAR.EQ.2) GO TO 1888
      CALL ERROR (1811,IBLK,IBLK,IBLK)
      GO TO 2940
C
C     LOAD THE NEW PROBLEM ON THE MASTER AND INTERMEDIATE FILES
C
 1820 READ (INPT,1822,END=2922) (NODET(I), I=1,21)
 1822 FORMAT (3A4, A1, A3, 16A4)
      DO 1825 I=1,21
      IF(NODET(I).NE.IBLK) GO TO 1830
 1825 CONTINUE
      GO TO 1820
 1830 IF(NODET(1).EQ.LASTA.AND.NODET(2).EQ.LASTB) GO TO 1844
      IF(L8080.EQ.0) WRITE (IOUT,1711)
      LASTA = NODET(1)
      LASTB = NODET(2)
 1844 IF(L8080.EQ.0) WRITE (IOUT,1855) (NODET(I), I=1,21)
 1855 FORMAT (1H , 3A4, A1, A3, 16A4)
      WRITE (IWF1,1822) (NODET(I), I=1,21)
      WRITE (IWF2,1822) (NODET(I), I=1,21)
      IF(NODET(1).EQ.IDX(1)) ISAVE = 1
      IF(NODET(1).NE.IDX(2)) GO TO 1820
      IF(ISAVE.EQ.1) GO TO 2011
 1866 CALL ERROR (1877,IBLK,IBLK,IBLK)
      GO TO 2940
C
C     STORE CHANGES IN INTERNAL PROGRAM STORAGE AREAS PRIOR TO LOADING
C        THE INTERMEDIATE FILE AND PERHAPS CREATING A NEW MASTER FILE
C
 1888 ICOUNT = 0
 1899 ICOUNT = ICOUNT + 1
      IF(ICOUNT.LE.MARC.AND.ICOUNT.LE.MNODE) GO TO 1911
      CALL ERROR (1900,IBLK,IBLK,IBLK)
      GO TO 2940
 1911 READ (INPT,1322,END=2922) NODEI(ICOUNT), NODEO(ICOUNT),
     1ICRITA(ICOUNT), ISTATE(ICOUNT), IPOINT(ICOUNT), JPOINT(ICOUNT),
     2UTIMEA(ICOUNT), TIMEA(ICOUNT), UCOSTA(ICOUNT), COSTA(ICOUNT),
     3UPERFA(ICOUNT), PERFA(ICOUNT), KEEPC(ICOUNT), KEEPP(ICOUNT),
     4LOGI(ICOUNT), LOGO(ICOUNT), ISTAT(ICOUNT), ICRITN(ICOUNT),
     5NARCI(ICOUNT), NARCO(ICOUNT), NPOINT(ICOUNT)
      IF(NODEI(ICOUNT).NE.IBLK) GO TO 1920
      IF(NODEO(ICOUNT).NE.IBLK) GO TO 1920
      IF(ICRITA(ICOUNT).NE.IBLK) GO TO 1920
      IF(ISTATE(ICOUNT).NE.IBLK) GO TO 1920
      IF(IPOINT(ICOUNT).NE.IBLK) GO TO 1920
      IF(JPOINT(ICOUNT).NE.IBLK) GO TO 1920
      IF(UTIMEA(ICOUNT).NE.IBLK) GO TO 1920
      IF(TIMEA(ICOUNT).NE.IBLK) GO TO 1920
      IF(UCOSTA(ICOUNT).NE.IBLK) GO TO 1920
      IF(COSTA(ICOUNT).NE.IBLK) GO TO 1920
      IF(UPERFA(ICOUNT).NE.IBLK) GO TO 1920
      IF(PERFA(ICOUNT).NE.IBLK) GO TO 1920
      IF(KEEPC(ICOUNT).NE.IBLK) GO TO 1920
      IF(KEEPP(ICOUNT).NE.IBLK) GO TO 1920
      IF(LOGI(ICOUNT).NE.IBLK) GO TO 1920
      IF(LOGO(ICOUNT).NE.IBLK) GO TO 1920
      IF(ISTAT(ICOUNT).NE.IBLK) GO TO 1920
      IF(ICRITN(ICOUNT).NE.IBLK) GO TO 1920
      IF(NARCI(ICOUNT).NE.IBLK) GO TO 1920
      IF(NARCO(ICOUNT).NE.IBLK) GO TO 1920
```

```
      IF(NPOINT(ICOUNT).NE.IBLK) GO TO 1920
      GO TO 1911
 1920 IF(NODEI(ICOUNT).EQ.LASTA.AND.NODEO(ICOUNT).EQ.LASTB) GO TO 1922
      IF(L8080.EQ.0) WRITE (IOUT,1711)
      LASTA = NODEI(ICOUNT)
      LASTB = NODEO(ICOUNT)
 19220IF(L8080.EQ.0) WRITE (IOUT,1855) NODEI(ICOUNT), NODEO(ICOUNT),
     1ICRITA(ICOUNT), ISTATE(ICOUNT), IPOINT(ICOUNT), JPOINT(ICOUNT),
     2UTIMEA(ICOUNT), TIMEA(ICOUNT), UCOSTA(ICOUNT), COSTA(ICOUNT),
     3UPERFA(ICOUNT), PERFA(ICOUNT), KEEPC(ICOUNT), KEEPP(ICOUNT),
     4LOGI(ICOUNT), LOGO(ICOUNT), ISTAT(ICOUNT), ICRITN(ICOUNT),
     5NARCI(ICOUNT), NARCO(ICOUNT), NPOINT(ICOUNT)
      IF(NODEI(ICOUNT).EQ.IDX(2)) GO TO 1933
      IF(NODEI(ICOUNT).NE.IDX(1)) GO TO 1899
      JCOUNT = ICOUNT
      ISAVE = 1
      GO TO 1899
 1933 IF(ISAVE.EQ.0) GO TO 1866
C
C     LOAD CHANGES AND UNCHANGED ARCS(FIRST PASS) AND NODES(SECOND PASS)
C
      REWIND IWF1
      REWIND IWF2
      ISAVE = 1
 1944 NODESM = IDX(7)
      MAXTAG = IDX(7)
 1955 READ (IWF1,1822) (NODET(I), I=1,21)
      IF(NODET(1).EQ.NODESM.AND.NODET(2).EQ.MAXTAG) GO TO 1955
      IF(NODET(1).EQ.IDX(1).OR.NODET(1).EQ.IDX(2)) GO TO 1999
      M = 1
      DO 1977 I=ISAVE,JCOUNT
      IF(NODET(1).NE.NODEI(I).OR.NODET(2).NE.NODEO(I)) GO TO 1977
      NODESM = NODET(1)
      MAXTAG = NODET(2)
      M = 2
      IF(ICRITA(I).EQ.IDX(3)) GO TO 1966
     0WRITE (IWF2,1822) NODEI(I), NODEO(I), ICRITA(I), ISTATE(I),
     1IPOINT(I), JPOINT(I), UTIMEA(I), TIMEA(I), UCOSTA(I), COSTA(I),
     2UPERFA(I), PERFA(I), KEEPC(I), KEEPP(I), LOGI(I), LOGO(I),
     3ISTAT(I), ICRITN(I), NARCI(I), NARCO(I), NPOINT(I)
 1966 NODEI(I) = IDX(7)
 1977 CONTINUE
      GO TO (1988,1955),M
 1988 WRITE (IWF2,1822) (NODET(I), I=1,21)
      GO TO 1955
C
C     LOAD ADD ON ARCS(FIRST PASS) AND NODES(SECOND PASS)
C
 1999 DO 2000 I=ISAVE,JCOUNT
      IF(NODEI(I).EQ.IDX(7)) GO TO 2000
     0WRITE (IWF2,1822) NODEI(I), NODEO(I), ICRITA(I), ISTATE(I),
     1IPOINT(I), JPOINT(I), UTIMEA(I), TIMEA(I), UCOSTA(I), COSTA(I),
     2UPERFA(I), PERFA(I), KEEPC(I), KEEPP(I), LOGI(I), LOGO(I),
     3ISTAT(I), ICRITN(I), NARCI(I), NARCO(I), NPOINT(I)
 2000 CONTINUE
      IF(ICOUNT.EQ.JCOUNT) GO TO 2011
      ISAVE = JCOUNT + 1
      JCOUNT = ICOUNT
      GO TO 1944
C
C     LIST THE INTERMEDIATE FILE, LOAD THE TRANSPORT FILE NO. 1, LOAD
C         ARC AND NODE NAMES AND CREATE A NEW MASTER FILE IF REQUESTED
```

```
C
 2011 REWIND IWF1
      REWIND IWF2
      REWIND IWF3
      MAXTAG = IDX(7)
      NODESM = IDX(7)
      NARC = 0
      NNODE= 0
      ISAVE = 1
      IF(JSTAR.EQ.0) GO TO 2022
      LASTA = IDX(7)
      LASTB = IDX(7)
      IF(L8080.EQ.0) WRITE (IOUT,2020)
 2020 FORMAT (37H1LISTING CF THE RECONSTITUTED PROBLEM)
 2022 ICOUNT = 0
 2033 ICOUNT = ICOUNT + 1
      IF(ICOUNT.LE.MARC.AND.ICOUNT.LE.MNODE) GO TO 2055
      CALL ERROR (2044,IBLK,IBLK,IBLK)
      GO TO 2940
 20550READ (IWF2,1822) NODEI(ICOUNT), NODEO(ICOUNT), ICRITA(ICOUNT),
     1ISTATE(ICOUNT), IPOINT(ICOUNT), JPOINT(ICOUNT), UTIMEA(ICOUNT),
     2TIMEA(ICOUNT), UCOSTA(ICOUNT), COSTA(ICOUNT), UPERFA(ICOUNT),
     3PERFA(ICOUNT), KEEPC(ICOUNT), KEEPP(ICOUNT), LOGI(ICOUNT),
     4LOGO(ICOUNT), ISTAT(ICOUNT), ICRITN(ICOUNT), NARCI(ICOUNT),
     5NARCO(ICOUNT), NPOINT(ICOUNT)
      IF(JSTAR.EQ.0) GO TO 2064
      IF(NODEI(ICOUNT).EQ.LASTA.AND.NODEO(ICOUNT).EQ.LASTB) GO TO 2060
      IF(L8080.EQ.0) WRITE (IOUT,1711)
      LASTA = NODEI(ICOUNT)
      LASTB = NODEO(ICOUNT)
 2060 IF(L8080.EQ.0) WRITE (IOUT,1855) NODEI(ICOUNT), NODEO(ICOUNT),
     1ICRITA(ICOUNT), ISTATE(ICOUNT), IPOINT(ICOUNT), JPOINT(ICOUNT),
     2UTIMEA(ICOUNT), TIMEA(ICOUNT), UCOSTA(ICOUNT), COSTA(ICOUNT),
     3UPERFA(ICOUNT), PERFA(ICOUNT), KEEPC(ICOUNT), KEEPP(ICOUNT),
     4LOGI(ICOUNT), LOGO(ICOUNT), ISTAT(ICOUNT), ICRITN(ICOUNT),
     5NARCI(ICOUNT), NARCO(ICOUNT), NPOINT(ICOUNT)
 2064 IF(ISTAR.NE.2) GO TO 2066
      0WRITE (IWF1,1822)NODEI(ICOUNT), NODEO(ICOUNT), ICRITA(ICOUNT),
     1ISTATE(ICOUNT), IPOINT(ICOUNT), JPOINT(ICOUNT), UTIMEA(ICOUNT),
     2TIMEA(ICOUNT), UCOSTA(ICOUNT), COSTA(ICOUNT), UPERFA(ICOUNT),
     3PERFA(ICOUNT), KEEPC(ICOUNT), KEEPP(ICOUNT), LOGI(ICOUNT),
     4LOGO(ICOUNT), ISTAT(ICOUNT), ICRITN(ICOUNT), NARCI(ICOUNT),
     5NARCO(ICOUNT), NPOINT(ICOUNT)
 2066 IF(ICOUNT.EQ.1) GO TO 2433
C
C     SAME FAMILY YET? IF NOT TAKE AN INVENTORY OF SATELLITE CARDS
C
      IF(NODEI(ICOUNT).NE.MAXTAG) GO TO 2077
      IF(NODEO(ICOUNT).EQ.NODESM) GO TO 2033
 2077 K = ICOUNT - 1
      IF(ISAVE.EQ.2) GO TO 2266
C
C     COUNT THE NUMBER OF THE VARIOUS TYPES OF SATELLITES THIS ARC HAS
C
      DO 2088 I=1,17
 2088 NODET(I) = 0
      IF(K.EQ.1) GO TO 2133
      DO 2122 J=2,K
      DO 2099 I=1,17
      L = I
      IF(IDA(L).NE.ICRITA(J)) GO TO 2099
      IF(IDB(L).EQ.ISTATE(J)) GO TO 2111
```

```
      IF(I.LE.11) GO TO 2099
      IF(IDX(11).EQ.ISTATE(J)) GO TO 2111
      IF(IDX(13).EQ.ISTATE(J)) GO TO 2111
      IF(IDX(14).EQ.ISTATE(J)) GO TO 2111
 2099 CONTINUE
      CALL ERROR (2100,NODEI(1),NODEO(1),IBLK)
      GO TO 2122
 2111 NODET(L) = NODET(L) + 1
 2122 CONTINUE
C
C     ENTER THE ARC NAME IN THE ARC NAME ARRAYS
C
 2133 IF(NARC.EQ.0) GO TO 2166
      DO 2155 I=1,NARC
      IF(IARC1(I).NE.NODEI(1).OR.IARC2(I).NE.NODEO(1)) GO TO 2155
      CALL ERROR (2144,NODEI(1),NODEO(1),IBLK)
 2155 CONTINUE
 2166 NARC = NARC + 1
      IF(NARC.LE.MARC) GO TO 2188
      CALL ERROR (2177,IBLK,IBLK,IBLK)
      GO TO 2940
 2188 IARC1(NARC) = NODEI(1)
      IARC2(NARC) = NODEO(1)
C
C     LOADING ARC DATA ON THE TRANSPORT FILE
C
      CWRITE (IWF3,2199) ICRITA(1), ISTATE(1), IPOINT(1), JPOINT(1),
     1UTIMEA(1), TIMEA(1), UCOSTA(1), (NODET(I), I=1,17)
 2199 FORMAT (A4, A1, A3, 4A4, 17I4)
      IF(K.EQ.1) GO TO 2255
      DO 2244 I=1,17
      L = 0
      DO 2233 J=2,K
      IF(IDA(I).NE.ICRITA(J)) GO TO 2233
      IF(IDB(I).EQ.ISTATE(J)) GO TO 2200
      IF(I.LE.11) GO TO 2233
      L = 1
      IF(IDX(11).EQ.ISTATE(J)) GO TO 2200
      L = 2
      IF(IDX(13).EQ.ISTATE(J)) GO TO 2200
      L = 3
      IF(IDX(14).NE.ISTATE(J)) GO TO 2233
 2200 WRITE (IWF3,2211)IPOINT(J),JPOINT(J),UTIMEA(J),TIMEA(J),UCOSTA(J),
     1COSTA(J),UPERFA(J),PERFA(J),KEEPC(J),KEEPP(J),LOGI(J),LOGO(J),
     2ISTAT(J), ICRITN(J), NARCI(J), NARCO(J), NPOINT(J), L
 2211 FORMAT (A3, 16A4, I1)
      IF(L.EQ.0) GO TO 2233
      IF(I.LT.12)GO TO 2222
      IF(I.GT.14)GO TO 2222
      IF(L.EQ.1) DISC = .FALSE.
      IF(L.EQ.2) INFC = .FALSE.
      IF(L.NE.3) GO TO 2222
      DISC = .FALSE.
      INFC = .FALSE.
 2222 IF(I.LT.15)GO TO 2233
      IF(L.EQ.1) DISP = .FALSE.
      IF(L.EQ.2) INFP = .FALSE.
      IF(L.NE.3) GO TO 2233
      DISP = .FALSE.
      INFP = .FALSE.
 2233 CONTINUE
 2244 CONTINUE
```

```
 2255 IF(NODEI(ICOUNT).NE.IDX(1)) GO TO 2422
      ISAVE = 2
      GO TO 2022
C
C     COUNT THE NUMBER OF THE VARIOUS TYPES OF SATELLITES THIS NODE HAS
C
 2266 DO 2277 I=1,3
 2277 NODET(1) = 0
      IF(K.EQ.1) GO TO 2322
      DO 2311 J=2,K
      DO 2288 I=1,3
      L = I
      IF(IDX(I+3).EQ.ICRITA(J)) GO TO 2300
 2288 CONTINUE
      CALL ERROR (2299,NODEI(1),NODEO(1),IBLK)
      GO TO 2311
 2300 NODET(L) = NODET(L) + 1
 2311 CONTINUE
C
C     ENTER THE NODE NAME IN THE NODE NAME ARRAYS
C
 2322 IF(NNODE.EQ.0) GO TO 2355
      DO 2344 I=1,NNODE
      IF(NODE1(I).NE.NODEI(1).OR.NODE2(I).NE.NODEO(1)) GO TO 2344
      CALL ERROR (2333,NODEI(1),NODEO(1),IBLK)
 2344 CONTINUE
 2355 NNODE = NNODE + 1
      IF(NNODE.LE.MNODE) GO TO 2377
      CALL ERROR (2366,IBLK,IBLK,IBLK)
      GO TO 2940
 2377 NODE1(NNODE) = NODEI(1)
      NODE2(NNODE) = NODEO(1)
C
C     LOADING NODE DATA ON THE TRANSPORT FILE
C
      WRITE (IWF3,2199) ICRITA(1), ISTATE(1), IPOINT(1), JPOINT(1),
     1UTIMEA(1), TIMEA(1), UCOSTA(1), (NODET(I), I=1,3)
      IF(K.EQ.1) GO TO 2411
      DO 2400 I=1,3
      DO 2399 J=2,K
      IF(IDX(I+3).NE.ICRITA(J)) GO TO 2399
      WRITE (IWF3,2388) ISTATE(J), IPOINT(J), JPOINT(J), UTIMEA(J),
     1TIMEA(J), UCOSTA(J), COSTA(J), UPERFA(J), PERFA(J), KEEPC(J),
     2KEEPP(J), LOGI(J), LOGO(J), ISTAT(J), ICRITN(J), NARCI(J),
     3NARCO(J), NPCINT(J)
 2388 FORMAT (A1, A3, 16A4)
 2399 CONTINUE
 2400 CONTINUE
 2411 IF(NODEI(ICOUNT).EQ.IDX(2)) GO TO 2444
C
C     PUT LAST RECORD READ IN INTERNAL STORAGE POSITION NO. 1
C
 2422 NODEI(1)  = NODEI(ICOUNT)
      NODEO(1)  = NODEO(ICOUNT)
      ICRITA(1) = ICRITA(ICOUNT)
      ISTATE(1) = ISTATE(ICOUNT)
      IPOINT(1) = IPOINT(ICOUNT)
      JPOINT(1) = JPOINT(ICOUNT)
      UTIMEA(1) = UTIMEA(ICOUNT)
      TIMEA(1)  = TIMEA(ICOUNT)
      UCOSTA(1) = UCOSTA(ICOUNT)
      COSTA(1)  = COSTA(ICOUNT)
```

```
      UPERFA(1)  =  UPERFA(ICOUNT)
      PERFA(1)   =  PERFA(ICOUNT)
      KEEPC(1)   =  KEEPC(ICOUNT)
      KEEPP(1)   =  KEEPP(ICOUNT)
      LOGI(1)    =  LOGI(ICOUNT)
      LOGO(1)    =  LOGO(ICOUNT)
      ISTAT(1)   =  ISTAT(ICOUNT)
      ICRITN(1)  =  ICRITN(ICOUNT)


      NARCI(1)   =  NARCI(ICOUNT)
      NARCO(1)   =  NARCO(ICOUNT)
      NPOINT(1)  =  NPOINT(ICOUNT)
      ICOUNT = 1
 2433 MAXTAG = NODEI(1)
      NODESM = NODEO(1)
      GO TO 2033
C
C     LOAD DATA ON TRANSPORT FILE NO. 2, FIRST TASK IS TO CHECK FOR
C        ALPHA IN THE NUMERIC FIELDS FOR ARC CARDS AND LOAD CONSTANT
C        OR PREVIOUS TRANSFORMATION DATA FOR FUNCTIONAL RELATIONSHIPS
C
 2444 REWIND IWF3
      REWIND IWF4
      ISAVE = 0
 2455 ISAVE = ISAVE + 1
      IF(ISAVE.GT.NARC) GO TO 2633
      M = 0
      OREAD (IWF3,2466) (ICRITN(I), I=1,4), (NODET(I), I=1,8),
     1(ISTATE(I), I=1,17)
 2466 FORMAT (4A4, 8A1, 17I4)
      OWRITE (IWF4,2477) (ICRITN(I), I=1,4), (NODET(I), I=1,5),
     1(ISTATE(I), I=1,17)
 2477 FORMAT (4A4, 5A1, 17I4)
      CALL CHECK (M,1,4)
      DO 2611 I=1,17
      IF(ISTATE(I).LE.0) GO TO 2611
      NODESM = ISTATE(I)
      DO 2600 J=1,NODESM
      READ (IWF3,2488) (NODET(K), K=1,68)
 2488 FORMAT (111A1)
      CALL CHECK (M,1,2)
      OGO TO (2533,2555,2555,2555,2533,2533,2599,2533,2533,2555,2499,
     12533,2555,2499,2533,2555,2499),I
C
C     CHECK FUNCTIONAL RELATIONSHIP DATA AND BOOT STRAP NUMERIC DATA
C
 2499 CALL CHECK (M, 3, 4)
      CALL CHECK (M,33,34)
      JCOUNT = 0
      DO 2500 K=1,48
 2500 JPOINT(K) = IBLK
      DO 2522 K=6,51,9
      JCOUNT = JCOUNT + 8
      ICOUNT = JCOUNT - 7
      L = K
      IF(L.GT.32) L = L + 3
      IF(NODET(L).NE.IBLK.AND.NODET(L).NE.IDX(8)) GO TO 2522
      LOW = L + 1
      IUP = L + 8
      CALL CHECK (M,LOW,IUP)
      DO 2511 N=ICOUNT,JCOUNT
```

```
         L = L + 1
 2511 JPOINT(N) = NODET(L)
 2522 CONTINUE
      GWRITE (IWF4,2488) (NODET(K), K=1,32), (JPOINT(K), K=1,24),
     1(NODET(K), K=33,62), (JPOINT(K), K=25,48.), NODET(68)
      GO TO 2600
C
C     CHECK MONTE CARLO, FILTER 1-2, DIST., HIST. AND SLACK DATA.
C         RECORD THE POSITION OF THE LAST DATA ELEMENT IN THE CARD.
C
 2533 IF(NODESM.GT.1) CALL ERROR (2544,IARC1(ISAVE),IARC2(ISAVE),IBLK)
 2555 CALL CHECK (M, 3,12)
      CALL CHECK (M,13,22)
      CALL CHECK (M,23,32)
      CALL CHECK (M,33,42)
      CALL CHECK (M,43,52)
      CALL CHECK (M,53,62)
      DO 2566 K=3,62
      L = 62 + 3 - K
      N = L - 2
      IF(NODET(L).NE.IBLK) GO TO 2577
 2566 CONTINUE
      N = 0
 2577 WRITE (IWF4,2588) (NODET(K), K=1,62), NODET(68), N
 2588 FORMAT (63A1, I5)
      GO TO 2600
 2599 WRITE (IWF4,2488) (NODET(K), K=1,68)
 2600 CONTINUE
 2611 CONTINUE
      IF(M.GT.0) CALL ERROR (2622,IARC1(ISAVE),IARC2(ISAVE),M)
      GO TO 2455
C
C     CHECK FOR ALPHA IN THE NUMERIC FIELDS OF THE NODE CARDS
C
 2633 ISAVE = 0
 2644 ISAVE = ISAVE + 1
      IF(ISAVE.GT.NNODE) GO TO 2711
      M = 0
      READ (IWF3,2655) (NODET(I), I=1,24), (ICRITN(L), L=1,3)
 2655 FORMAT (24A1, 3I4)
      WRITE (IWF4,2666) (NODET(I), I=1,21), (ICRITN(L), L=1,3)
 2666 FORMAT (21A1, 3I4)
      CALL CHECK (M, 1, 1)
      CALL CHECK (M, 2, 4)
      CALL CHECK (M, 5, 6)
      CALL CHECK (M, 7, 8)
      CALL CHECK (M, 9,12)
      CALL CHECK (M,13,16)
      CALL CHECK (M,17,20)
C
C     CHECK THE SATELLITES
C
      DO 2699 I=1,3
      L = ICRITN(I)
      IF(L.EQ.0) GO TO 2699
      IF(L.GT.1) CALL ERROR (2677,NODE1(ISAVE),NODE2(ISAVE),IBLK)
      DO 2688 K=1,L
      READ (IWF3,2488) (NODET(J), J=1,64)
 2688 WRITE (IWF4,2488) (NODET(J), J=1,64)
      IF(I.EQ.2) GO TO 2699
      CALL CHECK (M, 1, 8)
      CALL CHECK (M, 9,16)
```

```
      CALL CHECK (M,17,24)
      CALL CHECK (M,25,32)
      CALL CHECK (M,33,40)
      CALL CHECK (M,41,48)
      CALL CHECK (M,49,56)
      CALL CHECK (M,57,64)
 2699 CONTINUE
      IF(M.GT.0) CALL ERROR (2700,NODE1(ISAVE),NODE2(ISAVE),IBLK)
      GO TO 2644
C
C     INITIALIZE THE CORE STORAGE ARRAYS
C
 2711 REWIND IWF4
      NHIST = 0
      DO 2722 I=1,MHIST
      IOBS(I) = 0
      DO 2722 J=1,4
      XMIN(I,J) = SETUP
      HMIN(I,J) = SETUP
      HAVE(I,J) = 0.0
      XMAX(I,J) =-SETUP
 2722 HMAX(I,J) =-SETUP
      SLAK = .TRUE.
      IF(CTIME.EQ.1.0.AND.ITRACE.EQ.3) SLAK = .FALSE.
      NSLACK = 0
      DO 2733 I=1,MSLACK
      JOBS(I) = 0
      RMIN(I) = SETUP
      SMIN(I) = SETUP
      SAVE(I) = 0.0
      RMAX(I) =-SETUP
 2733 SMAX(I) =-SETUP
      DC 2744 I =1,MNODE
      NSLAK(I) = 0
      NARCI(I) = 0
      NARCO(I) = 0
 2744 NPOINT(I) = 0
      KNODE = 0
      DO 2755 I=1,MTAG
 2755 NSTORE(I) = 0
      DO 2766 I=1,MARC
      ISLAK(I) = 0
 2766 IPOINT(I) = 0
      KARC = 0
      DO 2777 I=1,LARC
 2777 ASTORE(I) = 0.0
      DO 2788 I=1,MITER
      NCDET(I) = 0
      DO 2788 J=1,4
 2788 STORET(I,J) = 0.0
C
C     CHECK THE DESIGNATED ARCS AND NODES FOR BEING VALID AND DECODE
C
      IF(ITRACE.EQ.3.AND.NTRIP.GT.0) CALL ERROR (2799,IBLK,IBLK,NWARN)
      IF(NTRIP.EQ.0) GO TO 2855
      NTRIP = 0
      DO 2844 I=1,10
      IF(ITRIP1(I).NE.IBLK) GO TO 2800
      IF(ITRIP2(I).NE.IBLK) GO TO 2800
      ITRIP1(I) = 0
      GO TO 2844
 2800 IF(I.GT.NTRIP) NTRIP = I
```

```
      DO 2811 J=1,NARC
      IF(ITRIP1(I).NE.IARC1(J)) GO TO 2811
      IF(ITRIP2(I).NE.IARC2(J)) GO TO 2811
      ITRIP1(I) = 1
      ITRIP2(I) = J
      GO TO 2844
 2811 CONTINUE
      DO 2822 J=1,NNODE
      IF(ITRIP1(I).NE.NODE1(J)) GO TO 2822
      IF(ITRIP2(I).NE.NODE2(J)) GO TO 2822
      ITRIP1(I) = 2
      ITRIP2(I) = J
      GO TO 2844
 2822 CONTINUE
      CALL ERROR (2833,ITRIP1(I),ITRIP2(I),IBLK)
 2844 CONTINUE
C
C     CHECK THE CORRELATION COMBINATIONS REQUESTED
C
 2855 IF(ICORR.EQ.C) RETURN
      ICORR = 10
      DO 2911 I=11,22
      IF(ITRIP1(I).EQ.IBLK.AND.ITRIP2(I).EQ.IBLK) GO TO 2911
      M = 0
      N = 0
      DO 2877 J=1,4
      IF(ITRIP1(I).NE.NUMB(J)) GO TO 2866
      M = J
 2866 IF(ITRIP2(I).NE.NUMB(J)) GO TO 2877
      N = J
 2877 CONTINUE
      IF(M.EQ.0) GO TO 2888
      IF(N.EQ.0) GC TO 2888
      IF(M.NE.N) GC TO 2900
 2888 CALL ERROR (2899,IBLK,IBLK,IBLK)
      RETURN
 2900 ICORR = ICORR + 1
      ITRIP1(ICORR) = M
      ITRIP2(ICORR) = N
 2911 CONTINUE
      RETURN
 2922 CALL ERROR (2933,IBLK,IBLK,IBLK)
      GC TO 2944
C
C     A STOP PROCESSING ERRCR HAS CCCURRED, DUMP THE REST OF THE PROBLEM
C
 2940 READ (INPT,1822,END=2922) (NODET(I), I=1,21)
      IF(NODET(1).EQ.IDX(2)) RETURN
      GO TO 2940
 2944 CALL EXIT
      END
      SUBROUTINE CHECK (M,LCW,IUP)
      IMPLICIT INTEGER*4(I-N)
     CCOMMON/TRIALS/STORET( 1000,4),TERM(10,8),KPOINT(10),NODET( 1000),
     1MTERM,NTERM,MITER,ITER
      DIMENSION ISYM(14)
      INTEGER*2LOW,IUP
      DATA ISYM/1H0,1H1,1H2,1H3,1H4,1H5,1H6,1H7,1H8,1H9,1H+,1H-,1H.,1H /
C
C     THIS SUBROUTINE CHECKS NUMERIC FIELD FOR ALPHA INFORMATION
C
      KEY  = 0
```

```fortran
      ITAL = 0
      DO 2999 I=LOW,IUP
      IF(NODET(I).NE.ISYM(14)) GO TO 2955
      IF(KEY.EQ.1) KEY = 2
      GO TO 2999
 2955 IF(KEY.EQ.2) GO TO 2988
      IF(KEY.EQ.1) GO TO 2966
      KEY = 1
      IF(NODET(I).EQ.ISYM(11).OR.NODET(I).EQ.ISYM(12)) GO TO 2999
 2966 DO 2977 J=1,10
      IF(NODET(I).EQ.ISYM(J)) GO TO 2999
 2977 CONTINUE
      IF(NODET(I).NE.ISYM(13))GO TO 2988
      ITAL = ITAL + 1
      GO TO 2999
 2988 M = M + 1
 2999 CONTINUE
      IF(ITAL.GT.1) M = M + 1
      RETURN
      END
      SUBROUTINE LOADT
      IMPLICIT INTEGER*4(I-N)
     0COMMON/TABLE/TAB1(3,3,3),TAB2(3,3,3),TAB3(3,3,3),TAB4(3,3,3),
     1TAB5(3,3,3),TAB6(3,3,3),TAB7(3,3,3),LXCK(7),MXCK(7),LYCK(7),
     2MYCK(7),LZCK(7),MZCK(7)
C
C     THIS SUBROUTINE IS FOR THE USER TO CODE INSTRUCTIONS FOR LOADING
C        DATA IN THE ARRAYS TAB1 - TAB7 FOR TABLE LOCK-UP (SEE DOARC).
C        THE UPPER BOUNDS OF THE ARRAYS TAB1 - TAB7 SHOULD BE LOADED IN
C        THE CHECK ARRAYS MXCK, MYCK AND MZCK.  THE LOWER BOUNDS OF THE
C        ARRAYS TAB1 - TAB7 SHOULD BE LOADED IN THE CHECK ARRAYS LXCK,
C        LYCK AND LZCK.
C
      DO 3000 I=1,7
      LXCK(I) = 1
      LYCK(I) = 1
      LZCK(I) = 1
      MXCK(I) = 3
      MYCK(I) = 3
 3000 MZCK(I) = 3
      DO 3005 I=1,3
      DO 3005 J=1,3
      DO 3005 K=1,3
      TAB1(I,J,K) =   I + J + K + 1
      TAB2(I,J,K) =   I + J + K + 2
      TAB3(I,J,K) =   I + J + K + 3
      TAB4(I,J,K) =   I + J + K + 4
      TAB5(I,J,K) =   I + J + K + 5
      TAB6(I,J,K) =   I + J + K + 6
 3005 TAB7(I,J,K) =   I + J + K + 7
      RETURN
      END
      SUBROUTINE LOADA
      IMPLICIT INTEGER*4(I-N)
     0COMMON EMARK(11),SETUP,FINT,DINT,FACTOR,STIME,SCOST,SPERF,CTIME,
     1CCOST,CPERF,BTIME,BCOST,BPERF,TIMESM,UNFORM,ITRIP1(22),ITRIP2(22),
     2KNT(6),INPT,IOUT,IPNH,IWF1,IWF2,IWF3,IWF4,IBLK,ITRACE,ICUT,ISEED,
     3IRROR,ISAVE,NODESM,MAXTAG,ICOUNT,JCOUNT,NTRIP,ICORR,LCOMP,NWARN,
     4NRK,NRKS,SLAK,MODE,INFC,INFP,DISC,DISP,CRIT,KGEN,MED,SINGLE,DOUBLE
     0COMMON/TRIALS/STORET( 1000,4),TERM(10,8),KPOINT(10),NODET( 1000),
     1MTERM,NTERM,MITER,ITER
     0COMMON/ARCS/ASTORE( 2800),UTIMEA( 350),TIMEA( 350),UCOSTA( 350),
```

```
     1COSTA( 350),UPERFA( 350),PERFA( 350),WORK( 350),ISTATE( 350),
     2NODEI( 350),NODEO( 350),ICRITA( 350),KEEPC( 350),KEEPP( 350),
     3IARC1( 350),IARC2( 350),IPOINT( 350),JPOINT( 350),ISLAK( 350),
     4KARC,LARC,MARC,NARC,ITALC,ITALP,ISTAR
     OCOMMON/NODES/TIMEN( 200),COSTN( 200),PERFN( 200),NSTORE( 5400),
     1NODE1( 200),NODE2( 200),LOGI( 200),LOGO( 200),NSTATE( 200),
     2NARCI( 200),NARCO( 200),ISTAT( 200),INSTAT( 200),ICRITN( 200),
     3NPOINT( 200),NSLAK( 200),JUMP( 200),KNODE,LNODE,MNODE,NNODE,MTAG,
     4NTAG
     OCOMMON/INTERN/XMIN(20,4),XMAX(20,4),HMIN(20,4),HMAX(20,4),
     1HAVE(20,4),ICBS(20),MHIST,NHIST
     OCOMMON/SLACK/RMIN(20),RMAX(20),SMIN(20),SMAX(20),SAVE(20),
     1JOBS(20),MSLACK,NSLACK
      DIMENSION TYPE(5), NF(4)
      LOGICAL SLAK,MODE,INFC,INFP,DISC,DISP,CRIT,KGEN,MED,SINGLE,DOUBLE
     ODATA TYPE,OMIT,SUM,BACKUP,IPOS,NEG,NF/1HT,1HC,1HP,1HK,1H ,1HO,1HS,
     11HB,1H+,1H-,4HNOFL,4HOW  ,4HDATA,4HGEN /
C
C     THIS SUBROUTINE LOADS THE BALANCE OF THE ARC AND NODE DATA INTO
C        THE STORAGE ARRAYS - FIRST TASK IS LOADING ARC DATA
C
      ISAVE = 0
 3011 ISAVE = ISAVE + 1
      IF(ISAVE.GT.NARC) RETURN
      IE = IARC1(ISAVE)
      JE = IARC2(ISAVE)
     OREAD (IWF4,3022) (NODET(I), I=1,4), UNFORM, M, (ISTATE(J), J=1,8),
     1(NSTATE(K), K=1,9)
 3022 FORMAT (4A4, F4.4, A1, 17I4)
      NODEI(ISAVE) = 0
      NODEO(ISAVE) = 0
     OIF(NODET(1).EQ.NF(1).AND.NODET(2).EQ.NF(2).AND.NODET(3).EQ.NF(3).
     1AND.NODET(4).EQ.NF(4)) GO TO 3088
C
C     STORE THE ADDRESSES OF THE INPUT AND OUTPUT NODES AND CHECK TO SEE
C        IF THIS ARC IS FUNCTIONING AS A TRANSPORTATION ARC
C
      DO 3033 I=1,NNODE
      IF(NODET(1).NE.NODE1(I).OR.NODET(2).NE.NODE2(I)) GO TO 3033
      NODEI(ISAVE) = I
      GO TO 3055
 3033 CONTINUE
      CALL ERROR (3044,IE,JE,IBLK)
 3055 DO 3066 I=1,NNODE
      IF(NODET(3).NE.NODE1(I).OR.NODET(4).NE.NODE2(I)) GO TO 3066
      NODEO(ISAVE) = I
      GO TO 3099
 3066 CONTINUE
      CALL ERROR (3077,IE,JE,IBLK)
 3088 UNFORM = 1.0
      M = IBLK
C
C     CHECK TO SEE IF THERE IS MORE THAN 1 MONTE CARLO OR FILTER OUTPUT
C        THEN COUNT THE NUMBER OF TIME, COST AND PERFORMANCE ENTRIES AND
C        LOAD REQUEST FOR A SLACK HISTOGRAM IF WANTED
C
 3099 ICOUNT = 0
      DO 3100 I=1,7
 3100 IF(ISTATE(I).GT.0) ICOUNT = ICOUNT + 1
      IF(ICOUNT.GT.1) CALL ERROR (3111,IE,JE,IBLK)
      JCOUNT = 0
      DO 3122 I=1,9
```

```
 3122 IF(NSTATE(I).GT.0) JCOUNT = JCOUNT + 1
      IF(M.EQ.IBLK.OR.CTIME.NE.1.0) GO TO 3166
      SLAK = .FALSE.
      NSLACK = NSLACK + 1
      IF(NSLACK.LE.MSLACK) GO TO 3155
      IF(ITRACE.EQ.3) CALL ERROR (3133,IE,JE,NWARN)
      IF(ITRACE.NE.3) CALL ERROR (3144,IE,JE,IBLK)
      NSLACK = 1
 3155 ISLAK(ISAVE) = NSLACK
 3166 IF(UNFORM.EQ.1.0.AND.ICOUNT.EQ.0.AND.JCOUNT.EQ.0) GO TO 3011
C
C     THIS ARC IS MORE THAN A TRANSPORTATION ARC, LOAD THE POINTER,
C         STORE PROBABILITY CF ARC COMPLETION AND FUN. REL. INDICATOR
C
      IPOINT(ISAVE) = KARC + 1
      KARC = KARC + 2
      IF(KARC.LE.LARC) GC TC 3199
 3177 CALL ERROR (3168,IBLK,IBLK,IBLK)
      RETURN
 3199 ASTORE(KARC-1) = UNFCRM
      ASTORE(KARC) = 0.0
C
C     READ IN THE MONTE CARLOS OR FILTERS 1 AND 2
C
      IF(ICOUNT.EQ.0) GO TC 3355
      DO 3222 I=1,6
      IF(ISTATE(I).EQ.0) GC TC 3222
      M = 1
      IF(I.GE.2.OR.I.LE.4) M = MARC/6
      IF(ISTATE(I).GT.M) CALL ERROR (3200,IE,JE,MARC)
      IF(I.GT.4) GC TO 3211
      KARC = KARC + 1
      IF(KARC.GT.LARC) GO TO 3177
      ASTORE(KARC) = I
 3211 CALL SEREAD (ISTATE(I),IE,JE,IADJ)
      IF(NODESM.LE.0.AND.I.LE.4) KARC = KARC - 1
      IF(KARC.GT.LARC) GC TO 3177
 3222 CCNTINUE
C
C     READ IN FILTER NO. 3
C
      IF(ISTATE(7).EQ.0) GC TC 3344
      MAXTAG = 0
      NODESM = ISTATE(7)
      IF(NODESM.GT.MARC) CALL ERROR (3233,IE,JE,MARC)
      DO 3266 I=1,NODESM
      MAXTAG = MAXTAG + 6
      IF(MAXTAG.LE.MARC) GC TO 3244
      MAXTAG = 6
 3244 M = MAXTAG - 5
      READ (IWF4,3255) J,(JPOINT(K),KEEPC(K),KEEPP(K),K=M,MAXTAG),IADJ
 3255 FORMAT (I2, 6(1X,A1,2A4), I1)
 3266 IF(J.NE.I) CALL ERROR (3277,IE,JE,IBLK)
      DO 3333 I=1,MAXTAG
      IF(KEEPC(I).EQ.IBLK.AND.KEEPP(I).EQ.IBLK) GO TO 3333
      IF(JPOINT(I).EQ.IPCS.CR.JPOINT(I).EQ.NEG) GC TO 3299
      CALL ERROR (3288,IE,JE,IBLK)
      GC TO 3333
 3299 DO 3300 J=1,NARC
      K = J
      IF(KEEPC(I).EQ.IARC1(K).AND.KEEPP(I).EQ.IARC2(K)) GO TO 3322
 3300 CONTINUE
```

```
      CALL ERROR (3311,KEEPC(I),KEEPP(I),IBLK)
      GO TO 3333
 3322 KARC = KARC + 1
      IF(KARC.GT.LARC) GC TC 3177
      IF(JPOINT(I).EQ.NEG) K = -K
      ASTORE(KARC) = K
 3333 CONTINUE
C
C     ENTER THE END OF FIRST SEGMENT INDICATOR
C
 3344 KARC = KARC + 1
      IF(KARC.GT.LARC) GC TC 3177
      ASTORE(KARC) = EMARK(1)
C
C     IF PRESENT, READ IN A SLACK HISTOGRAM MIN-MAX CARD
C
 3355 IF(ISTATE(8).LE.0) GC TO 3388
      IF(ISLAK(ISAVE).EQ.0) CALL ERROR (3366,IE,JE,IBLK)
      IF(ISTATE(8).NE.1) CALL ERROR (3377,IE,JE,IBLK)
      ICOUNT = -9999
      CALL SEREAD (ISTATE(8),IE,JE,IADJ)
      SMIN(NSLACK) = TIMEA(1)
      SMAX(NSLACK) = TIMEA(2)
 3388 IF(JCOUNT.EQ.0) GO TO 4466
C
C     READ IN TIME, COST AND PERFORMANCE GENERATES FOR THIS ARC
C
      DO 4388 I=3,9,3
      N = I/3
      LOGI(N) = 0
      LOGO(N) = 0
      TIMESM = 2*N
      M = I - 2
      J = I - 1
      ISTAT(M) = 0
      ISTAT(J) = 0
      ISTAT(I) = 0
      IF(NSTATE(M).GT.0.AND.NSTATE(J).GT.0) CALL ERROR (3399,IE,JE,IBLK)
      M1 = MARC/6
      M2 = MNODE
      IF(MARC.LT.MNODE) M2 = MARC
      IF(NSTATE(J).GT.M1) CALL ERROR (3400,IE,JE,MARC)
      IF(NSTATE(I).GT.M2) CALL ERROR (3411,IE,JE,M2)
C
C     THIS SECTION MODIFIES AND STORES STANDARD PROBABILITY DISTRIBUTION
C
      IF(NSTATE(M).EQ.0) GC TC 3922
      CALL SEREAD (NSTATE(M),IE,JE,IADJ)
      IF(NODESM.LE.0) GO TC 3922
      IF(KARC.GT.LARC) GC TC 3177
C
C     CHECK THE DISTRIBUTION INDICATOR
C
      M = KARC - NODESM + 1
      J = ASTORE(M) + 0.001
      IF(J.GE.1.AND.J.LE.14) GO TO 3455
      CALL ERROR (3422,IE,JE,J)
      GO TO 3911
C
C     PERFORM THE CHECKS AND CONVERSIONS REQUIRED OF EACH DISTRIBUTION -
C         FIRST SET UP THE ERROR MESSAGE FOR AN INCORRECT NO. OF ENTRIES
C
```

```
  3433 CALL ERROR (3444,IE,JE,J)
       GO TO 3911
  3455 M1 = M + 1
       M2 = M + 2
       M3 = M + 3
       M4 = M + 4
      CGO TO (3466,3477,3488,3499,3511,3544,3588,3611,3644,3677,3699,
     13733,3777,3811),J
C
C      CONSTANT
C
  3466 IF(NODESM.NE.2) GC TC 3433
       GO TO 3911
C
C      UNIFORM
C
  3477 IF(NODESM.NE.3) GC TC 3433
       GO TO 3899
C
C      TRIANGULAR
C
  3488 IF(NODESM.NE.4) GO TC 3433
       X = ASTORE(M3)
       GC TO 3877
C
C      NCRMAL
C
  3499 IF(NODESM.NE.5) GC TC 3433
  3500 X = ASTORE(M3)
       Y = ASTORE(M4)
       GO TO 3855
C
C      LCGNORMAL
C
  3511 IF(NODESM.NE.5) GO TC 3433
       X = ASTORE(M3)
       Y = ASTORE(M4)
       IF(ASTORE(M1).GT.C.O.AND.X.GT.O.O) GO TO 3533
       CALL ERROR (3522,IE,JE,IBLK)
       GC TO 3911
  3533 UNFORM = ALCG((Y*Y)/(X*X) + 1.C)
       ASTORE(M3) = ALOG(X) - C.5*UNFORM
       ASTORE(M4) = SQRT(UNFORM)
       GC TO 3855
C
C      GAMMA
C
  3544 IF(NODESM.NE.5) GO TC 3433
       X = ASTORE(M3)
       Y = ASTORE(M4)
       IF(ASTORE(M1).GT.O.O.AND.X.GT.O.C.AND.ASTORE(M4).GT.O.O)GO TO 3566
       CALL ERROR (3555,IE,JE,IBLK)
       GO TO 3911
  3566 UNFORM = Y*Y
       ASTORE(M3) = X/UNFCRM
       ASTORE(M4) = X*ASTORE(M3)
       IF(ASTCRE(M4).LE.C.O) CALL ERROR (3577,IE,JE,IBLK)
       GC TO 3855
C
C      WEIBULL
C
  3588 IF(NODESM.NE.5) GC TC 3433
```

```
       OIF(ASTORE(M1).GE.0.0.AND.ASTORE(M3).GT.0.0.AND.ASTORE(M4).GT.0.0)
      1GO TO 3600
       CALL ERROR (3599,IE,JE,IBLK)
       GO TO 3911
 3600 ASTORE(M4) = 1.0/ASTORE(M4)
       GO TO 3899
C
C      ERLANG
C
 3611 IF(NODESM.NE.5) GO TC 3433
       X = ASTORE(M3)
       K = ASTORE(M4) + 0.001
       UNFORM = K
       OIF(ASTORE(M1).GT.0.0.AND.X.GT.0.0.AND.ASTORE(M4).GT.0.0.AND.
      1ASTORE(M4).EQ.UNFORM) GO TO 3633
       CALL ERROR (3622,IE,JE,IBLK)
       GO TO 3911
 3633 ASTORE(M3) = ASTORE(M4)/X
       Y = ASTORE(M4)/(ASTORE(M3)*ASTORE(M3))
       GO TO 3855
C
C      CHI SQUARE
C
 3644 IF(NODESM.NE.4) GO TC 3433
       X = ASTORE(M3)
       K = X + 0.001
       UNFORM = K
       IF(ASTORE(M1).GT.0.0.AND.UNFORM.EQ.X.AND.X.GT.0.0) GO TO 3666
       CALL ERROR (3655,IE,JE,IBLK)
       GO TO 3911
 3666 Y = 2.0*X
       L = K/2
       ICCUNT = L*2
       IF(K.NE.ICOUNT) L = -L
       ASTORE(M3) = L
       GO TO 3855
C
C      BETA
C
 3677 IF(NODESM.NE.5) GO TC 3433
       OIF(ASTORE(M1).GT.0.0.AND.ASTORE(M3).GT.0.0.AND.ASTORE(M4).GT.0.0)
      1GO TO 3899
       CALL ERROR (3688,IE,JE,IBLK)
       GO TO 3911
C
C      POISSON
C
 3699 IF(NODESM.NE.4) GO TC 3433
       L = ASTORE(M2) - ASTORE(M1) + 0.001
       IF(ASTORE(M1).GE.0.0.AND.ASTORE(M3).GT.0.0.AND.L.GE.1) GO TO 3711
       CALL ERROR (3700,IE,JE,IBLK)
       GO TO 3911
 3711 IF(ASTORE(M3).LE.10.0) GO TO 3722
       KARC = KARC + 1
       IF(KARC.GT.LARC) GO TC 3177
       ASTORE(M) = 4.0
       ASTORE(M4) = SQRT(ASTCRE(M3))
       GO TO 35C0
 3722 X = ASTCRE(M3)
       ASTORE(M3) = EXP(-ASTCRE(M3))
       GO TO 3877
C
```

```
C      PASCAL
C
 3733 IF(NODESM.NE.5) GO TO 3433
      IF(ASTORE(M3).LE.0.0.OR.ASTORE(M3).GE.1.0) GO TO 3744
      K = ASTORE(M4) + 0.001
      UNFORM = K
      IF(UNFORM.EQ.ASTORE(M4).AND.K.GT.0.AND.ASTORE(M1).GE.0.)GO TO 3766
 3744 CALL ERROR (3755,IE,JE,IBLK)
      GO TO 3911
 3766 X = (ASTORE(M4)*(1.0 - ASTORE(M3)))/ASTORE(M3)
      Y = X/ASTORE(M3)
      ASTORE(M3) = -ALOG(1.0 - ASTORE(M3))
      GO TO 3855
C
C      BINOMIAL
C
 3777 IF(NODESM.NE.5) GO TO 3433
      IF(ASTORE(M3).LE.0.0.OR.ASTORE(M3).GE.1.0) GO TO 3788
      K = ASTORE(M4) + 0.001
      UNFORM = K
      IF(UNFORM.EQ.ASTORE(M4).AND.K.GT.0.AND.ASTORE(M1).GE.0.)GO TO 3800
 3788 CALL ERROR (3799,IE,JE,IBLK)
      GO TO 3911
 3800 X = ASTORE(M4)*ASTORE(M3)
      Y = X*(1.0 - ASTORE(M3))
      GO TO 3855
C
C      HYPERGEOMETRIC
C
 3811 IF(NODESM.NE.6) GO TO 3433
      IF(ASTORE(M3).GT.0.0.AND.ASTORE(M3).LT.1.0) GO TO 3844
 3822 CALL ERROR (3833,IE,JE,IBLK)
      GO TO 3911
 3844 K = ASTORE(M4) + 0.001
      UNFORM = K
      M5 = M+5
      L = ASTORE(M5) + 0.001
      TIMESM = L
     0IF(UNFORM.NE.ASTORE(M4).OR.TIMESM.NE.ASTORE(M5).OR.K.LE.1.OR.
     1L.LE.0.OR.K.LE.L.OR.ASTORE(M1).LT.0.0) GO TO 3822
      X = ASTORE(M5)*ASTORE(M3)
      UNFORM = (ASTORE(M4) - ASTORE(M5))/(ASTORE(M4) - 1.0)
      Y = X*(1.0 - ASTORE(M3))*UNFORM
C
C      CHECK FOR POSITIVE VARIANCE, MEAN WITHIN MIN-MAX, MIN-MAX OKAY
C
 3855 IF(Y.LE.0.0) CALL ERROR (3866,IE,JE,IBLK)
 3877 IF(X.LT.ASTORE(M1).OR.X.GT.ASTORE(M2)) CALL ERROR(3888,IE,JE,IBLK)
 3899 IF(ASTORE(M1).GE.ASTORE(M2)) CALL ERROR (3900,IE,JE,IBLK)
C
C      ENTER THE END OF SEGMENT INDICATOR FOR DISTRIBUTION INPUT
      NP = NPOINT(I)
      IE = NODE1(I)
      JE = NODE2(I)
      IF(IE.EQ.IBLK.AND.JE.EQ.IBLK) CALL ERROR (5299,IBLK,IBLK,I)
      DO 5300 J=1,NARC
 5300 IF(IARC1(J).EQ.IE.AND.IARC2(J).EQ.JE) CALL ERROR (5311,IE,JE,IBLK)
      LOGII = LOGI(I)
      LOGOO = LOGO(I)
      IF(LOGII.GT.4) GO TO 5322
      IF(LOGOO.GE.11) LOGOO = 1
 5322 IF(LOGII.LT.1.OR.LOGII.GT.8) CALL ERROR (5333,IE,JE,IBLK)
```

```
      IF(LOGII.GT.4) GO TO 5477
      IF(LOGOO.LT.1) GO TO 5355
      IF(LOGOO.LE.6) GO TO 5377
      K = -9
      L = 0
      M = NNODE - 1
      IF(M.GT.99) M = 99
 5344 K = K + 10
      L = L + 1
      IF(L.GT.M) GO TO 5355
      IF(K - LOGO(I))5344,5377,5344
 5355 CALL ERROR (5366,IE,JE,IBLK)
 5377 IF(LOGOO.LT.3) GO TO 5399
      IF(NO.LE.1) CALL ERROR (5368,IE,JE,IBLK)
C
C     CHECK TERMINAL NODES
C
 5399 IF(LOGOO.NE.1) GO TO 5477
      IF(ISTAT(I).LE.0.AND.ISTAT(I).GE.-1) GO TO 5411
      CALL ERROR (5400,IE,JE,IBLK)
 5411 JCOUNT = JCOUNT + 1
      IF(NI.EQ.0) CALL ERROR (5422,IE,JE,IBLK)
      IF(NO.NE.0) CALL ERROR (5433,IE,JE,IBLK)
      IF(LOGII.LT.2.OR.LOGII.GT.4) CALL ERROR (5444,IE,JE,IBLK)
      DO 5466 J=1,NI
      M = NP + J - 1
      M = NSTORE(M)
      L = IPOINT(M)
      IF(L.EQ.0) GO TO 5466
      IF(ASTORE(L).NE.1.0) CALL ERROR (5455,IARC1(M),IARC2(M),IBLK)
 5466 CONTINUE
      GO TO 5688
C
C     CHECK FOR PROPER USE OF INTERNAL NODE STATISTICS OPERANDS
C
 5477 IF(ISTAT(I).EQ.0) GO TO 5500
      K = 0
      DO 5488 J=1,NNODE
      IF(I.EQ.J) GO TO 5488
      IF(ISTAT(I).NE.ISTAT(J)) GO TO 5488
      K = K + 1
 5488 CONTINUE
      IF(K.GT.2) CALL ERROR (5499,IE,JE,IBLK)
C
C     CHECK INITIAL NODES
C
 5500 IF(LOGII.NE.1) GO TO 5544
      ICOUNT = 1
      IF(NI.NE.0) CALL ERROR (5511,IE,JE,IBLK)
      IF(NO.EQ.0) CALL ERROR (5522,IE,JE,IBLK)
      IF(LOGOO.LT.2.OR.LOGOO.GT.3) CALL ERROR (5533,IE,JE,IBLK)
      GO TO 5688
C
C     CHECK INTERNAL NODES
C
 5544 IF(NI.EQ.0) CALL ERROR (5555,IE,JE,IBLK)
      IF(NO.EQ.0) CALL ERROR (5566,IE,JE,IBLK)
      DO 5588 J=1,NI
      L = NP + J - 1
      M = NSTORE(L)
      M = IPOINT(M)
      IF(M.EQ.0) GO TO 5577
```

```
       IF(ASTORE(M).NE.1.0) JUMP(I) = 1
 5577 DO 5588 K=1,NO
       M = NP + NI + K - 1
 5588 IF(NSTORE(L).EQ.NSTORE(M)) CALL ERROR (5599,IE,JE,IBLK)
C
C      CHECK QUE AND SORT NODES FOR INPUT - OUTPUT ARCS
C
       IF(LOGII.LT.7) GO TO 5633
       IF(JUMP(I).EQ.1) GO TO 5600
       IF(NI - NO)5688,5688,5611
 5600 IF(NI+1.EQ.NO) GO TO 5688
 5611 CALL ERROR (5622,IE,JE,IBLK)
       GO TO 5688
C
C      CHECK COMPARE AND PREFERRED NODES FOR INPUT - OUTPUT ARCS
C
 5633 IF(LOGII.LT.5) GO TO 5666
       IF(NI+1.EQ.NO) GO TO 5688
       IF(LOGCC.GT.1) GO TO 5644
       IF(NI.NE.NO) GO TO 5644
       IF(JUMP(I).EQ.0) GO TO 5688
 5644 CALL ERROR (5655,IE,JE,IBLK)
       GO TO 5688
C
C      CHECK AND, PAND, OR NODES FOR INPUT - OUTPUT ARCS
C
 5666 IF(JUMP(I).EQ.0) GO TO 5688
       IF(NO.GT.1) GO TO 5688
       CALL ERROR (5677,IE,JE,IBLK)
C
C      CHECK ALL OUTPUT LOGIC
C
 5688 IF(LOGCC.NE.2) GO TO 5711
       L = 0
       DO 5699 J=1,NO
       M = NSTORE(NP+NI+J-1)
       K = IPOINT(M)
       IF(K.EQ.0) GO TO 5699
       CALL SEEK (K,EMARK(1))
       IF(ITALP.NE.0) L = L + 1
 5699 CONTINUE
       IF(L.GT.0) CALL ERROR (5700,IE,JE,IBLK)
C
C      CHECK MONTE CARLO OUTPUT LOGIC
C
 5711 IF(LOGCC.NE.3) GO TO 5933
       DO 5722 J=1,MARC
 5722 TIMEA(J) = 0.0
       DO 5900 J=1,NO
       M = NSTORE(NP+NI+J-1)
       KE = IARC1(M)
       LE = IARC2(M)
       IF(IPOINT(M).GT.0) GO TO 5755
       IF(J.NE.NO) GO TO 5733
       IF(JUMP(I).EQ.1) GO TO 6066
 5733 CALL ERROR (5744,KE,LE,IBLK)
       GO TO 6066
 5755 CALL SEEK (IPOINT(M),EMARK(1))
       IF(J.LT.NO) GO TO 5777
       IF(JUMP(I).EQ.0) GO TO 5777
       IF(ITALP.NE.0) CALL ERROR (5766,KE,LE,IBLK)
       GO TO 6066
```

```
 5777 IF(ITALP.EQ.0) GO TO 5733
C
C     LOAD ARC NO. 1'S BOUNDARIES AND CHECK FOR AN ODD NO. OF ENTRIES
C
      IF(J.GT.1) GO TO 5811
      N = ITALP
      L = (ITALP/2)*2
      IF(N.EQ.L) GO TO 5799
      CALL ERROR (5788,KE,LE,IBLK)
      GO TO 6066
 5799 TIMESM = ASTORE(ITALC+1)
      IF(N.LE.2) GO TO 5866
      DO 5800 K=3,N,2
 5800 TIMEA(K) = ASTORE(ITALC+K)
      GO TO 5866
C
C     COMPARE THE NO. OF ENTRIES AND BOUNDARIES OF THE 1ST ARC AGAINST
C        ALL OTHER ARCS PLUS CHECK AND SUM THE PROBABILITIES
C
 5811 IF(N.EQ.L) GO TO 5833
      CALL ERROR (5822,KE,LE,IBLK)
      GO TO 6066
 5833 IF(N.LE.2) GO TO 5866
      DO 5844 K=3,N,2
 5844 IF(TIMEA(K).NE.ASTORE(ITALC+K)) CALL ERROR (5855,KE,LE,IBLK)
 5866 DO 5888 K=2,N,2
     0IF(ASTORE(ITALC+K).LT.0.0.OR.ASTORE(ITALC+K).GT.1.0)
     1CALL ERROR (5877,KE,LE,IBLK)
 5888 TIMEA(K) = TIMEA(K) + ASTORE(ITALC+K)
      IF(TIMESM.NE.ASTORE(ITALC+1)) CALL ERROR (5899,KE,LE,IBLK)
 5900 CONTINUE
C
C     DO EACH OF THE DISTRIBUTIONS SUM TO ONE
C
      DO 5911 K=2,N,2
 5911 IF(TIMEA(K).LT.0.99.OR.TIMEA(K).GT.1.01)CALLERROR(5922,IE,JE,IBLK)
C
C     CHECK FILTER OUTPUT LOGIC
C
 5933 IF(LOGOO.LT.4) GO TO 6066
      NODESM = 0
      DO 6044 J=1,NO
      M = NSTORE(NP+NI-1+J)
      IF(IPOINT(M).EQ.0) GO TO 5944
      KE = IARC1(M)
      LE = IARC2(M)
      CALL SEEK (IPOINT(M),EMARK(1))
      IF(ITALP.GT.0) GO TO 5955
 5944 NODESM = NODESM + 1
      GO TO 6044
 5955 IF(LOGOO.EQ.6) GO TO 6044
      N = ITALP
      L = (ITALP/2)*2
      IF(N.EQ.L) GO TO 5977
      CALL ERROR (5966,KE,LE,IBLK)
      GO TO 6044
 5977 DO 5988 K=2,ITALP,2
 5988 IF(ASTORE(ITALC+K-1).GT.ASTORE(ITALC+K))CALLERROR(5999,KE,LE,IBLK)
C
C     CHECK FILTER NO. 2 OUTPUT LOGIC
C
      IF(LOGOO.EQ.4) GO TO 6044
```

```
      IF(LOGII.NE.3) CALL ERROR (6000,IE,JE,IBLK)
      IF(L.NE.2) CALL ERROR (6011,KE,LE,IBLK)
      IF(ASTORE(ITALC+1).LT.0.0) CALL ERROR (6022,KE,LE,IBLK)
      IF(ASTORE(ITALC+2).GT.FLOAT(NI)) CALL ERROR (6033,KE,LE,IBLK)
 6044 CONTINUE
      IF(NODESM.NE.1) CALL ERROR (6055,IE,JE,IBLK)
 6066 CONTINUE
C
C     ANY INITIAL AND TERMINAL NODES
C
      IF(ICOUNT.EQ.0) CALL ERROR (6077,IBLK,IBLK,IBLK)
      IF(JCOUNT.EQ.0) CALL ERROR (6088,IBLK,IBLK,IBLK)
      IF(JCOUNT*4.GT.MARC) CALL ERROR (6099,IBLK,IBLK,IBLK)
C
C     CHECK ARCS ERRORS
C
      DO 6133 I=1,NARC
      KE = IARC1(I)
      LE = IARC2(I)
      IF(NODEI(I).GT.NODEO(I)) CALL ERROR (6100,KE,LE,NWARN)
      IF(KE.EQ.IBLK.AND.LE.EQ.IBLK) CALL ERROR (6111,IBLK,IBLK,I)
      J = IPOINT(I)
      IF(J.LE.0) GO TO 6133
      IF(ASTORE(J).LE.0.0.OR.ASTORE(J).GT.1.0)CALLERROR(6122,KE,LE,IBLK)
 6133 CONTINUE
      IF(IRROR.GT.0) RETURN
C
C     LIST THE COMPUTER TIME CONSUMPTION WARNING MESSAGES
C
      IF(MED) GO TO 6155
      CALL ERROR (6144,IBLK,IBLK,NWARN)
 6155 IF(ICPGAP.EQ.0) GO TO 6177
      CALL ERROR (6166,IBLK,IBLK,NWARN)
 6177 IF(MODE) GO TO 6199
      IF(ITRACE.EQ.0) GO TO 6199
      CALL ERROR (6188,IBLK,IBLK,NWARN)
 6199 CRIT = .TRUE.
      IF(CTIME.EQ.0.0.AND.CCOST.EQ.0.0.AND.CPERF.EQ.0.0) GO TO 6222
      CRIT = .FALSE.
      CALL ERROR (6200,IBLK,IBLK,NWARN)
      IF(SLAK) GO TO 6222
      CALL ERROR (6211,IBLK,IBLK,NWARN)
 6222 IF(ITRACE.EQ.3) WRITE (IOUT,6233)
 6233 FORMAT (47HONOTE--- * INDICATES BEING ON THE CRITICAL PATH)
      IF(ITRACE.NE.2) GO TO 6344
      WRITE (IOUT,6244)
 6244 FORMAT (1H1, 6X, 8HTERMINAL, 4X, 9HBEGINNING)
      IF(INFC) GO TO 6266
      WRITE (IOUT,6255)
 6255 FORMAT (1H+, 30X, 13HCOST INFLATED)
 6266 IF(DISC) GO TO 6288
      WRITE (IOUT,6277)
 6277 FORMAT (1H+, 45X, 15HCOST DISCOUNTED)
 6288 IF(INFP) GO TO 6300
      WRITE (IOUT,6299)
 6299 FORMAT (1H+, 62X, 13HPERF INFLATED)
 6300 IF(DISP) GO TO 6322
      WRITE (IOUT,6311)
 6311 FORMAT (1H+, 77X, 15HPERF DISCOUNTED)
 6322 WRITE (IOUT,6333)
 6333 FORMAT (6H COUNT, 5X, 4HNODE, 9X, 4HSEED, 22X, 4HTIME, 17X, 9HPATH
     1 COST, 14X, 12HOVERALL COST, 15X, 11HPERFORMANCE)
```

```
6344 DO 6355 I=1,MNODE
6355 ICRITN(I) = 0
     DO 6366 I=1,MARC
6366 ICRITA(I) = 0
     RETURN
     END
     SUBROUTINE DCNET
     IMPLICIT INTEGER*4(I-N)
     0COMMON EMARK(11),SETUP,FINT,DINT,FACTOR,STIME,SCOST,SPERF,CTIME,
    1CCOST,CPERF,BTIME,BCCST,BPERF,TIMESM,UNFORM,ITRIP1(22),ITRIP2(22),
    2KNT(6),INPT,IOUT,IPNH,IWF1,IWF2,IWF3,IWF4,IBLK,ITRACE,ICUT,ISEED,
    3IRROR,ISAVE,NODESM,MAXTAG,ICOUNT,JCOUNT,NTRIP,ICORR,LCOMP,NWARN,
    4NRK,NRKS,SLAK,MODE,INFC,INFP,DISC,DISP,CRIT,KGEN,MED,SINGLE,DOUBLE
     0COMMON/TRIALS/STORET( 1000,4),TERM(10,8),KPOINT(10),NODET( 1000),
    1MTERM,NTERM,MITER,ITER
     0COMMON/ARCS/ASTORE( 2800),UTIMEA( 350),TIMEA( 350),UCOSTA( 350),
    1COSTA( 350),UPERFA( 350),PERFA( 350),WORK( 350),ISTATE( 350),
    2NODEI( 350),NODEO( 350),ICRITA( 350),KEEPC( 350),KEEPP( 350),
    3IARC1( 350),IARC2( 350),IPOINT( 350),JPOINT( 350),ISLAK( 350),
    4KARC,LARC,MARC,NARC,ITALC,ITALP,ISTAR
     0COMMON/NODES/TIMEN( 200),COSTN( 200),PERFN( 200),NSTORE( 5400),
    1NODE1( 200),NODE2( 200),LOGI( 200),LOGO( 200),NSTATE( 200),
    2NARCI( 200),NARCO( 200),ISTAT( 200),INSTAT( 200),ICRITN( 200),
    3NPOINT( 200),NSLAK( 200),JUMP( 200),KNODE,LNODE,MNODE,NNODE,MTAG,
    4NTAG
     0COMMON/INTERN/XMIN(20,4),XMAX(20,4),HMIN(20,4),HMAX(20,4),
    1HAVE(20,4),ICBS(20),MHIST,NHIST
     0COMMON/SLACK/RMIN(20),RMAX(20),SMIN(20),SMAX(20),SAVE(20),
    1JOBS(20),MSLACK,NSLACK
     0COMMON/CPGAP/T1(10),T2(10),CCNFL(10),CONFS(10),CAVE(10),CSMIN(10),
    1CSMAX(10),CHMIN(10),CHMAX(10),PAVE(10),PSMIN(10),PSMAX(10),
    2PHMIN(10),PHMAX(10),KCOBS(10),KPOBS(10),MCPGAP,NCPGAP,ICPGAP
     DIMENSION SLACKA( 350), SLACKN( 200), IPATHA( 350), IPATHN( 200)
     0LOGICAL SLAK,MODE,INFC,INFP,DISC,DISP,CRIT,KGEN,MED,SINGLE,DOUBLE,
    1ICHECK,JCHECK
     DATA KSTAR/1H*/
C
C
C    DONET CONDUCTS SIMULATION, FINDS CRIT-OPT PATH, STORES STATISTICS
C      ARC STATUS - THIS IS CARRIED BY THE VARIABLE ISTATE
C        -1 - EQUIVALENT TO 0, THIS OPTION IS USED FOR BACKTRACKING
C         0 - LOGICALLY ELIMINATED FROM THE NETWORK
C         1 - NONPROCESSED
C         2 - UNSUCCESSFULLY COMPLETED ARC
C         3 - SUCCESSFULLY COMPLETED ARC
C         4 - CRITICAL PATH CANDIDATE
C      NODE STATUS - THIS IS CARRIED BY THE VARIABLE NSTATE
C         0 - LOGICALLY ELIMINATED FROM THE NETWORK
C         1 - NONPROCESSED
C         2 - CANDIDATE FOR PROCESSING
C         3 - SUCCESSFULLY COMPLETED
C         4 - ON THE CRITICAL PATH
C         5 - ON THE CRITICAL PATH AFTER CALCULATION OF THE SLACKS
C
     NRK = 0
     NRKS = 0
     REWIND IWF2
     REWIND IWF3
     REWIND IWF4
     ICOUNT = 1
     JCOUNT = 0
     MAXTAG = 0
6377 JSEED = ISEED
```

```
       NTNODE = 0
       ISTAR = 0
       NTAG = KNODE
       DO 6380 I=1,6
 6380  KNT(I) = 0
       DO 6388 I=1,NARC
       SLACKA(I) = SETUP
       TIMEA(I) = SETUP
       COSTA(I) = SETUP
       PERFA(I) = SETUP
       UTIMEA(I) = SETUP
       UCOSTA(I) = SETUP
       UPERFA(I) = SETUP
       IPATHA(I) = IBLK
       ISTATE(I) = 1
       IF(NODEI(I).NE.0) GC TO 6388
       TIMEA(I) = 0.0
       COSTA(I) = 0.0
       PERFA(I) = C.0
       UTIMEA(I) = 0.0
       UCOSTA(I) = C.0
       UPERFA(I) = 0.0
       CALL DOARC(I)
 6388  CONTINUE
       DO 6399 I=1,NNODE
       SLACKN(I) = SETUP
       TIMEN(I) = SETUP
       COSTN(I) = SETUP
       PERFN(I) = SETUP
       NSTATE(I) = 1
       IPATHN(I) = IBLK
 6399  IF(LOGI(I).EQ.1) NSTATE(I) = 2
C
C      SIMULATE THE NETWORK VIA NODE REVIEW. CORRECT NODE ENTRY=1 REVIEW.
C          ICHECK = .TRUE. - NO VIRGIN NODES OR UNPROCESSED INPUT ARCS
C                            FOR CANDIDATES NODES.
C          ICHECK = .FALSE.- AT LEAST 1 OF THE 2 PRECEEDING OCCURRED.
C          JCHECK = .FALSE.- NO NODES PROCESSED THIS NODE REVIEW.
C          JCHECK = .TRUE. - AT LEAST 1 NODE WAS PROCESSED THIS REVIEW.
C
 6400  ICHECK = .TRUE.
       JCHECK = .FALSE.
       DO 7466 I=1,NNODE
       IF(NSTATE(I).EQ.2) GC TO 6411
       IF(NSTATE(I) - 1)7466,6433,7466
 6411  NI = NARCI(I)
       NP = NPOINT(I)
       IF(LOGI(I).EQ.1) GC TO 6455
       K = 0
 6422  L = NSTORE(NP+K)
       IF(ISTATE(L).NE.1) GC TO 6444
 6433  ICHECK = .FALSE.
       GC TO 7466
 6444  K = K + 1
       IF(K.NE.NI) GC TO 6422
 6455  JCHECK = .TRUE.
       NC = NARCC(I)
C
C      INITIATE THE CURRENT NODES OUTPUT ARCS AND INITIATE THOSE ARCS
C          OUTPUT NODES.
C
       IF(NO.LE.0) GC TO 6477
```

```
      DO 6466 K=1,NO
      L = NSTORE(NP+NI+K-1)
 6466 CALL INITAL(L)
C
C     CHECK AND, PAND, OR FOR LOGICAL ELIMINATIONS OR ESCAPES
C         AND (INFLUENCE OF THE INPUT ARC'S STATUS ON PROCESSING)
C         1. AT LEAST 1 ZERO = LOGICAL ELIMINATION
C         2. COMBINATION OF 2'S AND 3'S = GO OUT THE ESCAPE ARC
C         3. ALL 3'S = NORMAL PROCESSING
C         PAND - OR (INFLUENCE OF THE INPUT ARCS STATUS ON PROCESSING)
C         1. ALL ZEROS = LOGICAL ELIMINATION
C         2. COMBINATION OF 0'S AND 2'S = GO OUT THE ESCAPE ARC
C         3. AT LEAST ONE 3 = NORMAL PROCESSING
C
 6477 IF(LOGI(I).LT.2.OR.LOGI(I).GT.4) GO TO 6533
      KCOUNT = 0
      ITALC = 0
      ITALP = 0
      TSAVE = -SETUP
      CSAVE = 0.0
      M = 0
      N = 0
      J = 0
      DO 6500 K=1,NI
      L = NP + K - 1
      L = NSTORE(L)
      IF(ISTATE(L).GT.0) GO TO 6488
      M = M + 1
      GO TO 6500
 6488 IF(ISTATE(L).EQ.2) N = N + 1
      IF(ISTATE(L).EQ.3) J = J + 1
      IF(TIMEA(L).LT.TSAVE) GO TO 6499
      TSAVE = TIMEA(L)
 6499 KCOUNT = KCOUNT + 1
      IF(DOUBLE) CSAVE = CSAVE + CCSTA(L)
      IF(SINGLE) CALL DVECT (1,KCOUNT,L,CSAVE,SETUP)
      CSAVE = CSAVE - BCCST
 6500 CONTINUE
C
C     CHECK AND LOGIC
C
      IF(LOGI(I).NE.2) GO TO 6511
      IF(M.GT.0) GO TO 7455
      IF(N.GT.0) GO TO 6522
      GO TO 6533
C
C     CHECK PAND, OR LOGIC
C
 6511 IF(M.EQ.NI) GO TO 7455
      IF(J.GT.0) GO TO 6533
C
C     INITIALIZE THE ESCAPE ARC
C
 6522 TIMEN(I) = TSAVE
      COSTN(I) = CSAVE + BCCST
      PERFN(I) = BPERF
      NSTATE(I) = 3
      IF(SINGLE) CALL SVECT
      IF(IRROR.GT.0) RETURN
      N = NP + NI + NO - 1
      N = NSTORE(N)
      CALL DOARC(N)
```

```
          JPOINT(N) = ISTAR
          GC TO 7466
C
C         INITIALIZE FCR NORMAL PROCESSING
C
   6533 KCOUNT = 0
          MCOUNT = 0
          TSAVE = SETUP
          CSAVE = 0.0
          PSAVE = 0.0
          J = 0
          ITALC = 0
          ITALP = 0
          IF(LOGI(I).EQ.4) GC TC 6577
          TSAVE = -SETUP
          IF(LOGI(I).GE.5) GO TC 7111
          IF(LOGI(I).EQ.2.OR.LCGI(I).EQ.3) GO TO 6544
C
C         PROCESS NODES WITH 'INITIAL' INPUT LOGIC HERE
C
          TIMEN(I) = BTIME
          COSTN(I) = BCOST
          PERFN(I) = BPERF
          GC TO 6611
C
C         PROCESS NCDES WITH 'AND' AND 'PAND' INPUT LOGIC HERE
C
   6544 DO 6566 K=1,NI
          L = NP + K - 1
          L = NSTORE(L)
          IF(ISTATE(L).EQ.2) GC TO 6555
          IF(ISTATE(L).LE.0) GC TO 6566
          ISTATE(L) = 4
          KCOUNT = KCOUNT + 1
          IF(DOUBLE) PSAVE = PSAVE + PERFA(L)
          IF(SINGLE) CALL DVECT (2,KCOUNT,L,PSAVE,SETUP)
          PSAVE = PSAVE - BPERF
   6555 MCOUNT = MCOUNT + 1
          IF(DOUBLE) CSAVE = CSAVE + CCSTA(L)
          IF(SINGLE) CALL DVECT (1,MCOUNT,L,CSAVE,SETUP)
          CSAVE = CSAVE - BCOST
          IF(TIMEA(L).LT.TSAVE) GC TO 6566
          TSAVE = TIMEA(L)
   6566 CONTINUE
          GC TO 6600
C
C         PROCESS NODES WITH 'CR' INPUT LOGIC HERE
C
   6577 DO 6588 K=1,NI
          L = NP + K - 1
          L = NSTORE(L)
          IF(ISTATE(L).LE.0) GC TO 6588
          IF(TIMEA(L).GE.TSAVE) GO TO 6588
          TSAVE = TIMEA(L)
          J = L
   6588 CONTINUE
          ISAVE = 0
          DO 6599 K=1,NI
          L = NP + K - 1
          L = NSTORE(L)
          IF(ISTATE(L).LE.0) GC TO 6599
          KCOUNT = KCOUNT + 1
```

```
        IF(SINGLE) GO TO 6590
        CSAVE = CSAVE + COSTA(L)
        GO TO 6595
 6590 IF(L.EQ.J) X6 = SETUP
        IF(L.NE.J) X6 = TSAVE
        CALL DVECT (1,KCUNT,L,CSAVE,X6)
 6595 CSAVE = CSAVE - BCOST
 6599 CONTINUE
        IF(ISAVE.EQ.1) GO TO 6400
        IF(DOUBLE) PSAVE = PSAVE + PERFA(J)
        IF(SINGLE) CALL DVECT (2,1,J,PSAVE,SETUP)
        PSAVE = PSAVE - BPERF
        ISTATE(J) = 4
C
C       STORE THE NODE TIME, COST AND PERFORMANCE
C
 6600 TIMEN(I) = TSAVE
        COSTN(I) = CSAVE + BCOST
        PERFN(I) = PSAVE + BPERF
 6611 NSTATE(I) = 3
        IF(SINGLE) CALL SVECT
        IF(IRROR.GT.0) RETURN
        IF(LOGO(I).GE.11) GO TO 6622
        IF(LOGO(I).NE.1) GO TO 6633
 6622 NTNODE = NTNODE + 1
        MTNODE = I
        GO TO 7466
 6633 NCO = NO
        IF(JUMP(I).EQ.1) NCO = NO - 1
        IF(LOGO(I).GT.3) GO TO 6766
C
C       PROCESS NODES WITH 'ALL' OUTPUT LOGIC HERE
C
        IF(LOGO(I).NE.2) GO TO 6655
        DO 6644 K=1,NCO
        L = NP + NI + K - 1
        L = NSTORE(L)
        CALL DOARC(L)
 6644 JPOINT(L) = ISTAR
        GO TO 7466
C
C       PROCESS NODES WITH 'MONTE CARLO' OUTPUT LOGIC HERE
C
 6655 CALL RANDOM
        X2 = 0.0
        DO 6744 K=1,NOO
        L = NP + NI + K - 1
        L = NSTORE(L)
        CALL SEEK (IPOINT(L),EMARK(1))
C
C       IF THIS IS THE FIRST ARC, FIND THE CORRECT DISTRIBUTION
C
        IF(K.GT.1) GO TO 6733
        M = 2
        IF(ITALP.EQ.2) GO TO 6733
        LS = ITALC + 1
        X3 = 1.0
        X4 = 1.0
        IF(ASTORE(LS) - 3.0)6666,6677,6699
 6666 X5 = TIMEN(I)
        GO TO 6711
 6677 X5 = COSTN(I)
```

```
      IF(INFC) GO TO 6688
      X3 = FINT**((TIMEN(I) - BTIME)/FACTOR)
 6688 IF(DISC) GO TO 6711
      X4 = DINT**((TIMEN(I) - BTIME)/FACTOR)
      GO TO 6711
 6699 X5 = PERFN(I)
      IF(INFP) GO TO 6700
      X3 = FINT**((TIMEN(I) - BTIME)/FACTOR)
 6700 IF(DISP) GO TO 6711
      X4 = DINT**((TIMEN(I) - BTIME)/FACTOR)
 6711 DO 6722 ISAVE=3,ITALP,2
      M = ISAVE - 1
      LS = ITALC + ISAVE
      X6 = ASTORE(LS)*X3/X4
      IF(X6.GE.X5) GO TO 6733
 6722 CONTINUE
      M = ITALC + ITALP
C
C     DETERMINE WHICH OUTPUT ARC WILL BE PROCESSED
C
 6733 X1 = X2
      MS = ITALC + M
      X2 = X2 + ASTORE(MS)
      IF(UNFORM.GE.X1.AND.UNFORM.LT.X2) GO TO 6755
 6744 CONTINUE
 6755 CALL DOARC(L)
      JPOINT(L) = ISTAR
      GO TO 7466
C
C     PROCESS NODES WITH 'FILTER' OUTPUT LOGIC HERE
C
 6766 MCOUNT = 0
      DO 6977 K=1,NO
      L = NP + NI + K - 1
      L = NSTORE(L)
      N = IPOINT(L)
C
C     FIND THE ESCAPE ARC (NCOUNT POINTS TO THAT ARC)
C
      IF(N.EQ.0) GO TO 6777
      CALL SEEK (N,EMARK(1))
      IF(ITALP.GT.0) GO TO 6788
 6777 NCOUNT = L
      GO TO 6977
 6788 IF(LOGO(I) - 5)6799,6899,6911
C
C     PROCESS FILTER 1 LOGIC
C
 6799 N = NP + NI + NO
      N = NSTORE(N)
      TSAVE = 0.0
      CSAVE = 0.0
      PSAVE = 0.0
      IF(N.EQ.0) GO TO 6822
      IF(NSTATE(N).EQ.3) GO TO 6811
      CALL ERROR (6800,NODE1(I),NODE2(I),IBLK)
      GO TO 7499
 6811 TSAVE = TIMEN(N)
      CSAVE = COSTN(N)
      PSAVE = PERFN(N)
 6822 TSAVE = TIMEN(I) - TSAVE
      CSAVE = COSTN(I) - CSAVE
```

```
      PSAVE = PERFN(I) - PSAVE
      X1 = ASTORE(ITALC+1)
      X2 = ASTORE(ITALC+2)
      IF(X1.EQ.0.0.AND.X2.EQ.0.0) GO TO 6833
      IF(TSAVE.LE.X1.OR.TSAVE.GT.X2) GO TO 6977
 6833 IF(ITALP.EQ.2) GO TO 6966
      X3 = ASTORE(ITALC+3)
      X4 = ASTORE(ITALC+4)
      IF(X3.EQ.0.0.AND.X4.EQ.0.0) GO TO 6866
      IF(INFC) GO TO 6844
      X = FINT**((TIMEN(I) - BTIME)/FACTOR)
      X3 = X3*X
      X4 = X4*X
 6844 IF(DISC) GO TO 6855
      X = DINT**((TIMEN(I) - BTIME)/FACTOR)
      X3 = X3/X
      X4 = X4/X
 6855 IF(CSAVE.LE.X3.OR.CSAVE.GT.X4) GO TO 6977
 6866 IF(ITALP.EQ.4) GO TO 6966
      X5 = ASTORE(ITALC+5)
      X6 = ASTORE(ITALC+6)
      IF(X5.EQ.0.0.AND.X6.EQ.0.0) GO TO 6966
      IF(INFP) GO TO 6877
      X = FINT**((TIMEN(I) - BTIME)/FACTOR)
      X5 = X5*X
      X6 = X6*X
 6877 IF(DISP) GO TO 6888
      X = DINT**((TIMEN(I) - BTIME)/FACTOR)
      X5 = X5/X
      X6 = X6/X
 6888 IF(PSAVE.LE.X5.OR.PSAVE.GT.X6) GO TO 6977
      GO TO 6966
C
C     PROCESS FILTER #2 LOGIC HERE
C
 6899 X1 = 0.0
      DO 6900 J=1,NI
      N = NSTORE(NP+J-1)
 6900 IF(ISTATE(N).GT.2) X1 = X1 + 1.0
      IF(X1.LT.ASTORE(ITALC+1).OR.X1.GT.ASTORE(ITALC+2)) GO TO 6977
      GO TO 6966
C
C     PROCESS FILTER #3 LOGIC HERE
C
 6911 DO 6955 J=1,ITALP
      X1 = ASTORE(ITALC+J)
      N = ABS(X1) + 0.001
      IF(ISTATE(N).NE.1) GO TO 6933
      CALL ERROR (6922,IARC1(N),IARC2(N),IBLK)
      GO TO 7499
 6933 IF(X1.GT.0.0) GO TO 6944
      IF(ISTATE(N).EQ.3.OR.ISTATE(N).EQ.4) GO TO 6977
      GO TO 6955
 6944 IF(ISTATE(N).LE.0.OR.ISTATE(N).EQ.2) GO TO 6977
 6955 CONTINUE
C
C     PROCESS THIS CONSTRAINED ARC
C
 6966 MCCUNT = MCOUNT + 1
      CALL DOARC (L)
      JPOINT(L) = ISTAR
 6977 CONTINUE
```

```
      IF(MCOUNT.GT.0) GO TC 7466
      CALL DOARC (NCOUNT)
      JPOINT(NCOUNT) = ISTAR
      GO TO 7466
C
C     PROCESS NODES WITH SCRT LOGIC
C
 6988 NCOUNT = 0
      N = NP + NI + NO
 6999 J = 0
      X1 = 0.0
      X2 = 0.0
      X3 = 0.0
      DO 7000 K=1,NI
      L = NP + K - 1
      L = NSTCRE(L)
      IF(ISTATE(L).NE.3) GC TO 7000
      J = L
      X1 = X1 + TIMEA(L)
      X2 = X2 + CCSTA(L)
      X3 = X3 + PERFA(L)
 7000 CCNTINUE
      IF(J.EQ.0) GC TO 7433
      X = SETUP
      DO 7011 K=1,NI
      L = NP + K - 1
      L = NSTORE(L)
      IF(ISTATE(L).NE.3) GC TO 7011
      X4 = 0.0
      X5 = 0.0
      X6 = 0.0
      IF(X1.NE.0.0) X4 = TIMEA(L)/X1
      IF(X2.NE.0.0) X5 = CCSTA(L)/X2
      IF(X3.NE.0.0) X6 = PERFA(L)/X3
      TIMESM = X4*FLOAT(NSTORE(N)) + X5*FLOAT(NSTCRE(N+1)) -
     1X6*FLOAT(NSTCRE(N+2))
      IF(TIMESM.GE.X) GO TC 7011
      X = TIMESM
      J = L
 7011 CCNTINUE
      ISTATE(J) = 4
      ITALC = 0
      ITALP = 0
      IF(DOUBLE) GC TO 7020
      CALL DVECT (1,1,J,UNFORM,SETUP)
      CALL DVECT (2,1,J,UNFCRM,SETUP)
      CALL SVECT
      IF(IRROR.GT.0) RETURN
 7020 M = NP + NI + NCOUNT
      NCCUNT = NCCUNT + 1
      M = NSTORE(M)
      TIMEA(M) = TIMEA(J)
      CALL DCARC(M)
      JPCINT(M) = ISTAR
      COSTA(M) = CCSTA(J) + UCOSTA(M)
      PERFA(M) = PERFA(J) + UPERFA(M)
      GC TO 6999
C
C     PROCESS NODES WITH QLEUE LOGIC, FIRST INITIALIZE
C
 7022 IF(LOGI(I).EC.8) GC TC 6988
      TIMEN(I) = 0.0
```

```
      COSTN(I) = 0.0
      PERFN(I) = 0.0
      N = 0
 7033 LS = 0
      X = SETUP
C
C     ARRANGE THE ARCS IN THE QUEUE FROM EARLIEST TO LATEST, FIRST FIND
C        MIN OF CURRENT CANDIDATE SET OF INPUT ARCS
C
      DO 7044 K=1,NI
      M = NP + K - 1
      L = NSTORE(M)
      IF(ISTATE(L).NE.3) GO TO 7044
      IF(TIMEA(L).GE.X) GO TO 7044
      X = TIMEA(L)
      LS = L
      MS = NSTORE(M+NI)
 7044 CONTINUE
C
C     IF THERE WAS A SELECTION, RECORD IT
C
      IF(LS.EQ.0) GO TO 7055
      ISTATE(LS) = 4
      ITALC = 0
      ITALP = 0
      IF(DOUBLE) GO TO 7050
      CALL DVECT (1,1,LS,UNFORM,SETUP)
      CALL DVECT (2,1,LS,UNFORM,SETUP)
      CALL SVECT
      IF(IRROR.GT.0) RETURN
 7050 TIMEA(MS) = TIMEA(LS)
      CALL DOARC(MS)
      JPOINT(MS) = ISTAR
      COSTA(MS) = COSTA(LS) + UCOSTA(MS)
      PERFA(MS) = PERFA(LS) + UPERFA(MS)
      CSAVE = CSAVE + UCOSTA(MS)
      PSAVE = PSAVE + UPERFA(MS)
      IF(TSAVE.LT.TIMEA(MS)) TSAVE = TIMEA(MS)
      N = N + 1
      WORK(N) = LS
      SLACKA(N) = MS
      IF(N.LT.NI) GO TO 7033
C
C     ADJUST PROCESSING TIMES FOR ITEMS WAITING IN THE QUEUE PROVIDING
C        THERE ARE MORE ITEMS IN THE QUEUE THAN THERE ARE SERVERS
C
 7055 NSERVE = LOGC(I)
      IF(N.LE.NSERVE) GO TO 7099
      DO 7066 K=1,NSERVE
 7066 IPATHA(K) = K
      MCOUNT = NSERVE + 1
      DO 7088 K=MCOUNT,N
      X = SETUP
      NCOUNT = K - 1
C
C     FIND THE SERVER FINISHED FIRST
C
      DO 7077 ISAVE=1,NCOUNT
      IF(IPATHA(ISAVE).EQ.0) GO TO 7077
      M = SLACKA(ISAVE) + 0.1
      IF(TIMEA(M).GE.X) GO TO 7077
      X = TIMEA(M)
```

```
      L = ISAVE
 7077 CONTINUE
C
C     PUT THE SERVER ON A NEW ARC AND COMPUTE THE DELAY TIME FOR THAT
C         NEW ARC IF THERE IS DELAY TIME
C
      IPATHA(K) = IPATHA(L)
      IPATHA(L) = 0
      L = WORK(L) + 0.1
      IF(TIMEA(L).GE.X) GC TO 7088
      M = SLACKA(K) + 0.1
      TIMEA(M) = TIMEA(M) + X - TIMEA(L)
      IF(TSAVE.LT.TIMEA(M)) TSAVE = TIMEA(M)
 7088 CONTINUE
 7099 DO 7100 N=1,NARC
      SLACKA(N) = SETUP
 7100 IPATHA(N) = IBLK
      GO TO 7433
C
C     ARE THERE ENOUGH SUCCESSFULLY PROCESSED INPUT ARCS TO SATISFY THE
C         OUTPUT ARC SET REQUIREMENT
C
 7111 MCOUNT = 0
      M = 1
      DO 7144 K=1,NI
      L = NSTORE(NP+K-1)
      IF(ISTATE(L).LE.0.CR.ISTATE(L).EQ.2) GO TO 7122
      IF(DOUBLE) PSAVE = PSAVE + PERFA(L)
      IF(SINGLE) CALL DVECT (2,KCOUNT,L,PSAVE,SETUP)
      PSAVE = PSAVE - BPERF
      KCOUNT = KCOUNT + 1
 7122 IF(ISTATE(L).LE.0) GO TO 7144
      MCOUNT = MCOUNT + 1
      IF(DOUBLE) CSAVE = CSAVE + COSTA(L)
      IF(SINGLE) CALL DVECT (1,MCOUNT,L,CSAVE,SETUP)
      CSAVE = CSAVE - BCOST
      GC TO (7133,7144),M
 7133 IF(LOGU(I).GT.0.AND.LOGI(I).EQ.6.AND.KCOUNT.EQ.LOGO(I)) M = 2
      IF(TSAVE.LT.TIMEA(L)) TSAVE = TIMEA(L)
 7144 CONTINUE
      IF(MCOUNT.EQ.0) GO TC 7455
C
C     IF THERE ARE NOT ANY SUCCESSFUL INPUT ARCS - OR NOT ENOUGH, FIRE
C         THE ESCAPE OUTPUT ARC - FIRST FIND THE TIME WHEN CONDITIONS
C         TURNED INFEASIBLE IF THERE WASN'T ENOUGH SUCCESSFUL INPUT ARCS
C
      IF(KCOUNT.LE.0) GO TC 7155
      IF(LOGI(I) - 6)7200,7200,7022
 7155 IF(LOGO(I).LT.0.OR.LCGI(I).GE.7) GO TO 7199
 7166 NCCUNT = LOGC(I)
 7177 TSAVE = SETUP
      DO 7188 K=1,NI
      M = NP + K - 1
      L = NSTORE(M)
      IF(ISTATE(L).LE.0) GC TC 7188
      M = NSTORE(M+NI)
      IF(TSAVE.LE.TIMEA(L)) GO TO 7188
      TSAVE = TIMEA(L)
      LS = L
      MS = M
 7188 CONTINUE
      TIMEA(MS) = 0.0
```

```
      MCOUNT = MCOUNT - 1
      IF(ISTATE(LS).GT.2) NCOUNT = NCOUNT - 1
      IF(MCOUNT.GE.NCOUNT) GO TO 7177
 7199 NSTATE(I) = 3
      TIMEN(I) = TSAVE
      CCSTN(I) = CSAVE + BCCST
      PERFN(I) = BPERF
      IF(SINGLE) CALL SVECT
      IF(IRROR.GT.0) RETURN
      M = NSTORE(NP+NI+NC-1)
      CALL DOARC(M)
      JPOINT(M) = ISTAR
      GO TO 7466
 7200 N = NP + NI + NC
      NCOUNT = -LOGO(I)
      IF(LOGO(I).LT.0) GO TO 7211
      IF(KCOUNT.LT.LOGO(I)) GO TO 7166
      IF(LOGI(I) - 5)7166,7266,7388
 7211 IF(LOGI(I).EQ.6) GO TO 7377
C
C     CALCULATE OUTPUT ARC TIMES (DESIRED CONDITIONS) WHEN THERE IS NOT
C         A FULL SUBSET AND LESS THAN 100% WEIGHT ON TIME
C         MCOUNT - CURRENT NUMBER OF REMAINING OUTPUT ARC CANDIDATES
C         NCOUNT - CURRENT NUMBER OF GOOD OUTPUT ARC'S STILL NEEDED
C
      IF(KCOUNT.GE.-LOGO(I).OR.IABS(NSTORE(N)).EQ.100) GO TO 7266
 7222 X1 = SETUP
      DO 7233 K=1,NI
      M = NP + K - 1
      L = NSTORE(M)
      M = NSTORE(M+NI)
      IF(TIMEA(M).NE.SETUP) GO TO 7233
      IF(X1.LE.TIMEA(L)) GO TO 7233
      X1 = TIMEA(L)
      LS = L
      MS = M
 7233 CONTINUE
      TIMEA(MS) = 0.0
      MCOUNT = MCOUNT - 1
      IF(ISTATE(LS).GT.2) NCOUNT = NCOUNT - 1
      IF(MCOUNT.GT.NCOUNT) GO TO 7222
C
C     THE PIVOT TIME HAS BEEN FOUND, ASSIGN TIME VALUES TO OUTPUT ARCS
C
      DO 7255 K=1,NI
      M = NP + K - 1
      L = NSTORE(M)
      M = NSTORE(M+NI)
      IF(TIMEA(M).EQ.SETUP) GO TO 7244
      TIMEA(M) = X1
      GO TO 7255
 7244 TIMEA(M) = TIMEA(L)
 7255 CONTINUE
C
C     PROCESSING POINT FOR THE COMPARE NODE - COMPUTE NORMALIZATION BASE
C
 7266 MCOUNT = 0
      IF(IABS(NSTORE(N)).NE.100) GO TO 7288
      IF(LOGO(I).GT.0) GO TO 7277
      TSAVE = SETUP
      GO TO 7288
 7277 TSAVE =-SETUP
```

```
 7288 NCOUNT = 0
      X1 = 0.0
      X2 = 0.0
      X3 = 0.0
      DO 7299 K=1,NI
      L = NSTORE(NP+K-1)
      IF(ISTATE(L).LE.0.OR.ISTATE(L).EQ.2.OR.ISTATE(L).EQ.4) GO TO 7299
      X1 = X1 + TIMEA(L)
      X2 = X2 + COSTA(L)
      X3 = X3 + PERFA(L)
      NCOUNT = 1
 7299 CONTINUE
      IF(NCOUNT.EQ.0) GO TO 7333
C
C     CHOOSE THE WINNING INPUT ARC
C
      UNFORM = SETUP
      DO 7300 K=1,NI
      L = NSTORE(NP+K-1)
      IF(ISTATE(L).LE.0.OR.ISTATE(L).EQ.2.OR.ISTATE(L).EQ.4) GO TO 7300
      X4 = 0.0
      X5 = 0.0
      X6 = 0.0
      IF(X1.NE.0.0) X4 = TIMEA(L)/X1
      IF(X2.NE.0.0) X5 = COSTA(L)/X2
      IF(X3.NE.0.0) X6 = PERFA(L)/X3
      TIMESM = X4*FLOAT(NSTORE(N)) + X5*FLOAT(NSTORE(N+1)) -
     1X6*FLOAT(NSTORE(N+2))
      IF(TIMESM.GE.UNFORM) GO TO 7300
      UNFORM = TIMESM
      J = L
      M = K
 7300 CONTINUE
C
C     ANOTHER WINNING INPUT ARC HAS BEEN DETERMINED - UPDATE
C
      ISTATE(J) = 4
      ITALC = 0
      ITALP = 0
      IF(DOUBLE) GO TO 7310
      CALL DVECT (1,1,J,UNFORM,SETUP)
      CALL DVECT (2,1,J,UNFORM,SETUP)
      CALL SVECT
      IF(IRROR.GT.0) RETURN
 7310 M = NSTORE(NP+NI+M-1)
      JPOINT(M) = ISTAR
      IF(IABS(NSTORE(N)).NE.100) GO TO 7322
      IF(LOGO(I).GT.0) GO TO 7311
      IF(TSAVE.GT.TIMEA(J)) TSAVE = TIMEA(J)
      TIMEA(M) = TIMEA(J)
      GO TO 7322
 7311 IF(TSAVE.LT.TIMEA(J)) TSAVE = TIMEA(J)
 7322 MCOUNT = MCOUNT + 1
      IF(MCOUNT.LT.IABS(LOGO(I))) GO TO 7288
C
C     FOR THE SPECIAL CASE OF TIME ALONE, TRIM THOSE INPUT ARCS WHOSE
C         TIME EXCEEDS THE NODE TIME
C
 7333 IF(IABS(NSTORE(N)).NE.100) GO TO 7433
      KCOUNT = 0
      ITALC = 0
      ITALP = 0
```

```
      ISAVE = 0
      DO 7355 K=1,NI
      L = NSTORE(NP+K-1)
      M = ISTATE(L)
      IF(M.LE.0.OR.M.EQ.2) GO TO 7344
      KCOUNT = KCOUNT + 1
      IF(SINGLE) GO TO 7335
      PSAVE = PSAVE + PERFA(L)
      GO TO 7340
 7335 IF(M.EQ.4) X6 = SETUP
      IF(M.NE.4) X6 = TSAVE
      CALL DVECT (2,KOUNT,L,PSAVE,X6)
 7340 PSAVE = PSAVE - BPERF
 7344 IF(M.LE.0) GO TO 7355
      IF(SINGLE) GO TO 7345
      CSAVE = CSAVE + COSTA(L)
      GO TO 7350
 7345 IF(M.EQ.4) X6 = SETUP
      IF(M.NE.4) X6 = TSAVE
      CALL DVECT (1,KOUNT,L,CSAVE,X6)
 7350 CSAVE = CSAVE - BCCST
 7355 CONTINUE
      IF(ISAVE.EQ.0) GO TO 7433
      DO 7366 K=1,NI
      L = NSTORE(NP+K-1)
 7366 IF(ISTATE(L).EQ.4) ISTATE(L) = 3
      GO TO 6400
C
C     PROCESSING POINT FOR THE PREFERRED NODE
C
 7377 TSAVE = -SETUP
 7388 MCOUNT = 0
      K = 0
 7399 K = K + 1
      IF(K.GT.NI.OR.MCOUNT.EQ.IABS(LOGO(I))) GO TO 7433
      MS = NP + K - 1
      LS = NSTORE(MS)
      MS = NSTORE(MS+NI)
      IF(LOGO(I).GT.0.AND.ISTATE(LS).LE.0) GO TO 7399
      IF(LOGO(I).GT.0) GO TO 7422
      X1 = SETUP
      DO 7400 J=1,NCOUNT
      M = NP + J - 1
      L = NSTORE(M)
      M = NSTORE(M+NI)
      IF(ISTATE(L).LE.0.OR.TIMEA(M).NE.SETUP) GO TO 7400
      IF(TIMEA(L).GE.X1) GO TO 7400
      X1 = TIMEA(L)
      LS = L
      MS = M
 7400 CONTINUE
      IF(X1.NE.SETUP) GO TO 7411
      NCOUNT = NCOUNT + 1
      IF(NCOUNT.GT.NI) NCOUNT = NI
      GO TO 7399
 7411 IF(ISTATE(LS).EQ.2) NCOUNT = NCOUNT + 1
      IF(NCOUNT.GT.NI) NCOUNT = NI
      IF(TSAVE.LT.TIMEA(LS)) TSAVE = TIMEA(LS)
      IF(ISTATE(LS).LE.2) GO TO 7399                                        GLM30
      IF(LOGO(I).GT.0) TIMEA(MS) = TSAVE
      ISTATE(LS) = 4
      ITALC = 0
```

```
       ITALP = 0
       IF(DOUBLE) GC TO 7420
       CALL DVECT (1,1,LS,UNFORM,SETUP)
       CALL DVECT (2,1,LS,UNFORM,SETUP)
       CALL SVECT
       IF(IRROR.GT.0) RETURN
  7420 JPOINT(MS) = ISTAR
       MCOUNT = MCOUNT + 1
       GO TO 7399
C
C      SUCCESSFUL COMPLETICN - CALCULATE NODE AND OUTPUT ARC STATISTICS
C
  7433 NSTATE(I) = 3
       TIMEN(I) = TSAVE
       COSTN(I) = CSAVE + BCCST
       PERFN(I) = PSAVE/FLOAT(KCOUNT) + BPERF
       IF(LOGI(I).GE.7) GC TC 7456
       DO 7444 K=1,NI
       M = NP + K - 1
       L = NSTORE(M)
       IF(ISTATE(L).NE.4) GC TO 7444
       M = NSTORE(M+NI)
       CALL DCARC(M)
       COSTA(M) = UCOSTA(M) + COSTA(L)
       PERFA(M) = UPERFA(M) + PERFA(L)
  7444 CONTINUE
       GC TO 7466
C
C      PRESENT NODE UNDER CCNSIDERATION IS BEING LOGICALLY ELIMINATED
C
  7455 NSTATE(I) = C
  7466 CONTINUE
C
C      IS THIS LAST NODE REVIEW NEEDED FOR THIS ITERATION
C
       IF(IRROR.GT.0) GO TC 7499
       IF(ICHECK) GO TO 75CC
       IF(JCHECK) GC TO 64CC
  7477 CALL ERRCR (7488,IBLK,IBLK,IBLK)
  7499 NODESM = MNCDE
       GO TO 8033
C
C      DETERMINE THE TERMINAL NODE FOR THIS ITERATION
C
  7500 IF(NTNCDE - 1)7477,7511,7522
  7511 NCDESM = MTNCDE
       TIMESM = TIMEN(MTNODE)
       GC TO 7588
  7522 K = -9
  7533 K = K + 1C
       UNFORM = SETUP
       TSAVE = 0.0
       CSAVE = 0.0
       PSAVE = C.0
       X1 = -SETUP
       NODESM = 0
       L = 0
       DO 7544 I=1,NNCDE
       IF(NSTATE(I).EQ.1.OR.NSTATE(I).EQ.2) GO TC 7477
       IF(LOGI(I).GE.5) GO TO 7544
       IF(LOGC(I).NE.K.CR.NSTATE(I).NE.3) GO TO 7544
       IF(TIMEN(I).GT.X1) X1 = TIMEN(I)
```

```
      L = L + 1
      NODESM = I
      TSAVE = TSAVE + TIMEN(I)
      CSAVE = CSAVE + COSTN(I)
      PSAVE = PSAVE + PERFN(I)
 7544 CONTINUE
      IF(L - 1)7533,7577,7555
 7555 DO 7566 I=1,NNODE
      IF(LOGI(I).GE.5) GO TO 7566
      IF(LOGO(I).NE.K.OR.NSTATE(I).NE.3) GO TO 7566
      X4 = 0.0
      X5 = 0.0
      X6 = 0.0
      IF(TSAVE.NE.0.0.AND.STIME.NE.0.0) X4 = STIME*(TIMEN(I)/TSAVE)
      IF(CSAVE.NE.0.0.AND.SCOST.NE.0.0) X5 = SCOST*(COSTN(I)/CSAVE)
      IF(PSAVE.NE.0.0.AND.SPERF.NE.0.0) X6 = SPERF*(PERFN(I)/PSAVE)
      X2 = X4 + X5 - X6
      IF(X2.GE.UNFORM) GO TO 7566
      UNFORM = X2
      NODESM = I
 7566 CONTINUE
 7577 TIMESM = TIMEN(NODESM)
      IF(ABS(STIME).NE.1.0) TIMESM = X1
C
C     CHECK MAX STORAGE REQUIRED OF COST PERFORMANCE VECTOR STORAGE AND
C        STORE THE ITERATION RESULTS AND LIST SPECIAL PRINTOUT IF WANTED
C
 7588 IF(NTAG.GT.MAXTAG) MAXTAG = NTAG
      NODET(ICOUNT) = NODESM
      STORET(ICOUNT,1) = TIMESM
      STORET(ICOUNT,2) = COSTN(NODESM)
      CALL OCOST (TIMESM)
      STORET(ICOUNT,3) = UNFORM
      STORET(ICOUNT,4) = PERFN(NODESM)
      IF(ITRACE.EQ.2) WRITE (IOUT,7599) ICOUNT, NODE1(NODESM),
     1NODE2(NODESM), JSEED, (STORET(ICOUNT,K), K=1,4)
 7599 FORMAT (1H , I5, 1X, 2A4, I13, 4G26.12)
C
C     FIND THE CRITICAL-OPTIMUM PATH FOR THIS ITERATION, IF WANTED
C
      IF(CRIT) GO TO 7922
      IF(ISTAT(NODESM).EQ.-1) GO TO 7922
      ISAVE = NODESM
      JCOUNT = JCOUNT + 1
 7600 ICRITN(ISAVE) = ICRITN(ISAVE) + 1
      IF(CTIME.EQ.1.0) SLACKN(ISAVE) = 0.0
      IPATHN(ISAVE) = KSTAR
      NSTATE(ISAVE) = 4
      L = NARCI(ISAVE)
      IF(L.LE.0) GO TO 7655
      M = 4
 7611 TSAVE = 0.0
      CSAVE = 0.0
      PSAVE = 0.0
      DO 7622 I=1,L
      K = NPOINT(ISAVE) + I - 1
      K = NSTORE(K)
      IF(ISTATE(K).NE.M) GO TO 7622
      TSAVE = TSAVE + TIMEA(K)
      CSAVE = CSAVE + COSTA(K)
      PSAVE = PSAVE + PERFA(K)
 7622 CONTINUE
```

```
      J = 0
      UNFORM = -SETUP
      DO 7633 I=1,L
      K = NPOINT(ISAVE) + I - 1
      K = NSTORE(K)
      IF(ISTATE(K).NE.M) GC TO 7633
      X4 = 0.0
      X5 = 0.0
      X6 = 0.0
      IF(TSAVE.NE.0.0.AND.CTIME.NE.0.0) X4 = CTIME*(TIMEA(K)/TSAVE)
      IF(CSAVE.NE.0.0.AND.CCOST.NE.0.0) X5 = CCOST*(COSTA(K)/CSAVE)
      IF(PSAVE.NE.0.0.AND.CPERF.NE.0.0) X6 = CPERF*(PERFA(K)/PSAVE)
      X1 = X4 + X5 - X6
      IF(X1.LE.UNFCRM) GC TC 7633
      UNFORM = X1
      J = K
 7633 CONTINUE
      M = 2
      IF(J.EQ.C) GC TO 7611
      ICRITA(J) = ICRITA(J) + 1
      IPATHA(J) = KSTAR
      IF(CTIME.NE.1.C) GC TC 7644
      SLACKA(J) = 0.0
      WORK(J) = TIMEA(J) - UTIMEA(J)
 7644 ISAVE = NODEI(J)
      GO TO 7600
C
C     COMPUTE THE SLACKS, START WITH THE WINNING TERMINAL NODE AND FIND
C         THE SLACKS FOR THE INPUT ARCS AND THEIR LATEST STARTING TIME
C
 7655 IF(SLAK) GO TO 7922
      NSTATE(NCDESM) = 5
      NI = NARCI(NCDESM)
      NP = NPOINT(NODESM)
      DO 7677 K=1,NI
      L = NSTORE(NP+K-1)
      IF(ISTATE(L).GT.1) GC TO 7666
      SLACKA(L) = -SETUP
      GO TC 7677
 7666 IF(SLACKA(L).NE.SETUP) GO TO 7677
      SLACKA(L) = TIMEN(NODESM) - TIMEA(L)
      WORK(L) = TIMEA(L) - UTIMEA(L) + SLACKA(L)
 7677 CONTINUE
C
C     COMPUTE THE SLACKS FOR THE REST OF THE NODES AND ARCS
C
      KCOUNT = 0
 7688 ICHECK = .TRUE.
      DO 7844 ISAVE=1,NNCCE
      I = NNODE + 1 - ISAVE
C
C     IF THIS NODE IS ACTIVE AND HAS NOT BEEN PROCESSED, PROCESS IT NOW
C
      IF(NSTATE(I).EQ.5) GC TC 7844
      IF(NSTATE(I).NE.4.AND.SLACKN(I).NE.SETUP) GO TO 7844
      ICHECK = .FALSE.
      NP = NPOINT(I)
      NI = 0
      IF(LOGI(I).NE.1) NI = NARCI(I)
C
C     IF THIS NODE IS A NCN-CRITICAL TERMINAL NODE OR HAS BEEN ELIMINAT-
C         ED OR IS OF NO CCNSEQUENCE TO THE CRITICAL PATH ANALYSIS, WIPE
```

```
C          IT OUT AND ITS INPUT NODES
C
      IF(NSTATE(I).EQ.0) GO TO 7699
      IF(LOGI(I).GE.5) GO TO 7711
      IF(LOGO(I).GE.11)GO TO 7699
      IF(LOGO(I).NE.1) GO TO 7711
 7699 SLACKN(I) = -SETUP
      IF(NI.EQ.0) GO TO 7844
      DO 7700 K=1,NI
      L = NSTORE(NP+K-1)
 7700 SLACKA(L) = -SETUP
      GO TO 7844
C
C     IF THE ACTIVE OUTPUT ARCS OF THIS NODE ALL HAVE NON-INITIAL VALUES
C         FOR THEIR SLACKS, COMPUTE THE NODE SLACK
C
 7711 NO = NARCO(I)
      UNFORM = SETUP
      DO 7722 K=1,NO
      M = NSTORE(NP+NI+K-1)
      IF(SLACKA(M).EQ.SETUP) GO TO 7844
      IF(SLACKA(M).EQ.-SETUP.OR.NSTATE(I).EQ.4) GO TO 7722
      IF(WORK(M).LT.UNFORM) UNFORM = WORK(M)
 7722 CONTINUE
      IF(NSTATE(I).NE.4) GO TO 7733
      UNFORM = TIMEN(I)
      GO TO 7744
 7733 IF(UNFORM.EQ.SETUP) GO TO 7699
      SLACKN(I) = UNFORM - TIMEN(I)
C
C     COMPUTE SLACKS FOR ARCS FLOWING INTO NODES HAVING SPLIT NODE LOGIC
C
 7744 IF(LOGI(I).EQ.1) GO TO 7833
      IF(LOGI(I).GE.5) GO TO 7800
 7755 DO 7799 K=1,NI
      L = NSTORE(NP+K-1)
      IF(ISTATE(L).GT.1) GO TO 7777
 7766 SLACKA(L) = -SETUP
      GO TO 7799
 7777 IF(SLACKA(L).NE.SETUP) GO TO 7799
      IF(LOGI(I).NE.4) GO TO 7788
      IF(TIMEA(L).NE.TIMEN(I)) GO TO 7766
 7788 SLACKA(L) = UNFORM - TIMEA(L)
      WORK(L) = TIMEA(L) - UTIMEA(L) + SLACKA(L)
      IF(SLACKA(L).LT.0.0) SLACKA(L) = -SETUP
 7799 CONTINUE
      GO TO 7833
C
C     COMPUTE SLACKS FOR ARCS FLOWING INTO NODES HAVING SPECIAL NODE
C         LOGIC, EXCEPT IF THE ESCAPE ARC WAS USED, THEN TREAT LIKE SPLIT
C
 7800 M = NSTORE(NP+NI+NC-1)
      IF(ISTATE(M).GT.1) GO TO 7755
      DO 7822 K=1,NI
      M = NP + K - 1
      L = NSTORE(M)
      IF(ISTATE(L).GT.1) GO TO 7811
      SLACKA(L) = -SETUP
      GO TO 7822
 7811 IF(SLACKA(L).NE.SETUP) GO TO 7822
      M = NSTORE(M+NI)
C
```

```
C         IF THE STARTING TIME CF THIS INPUT ARCS MATTING OUTPUT ARC IS THE
C            SAME AS THE NODE, USE NODE SLACK, ELSE USE OUTPUT ARC SLACK
C
          UNFORM = TIMEN(I)
          IF(ISTATE(M).GT.1) UNFCRM = TIMEA(M) - UTIMEA(M)
          X1 = UNFORM*C.C01
          X2 = UNFORM - X1
          X3 = UNFORM + X1
          X4 = SLACKA(M)
          IF(TIMEN(I).GE.X2.ANC.TIMEN(I).LE.X3) X4 = SLACKN(I)
          UNFORM = UNFCRM + X4
          SLACKA(L) = UNFORM - TIMEA(L)
          WCRK(L) = TIMEA(L) - UTIMEA(L) + SLACKA(L)
          IF(SLACKA(L).LT.0.0) SLACKA(L) = -SETUP
 7822 CONTINUE
 7833 IF(NSTATE(I).EQ.4) NSTATE(I) = 5
 7844 CCNTINUE
          IF(ICHECK) GC TO 7866
          KCOUNT = KCOUNT + 1
          IF(KCOUNT.LE.NNODE) GO TO 7688
          CALL ERRCR (7855,IBLK,IBLK,NWARN)
          SLAK = .TRUE.
          GC TO 7922
C
C         REVIEW MIN, MAX, AVERAGE, COUNT AND TEMPORARILY STORE SLACK DATA
C
 7866 IF(ITRACE.EQ.3) GO TC 7922
          K = 0
 7877 K = K + 1
          I = 0
 7888 I = I + 1
          GO TO (7899,7900),K
 7899 IF(I.GT.NARC) GO TC 7877
          M = ISLAK(I)
          X1 = SLACKA(I)
          GO TO 7911
 7900 IF(I.GT.NNODE) GO TO 7922
          M = NSLAK(I)
          X1 = SLACKN(I)
 7911 IF(M.LE.0.OR.X1.EQ.-SETUP.OR.X1.EQ.+SETUP) GC TO 7888
          IF(X1.LT.RMIN(M)) RMIN(M) = X1
          IF(X1.GT.RMAX(M)) RMAX(M) = X1
          SAVE(M) = SAVE(M) + X1
          JOBS(M) = JCBS(M) + 1
          WRITE (IWF4) M, X1
          NRKS = NRKS + 1
          GO TO 7888
C
C         IF REQUESTED, COMPUTE THE COST-PERFORMANCE TIME INTERVALS
C
 7922 IF(ITRACE.EQ.3.OR.NCPGAP.EQ.0) GO TO 7999
          L = 2*NCPGAP
          DO 7988 I=1,MCPGAP
          K = NCPGAP + I
          WORK(I) = 0.0
          WORK(K) = 0.0
          IF(T1(I) - BTIME)7933,7944,7955
 7933 X1 = C.0
          X2 = 0.0
          GO TO 7966
 7944 X1 = BCCST
          X2 = BPERF
```

```
         GO TO 7966
    7955 CALL OCOST (T1(I))
         X1 = UNFORM
         X2 = TIMESM
    7966 CALL OCOST (T2(I))
         X1 = UNFORM - X1
         X2 = TIMESM - X2
         IF(X1.LE.0.0) GO TO 7977
         IF(X1.LT.CSMIN(I)) CSMIN(I) = X1
         IF(X1.GT.CSMAX(I)) CSMAX(I) = X1
         CAVE(I) = CAVE(I) + X1
         KCOBS(I) = KCOBS(I) + 1
         WORK(I) = X1
    7977 IF(X2.LE.0.0) GO TO 7988
         IF(X2.LT.PSMIN(I)) PSMIN(I) = X2
         IF(X2.GT.PSMAX(I)) PSMAX(I) = X2
         PAVE(I) = PAVE(I) + X2
         KPOBS(I) = KPOBS(I) + 1
    7988 WORK(K) = X2
         WRITE (IWF2) (WORK(I), I=1,L)
   C
   C     IF REQUESTED, MAKE A TRACE OF THIS ITERATION, ARCS FIRST
   C
    7999 IF(ITRACE.EQ.3) GO TO 8033
         IF(NTRIP.LE.0) GO TO 8233
         DO 8022 I=1,NTRIP
         J = ITRIP2(I)
         IF(ITRIP1(I) - 1)8022,8000,8011
    8000 IF(ISTATE(J) - 2)8022,8033,8033
    8011 IF(NSTATE(J).GE.3) GO TO 8033
    8022 CONTINUE
         GO TO 8233
    8033 WRITE (IOUT,8044) ICOUNT, NODE1(NODESM), NODE2(NODESM), JSEED
   8044 FORMAT (7HCITER =, I5, 20H OPT TERMINAL NODE =, 2A4, 12H INIT SEED
        1 =, I12)
         IF(INFC) GO TO 8066
         WRITE (IOUT,8055)
    8055 FORMAT (1H+, 66X, 13HCOST INFLATED)
    8066 IF(DISC) GO TO 8088
         WRITE (IOUT,8077)
    8077 FORMAT (1H+, 81X, 15HCOST DISCOUNTED)
    8088 IF(INFP) GO TO 8100
         WRITE (IOUT,8099)
    8099 FORMAT (1H+, 98X, 13HPERF INFLATED)
    8100 IF(DISP) GO TO 8122
         WRITE (IOUT,8111)
    8111 FORMAT (1H+, 113X, 15HPERF DISCOUNTED)
    8122 WRITE (IOUT,8133) (STORET(ICOUNT,K), K=1,4)
   8133 FORMAT (45H TIME, PATH-COST, OVERALL-COST, PERFORMANCE =, 4F21.4/
        1 1H , 48X, 4HTIME, 30X, 4HCOST, 26X, 11HPERFORMANCE/ 17H ARC NAME
        2STATUS, 8X, 5HSLACK, 3(10X,7HPRIMARY,7X,10HCUMULATIVE))
         DO 8188 I=1,NARC
         IF(ISTATE(I).LE.0.AND.IRROR.EQ.0) GO TO 8188
         IF(ISTATE(I).GT.1) GO TO 8155
         WRITE (IOUT,8144) IARC1(I), IARC2(I), IPATHA(I), ISTATE(I)
    8144 FORMAT (1H , 2A4, 1X, A1, 1X, I3, F15.2, 6F17.2)
         GO TO 8188
    8155 IF(SLACKA(I).EQ.+SETUP.OR.SLACKA(I).EQ.-SETUP.OR.SLAK) GO TO 8166
        0WRITE (IOUT,8144) IARC1(I),IARC2(I),IPATHA(I),ISTATE(I),SLACKA(I),
        1UTIMEA(I), TIMEA(I), UCOSTA(I), COSTA(I), UPERFA(I), PERFA(I)
         GO TO 8188
   8166 WRITE (IOUT,8177) IARC1(I), IARC2(I), IPATHA(I), ISTATE(I),
```

```
      1UTIMEA(I), TIMEA(I), UCOSTA(I), COSTA(I), UPERFA(I), PERFA(I)
 8177 FORMAT (1H , 2A4, 1X, A1, 1X, I3, 15X, 6G17.9)
 8188 CONTINUE
C
C     NOW LIST OUT THE NODES
C
      WRITE (IOUT,8199)
 8199 FORMAT (17H NODE NAME STATUS, 8X, 5HSLACK, 13X, 4HTIME, 13X,
     14HCOST, 6X, 11HPERFORMANCE)
      DO 8222 I=1,NNODE
      IF(NSTATE(I).EQ.0.AND.IRROR.EQ.0) GO TO 8222
      IF(NSTATE(I).GT.2) GO TO 8200
      WRITE (IOUT,8144) NODE1(I), NODE2(I), IPATHN(I), NSTATE(I)
      GO TO 8222
 8200 IF(SLACKN(I).EQ.+SETUP.OR.SLACKN(I).EQ.-SETUP.OR.SLAK) GO TO 8211
     0WRITE (IOUT,8144) NODE1(I), NODE2(I), IPATHN(I), NSTATE(I),
     1SLACKN(I), TIMEN(I), COSTN(I), PERFN(I)
      GO TO 8222
 8211 WRITE (IOUT,8177) NODE1(I), NODE2(I), IPATHN(I), NSTATE(I),
     1TIMEN(I), COSTN(I), PERFN(I)
 8222 CONTINUE
      IF(IRROR.GT.0) RETURN
C
C     IF REQUESTED, STORE INTERNAL NODE POINT AND INTERVAL STATISTICS
C         MINS AND MAXS AND ACCUMULATE FOR FINDING THE AVERAGES
C
 8233 IF(MODE) GO TO 8299
      DO 8288 M=1,NHIST
      I = 0
 8244 I = I + 1
      IF(I.GT.NNODE) GO TO 8288
      IF(ISTAT(I).NE.M) GO TO 8244
      IF(NSTATE(I).EQ.0)GO TO 8288
      CALL OCOST (TIMEN(I))
      WORK(1) = TIMEN(I)
      WORK(2) = COSTN(I)
      WORK(3) = UNFORM
      WORK(4) = PERFN(I)
C
C     CHECK FOR AND IF REQUESTED, COMPUTE INTERVAL (GAP) STATISTICS
C
 8255 I = I + 1
      IF(I.GT.NNODE) GO TO 8266
      IF(ISTAT(I).NE.M) GO TO 8255
      IF(NSTATE(I).EQ.0)GO TO 8288
      CALL OCOST (TIMEN(I))
      WORK(1) = TIMEN(I) - WORK(1)
      WORK(2) = COSTN(I) - WORK(2)
      WORK(3) = UNFORM   - WORK(3)
      WORK(4) = PERFN(I) - WORK(4)
C
C     REVIEW MIN, MAX, AVERAGE, COUNT AND TEMPORARILY STORE HISTO. DATA
C
 8266 DO 8277 K=1,4
      IF(WORK(K).LT.XMIN(M,K)) XMIN(M,K) = WORK(K)
      IF(WORK(K).GT.XMAX(M,K)) XMAX(M,K) = WORK(K)
 8277 HAVE(M,K) = HAVE(M,K) + WORK(K)
      IOBS(M) = IOBS(M) + 1
      IF(ITRACE.EQ.0.OR.ITRACE.EQ.3) GO TO 8288
      WRITE (IWF3) M, WORK(1), WORK(2), WORK(3), WORK(4)
      NRK = NRK + 1
 8288 CONTINUE
```

```
C
C          TALLY THE ITERATION JUST COMPLETED AND CHECK FOR IT BEING THE LAST
C
 8299 ICOUNT = ICOUNT + 1
      IF(ICOUNT.LE.ITER) GO TO 6377
      RETURN
      END
      SUBROUTINE DOARC (I)
      IMPLICIT INTEGER*4(I-N)
     0COMMON EMARK(11),SETUP,FINT,DINT,FACTOR,STIME,SCOST,SPERF,CTIME,
     1CCOST,CPERF,BTIME,BCCST,BP.ERF,TIMESM,UNFORM,ITRIP1(22),ITRIP2(22),
     2KNT(6),INPT,IOUT,IPNH,IWF1,IWF2,IWF3,IWF4,IBLK,ITRACE,ICUT,ISEED,
     3IRROR,ISAVE,NODESM,MAXTAG,ICOUNT,JCOUNT,NTRIP,ICORR,LCOMP,NWARN,
     4NRK,NRKS,SLAK,MODE,INFC,INFP,DISC,DISP,CRIT,KGEN,MED,SINGLE,DOUBLE
     0COMMON/ARCS/ASTORE( 2800),UTIMEA( 350),TIMEA( 350),UCOSTA( 350),
     1COSTA( 350),UPERFA( 350),PERFA( 350),WORK( 350),ISTATE( 350),
     2NODEI( 350),NODEO( 350),ICRITA( 350),KEEPC( 350),KEEPP( 350),
     3IARC1( 350),IARC2( 350),IPOINT( 350),JPOINT( 350),ISLAK( 350),
     4KARC,LARC,MARC,NARC,ITALC,ITALP,ISTAR
     0COMMON/NODES/TIMEN( 200),COSTN( 200),PERFN( 2C0),NSTORE( 5400),
     1NODE1( 2C0),NODE2( 2CC),LOGI( 200),LOGO( 200),NSTATE( 200),
     2NARCI( 200),NARCO( 2CC),ISTAT( 200),INSTAT( 200),ICRITN( 200),
     3NPCINT( 200),NSLAK( 200),JUMP( 200),KNODE,LNODE,MNODE,NNODE,MTAG,
     4NTAG
     0COMMON/TABLE/TAB1(3,3,3),TAB2(3,3,3),TAB3(3,3,3),TAB4(3,3,3),
     1TAB5(3,3,3),TAB6(3,3,3),TAB7(3,3,3),LXCK(7),MXCK(7),LYCK(7),
     2MYCK(7),LZCK(7),MZCK(7)
      DIMENSION IFIRE(3)
      LOGICAL SLAK,MODE,INFC,INFP,DISC,DISP,CRIT,KGEN,MED,SINGLE,DOUBLE
C
C         THIS SUBROUTINE CALCLLATES TIME COST PERFORMANCE FOR ARCS AND
C             DETERMINES THEIR SUCCESS-FAIL STATUS, INITIALIZE
C
      UTIMEA(I) = C.C
      UCOSTA(I) = C.C
      UPERFA(I) = 0.0
      J = IPOINT(I)
      IF(J.LE.C) GO TO 9333
      IF(ASTCRE(J+1).LE.C.C) GO TO 9333
      K = ASTCRE(J+1) + 0.CO1
      IFIRE(1) = K/100
      L = IFIRE(1)*100
      K = K - L
      IFIRE(2) = K/10
      IFIRE(3) = K - IFIRE(2)*10
C
C        GENERATE THE TIME COST PERFORMANCE VALUES FOR THIS ARC
C
      IDSC = 0
      IDSP = 0
      DO 9322 NN=1,3
      DO 9311 K=1,3
      IF(IFIRE(K).NE.NN) GC TO 9311
      ENDX = EMARK(10) - FLCAT(2*K)
      CALL SEEK (J,ENDX)
      RANVAR = 0.0
      IF(ITALP.EC.0) GO TO 8633
      IF(INFC.AND.INFP.AND.DISC.AND.DISP) GC TO 8311
      L = ASTORE(ITALC+ITALP) + 0.C01
      IF(ENDX.EC.EMARK(4)) IDSC = L
      IF(ENDX.EC.EMARK(6)) IDSP = L
C
```

```
C        GENERATE THE STOCHASTIC CONTRIBUTION
C
 8311 MMM = 0
      M = ASTORE(ITALC+1) + 0.001
      I1 = ITALC + 2
      I2 = ITALC + 3
      I3 = ITALC + 4
      I4 = ITALC + 5
 8322 MMM = MMM + 1
      IF(MMM.LT.1000) GO TO 8344
      CALL ERROR (8333,IARC1(I),IARC2(I),M)
      RETURN
 83440GO TO (8399,8400,8411,8444,8444,8455,8477,8455,8488,8500,8511,
     18455,8555,8577,8355),M
C
C        CREATE A VARIATE FROM A DISTRIBUTION ENTERED AS A HISTOGRAM
C
 8355 CALL RANDOM
      PROB1 = 0.0
      I3 = ITALC + ITALP - 1
      DO 8366 L=I2,I3,2
      MM = L
      PROB2 = PROB1 + ASTORE(MM)
      IF(UNFORM.LE.PROB2) GO TO 8377
 8366 PROB1 = PROB2
      GO TO 8388
 8377 PROB1 = (UNFORM - PROB1)/(PROB2 - PROB1)
 8388 UNFORM = ASTORE(MM-1) + (ASTORE(MM+1) - ASTORE(MM-1))*PROB1
      GO TO 8622
C
C        CREATE A CONSTANT VARIATE
C
 8399 UNFORM = ASTORE(I1)
      GO TO 8622
C
C        CREATE A UNIFORM VARIATE
C
 8400 CALL RANDOM
      UNFORM = ASTORE(I1) + (ASTORE(I2) - ASTORE(I1))*UNFORM
      GO TO 8622
C
C        CREATE A TRIANGULAR VARIATE
C
 8411 PROB1 = ASTORE(I2) - ASTORE(I1)
      TIMESM = (ASTORE(I3) - ASTORE(I1))/PROB1
      CALL RANDOM
      IF(UNFORM.GT.TIMESM) GO TO 8422
      TIMESM = SQRT(TIMESM*UNFORM)
      GO TO 8433
 8422 TIMESM = 1.0 - SQRT(1.0 - TIMESM - UNFORM + TIMESM*UNFORM)
 8433 UNFORM = ASTORE(I1) + TIMESM*PROB1
      GO TO 8622
C
C        CREATE A NORMAL OR LOGNORMAL-(M.EQ.5) VARIATE
C
 8444 CALL NORM
      UNFORM = UNFORM*ASTORE(I4) + ASTORE(I3)
      IF(M.EQ.5) UNFORM = EXP(UNFORM)
      GO TO 8611
C
C        CREATE A GAMMA(M.EQ.6), ERLANG(M.EQ.8) OR PASCAL(M.EQ.12)
C
```

```
 8455 ISAVE = ASTORE(I4) + C.001
      IF(M.GT.6) GO TO 8466
      TIMESM = ISAVE
      TIMESM = ASTORE(I4) - TIMESM
      CALL RANDOM
      IF(UNFORM.GT.TIMESM) GO TO 8466
      ISAVE = ISAVE + 1
 8466 CALL GAM
      UNFORM = TIMESM/ASTORE(I3)
      IF(M.EQ.12) GO TO 8544
      GO TO 8611
C
C     CREATE A WEIBULL VARIATE
C
 8477 CALL RANDOM
      UNFORM = ASTORE(I1) + ASTORE(I3)*((-ALOG(UNFORM))**ASTORE(I4))
      GO TO 8611
C
C     CREATE A CHI SQUARE VARIATE
C
 8488 PROB1 = 0.0
      IF(ASTORE(I3).GT.0.0) GO TO 8499
      CALL NORM
      PROB1 = UNFORM*UNFORM
 8499 ISAVE = ABS(ASTORE(I3)) + 0.001
      CALL GAM
      UNFORM = TIMESM/0.5 + PROB1
      GO TO 8611
C
C     CREATE A BETA VARIATE
C
 8500 ISAVE = ASTORE(I3) + C.5
      CALL GAM
      PROB1 = TIMESM
      ISAVE = ASTORE(I4) + C.5
      CALL GAM
      UNFORM = PROB1/(PROB1 + TIMESM)
      UNFORM = UNFORM*(ASTORE(I2) - ASTORE(I1)) + ASTORE(I1)
      GO TO 8611
C
C     CREATE A POISSON VARIATE
C
 8511 ISAVE = 0
      TIMESM = 1.0
 8522 CALL RANDOM
      TIMESM = TIMESM*UNFORM
      IF(TIMESM.LT.ASTORE(I3)) GO TO 8533
      ISAVE = ISAVE + 1
      GO TO 8522
 8533 UNFORM = ISAVE
      GO TO 8611
C
C     CREATE A PASCAL VARIATE
C
 8544 ISAVE = UNFORM
      UNFORM = ISAVE
      GO TO 8611
C
C     CREATE A BINOMIAL VARIATE
C
 8555 ISAVE = 0
      MM = ASTORE(I4) + 0.001
```

```
      DO 8566 L=1,MM
      CALL RANDOM
      IF(UNFORM.GT.ASTORE(I3)) GO TO 8566
      ISAVE = ISAVE + 1
 8566 CONTINUE
      UNFORM = ISAVE
      GO TO 8611
C
C     CREATE A HYPERGEOMETRIC VARIATE
C
 8577 PROB1 = ASTORE(I3)
      PROB2 = ASTORE(I4)
      MM = ASTORE(ITALC+6) + 0.001
      ISAVE = 0
      DO 8600 L=1,MM
      CALL RANDOM
      IF(UNFORM.GT.PROB1) GO TO 8588
      TIMESM = 1.0
      ISAVE = ISAVE + 1
      GO TO 8599
 8588 TIMESM = 0.0
 8599 PROB1 = (PROB2*PROB1 - TIMESM)/(PROB2 - 1.0)
 8600 PROB2 = PROB2 - 1.0
      UNFORM = ISAVE
C
C     PERFORM CHECKS THEN SUM NEW DEVIATE INTO APPROPRIATE TCP VECTOR
C
 8611 IF(UNFORM.LT.ASTORE(I1).OR.UNFORM.GT.ASTORE(I2)) GO TO 8322
 8622 IF(K.EQ.1) UTIMEA(I) = UTIMEA(I) + UNFORM
      IF(K.EQ.2) UCOSTA(I) = UCOSTA(I) + UNFORM
      IF(K.EQ.3) UPERFA(I) = UPERFA(I) + UNFORM
      RANVAR = UNFORM
 8633 ENDX = ENDX - 1.0
      CALL SEEK (J,ENDX)
      IF(ITALP.EQ.0) GO TO 9311
      IF(INFC.AND.INFP.AND.DISC.AND.DISP) GO TO 8644
      L = ASTORE(ITALC+ITALP) + 0.001
      IF(ENDX.EQ.EMARK(5)) IDSC = L
      IF(ENDX.EQ.EMARK(7)) IDSP = L
C
C     GENERATE THE VARIABLE CONTRIBUTION
C
 8644 TIMESM = 0.0
      ITALP = ITALC + ITALP - 1
      ITALC = ITALC + 9
      L = 0
      ISAVE = 0
      PROB1 = SETUP
      DO 9300 N=ITALC,ITALP,9
      IF(ISAVE.EQ.1) GO TO 8655
      IF(ASTORE(N-7).EQ.2.0) GO TO 9300
      IF(ASTORE(N-7).EQ.-2.0) GO TO 9300
 8655 DO 8744 I1=1,3
      I2 = ASTORE(N-7+2*I1) + 0.001
      I3 = N - 6 + 2*I1
      I4 = ABS(ASTORE(I3)) + 0.001
      IF(I2 - 4)8666,8722,8711
C
C     LOAD TIME, COST OR PERFORMANCE FROM A NODE OR AN ARC
C
 8666 IF(ASTORE(I3).GT.0.0) GO TO 8677
      IF(NSTATE(I4).NE.3) GO TO 8688
```

```
      IF(I2.EQ.1) UNFORM = TIMEN(I4)
      IF(I2.EQ.2) UNFORM = COSTV(I4)
      IF(I2.EQ.3) UNFORM = PERFN(I4)
      GO TO 8733
 8677 IF(I4.EQ.I.OR.ISTATE(I4).GE.2) GO TO 8700
 8688 ISAVE = 1
      IF(ASTORE(N-6).NE.0.C) GO TO 9300
      CALL ERROR (8699,IARC1(I),IARC2(I),IBLK)
      RETURN
 8700 IF(I2.EQ.1) UNFORM = UTIMEA(I4)
      IF(I2.EQ.2) UNFORM = UCC
      IF(I2.EQ.3) UNFORM = UPERFA(I4)
      GO TO 8733
C
C     LOAD THE RESULTS OF A PREVIOUS TRANSFORMATION OR LOAD A CONSTANT,
C         THEN LOAD TRANSFCRMATION VARIABLES X, Y AND Z
C
 8711 IF(I4.GT.0) GO TO 8720
      UNFORM = RANVAR
      GO TO 8733
 8720 UNFORM = WORK(I4)
      GC TO 8733
 8722 UNFORM = ASTCRE(I3)
 8733 IF(I1.EQ.1) X = UNFORM
      IF(I1.EQ.2) Y = UNFORM
 8744 IF(I1.EQ.3) Z = UNFORM
C
C     COMPUTE THE VALUE OF THIS TRANSFORMATION
C
      MM = ASTCRE(N-8) + C.001
      M = MM
      IF(M.LE.50) GO TO 8755
      M = M - 50
 8755 IF(M.GT.37) GO TO 9144
      GGO TO(8788,8799,88CO,8811,8822,8833,8844,8855,8866,8877,8888,8899,
     18900,8911,8911,8922,8933,8944,8955,8966,8977,8988,8999,9000,9011,
     29022,9C33,9044,9055,9C66,9077,9088,9C99,9100,9111,9122,9133),M
 8766 CALL ERROR (8777,IARC1(I),IARC2(I),MM)
      RETURN
 8788 UNFORM = X*Y*Z
      GO TO 9288
 8799 IF(Z.EQ.C.C) GO TO 8766
      UNFORM = (X*Y)/Z
      GO TO 9288
 88CO UNFORM = Y*Z
      IF(UNFORM.EQ.C.O) GO TO 8766
      UNFORM = X/UNFORM
      GO TO 9288
 8811 UNFORM = X*Y*Z
      UNFORM = 1.C/UNFORM
      GO TO 9288
 8822 UNFORM = X+Y+Z
      GC TO 9288
 8833 UNFORM = X+Y-Z
      GO TO 9288
 8844 UNFORM = X-Y-Z
      GC TO 9288
 8855 UNFORM = -X-Y-Z
      GO TO 9288
 8866 UNFORM = X*(Y+Z)
      GC TO 9288
 8877 UNFORM = X*(Y-Z)
```

```
      GC TO 9288
8888  UNFORM = Y+Z
      IF(UNFORM.EQ.0.0) GC TO 8766
      UNFORM = X/UNFORM
      GC TO 9288
8899  UNFORM = Y-Z
      IF(UNFORM.EQ.0.0) GO TO 8766
      UNFORM = X/UNFORM
      GC TO 9288
8900  IF(Y.LE.0.0) GO TC 8766
      UNFORM = X*(Y)**Z
      GC TO 9288
8911  UNFORM = Y*Z
      IF(UNFORM.LE.0.0) GO TO 8766
      IF(M.EQ.14) UNFORM = X*ALJG(UNFORM)
      IF(M.EQ.15) UNFORM = X*ALOG10(UNFORM)
      GO TO 9288
8922  UNFORM = X*SIN(Y*Z)
      GC TO 9288
8933  UNFORM = X*COS(Y*Z)
      GO TO 9288
8944  UNFORM = X*ATAN(Y*Z)
      GO TO 9288
8955  PROB1 = X
      PROB2 = Y
      UNFORM = 0.0
      GC TO 9288
8966  IF(PROB1.EQ.SETUP) GC TO 8766
      IF(PROB1.GE.PROB2) UNFORM = X
      IF(PROB1.LT.PROB2) UNFORM = Y
      GO TO 9288
8977  UNFORM = X
      IF(Y.LT.UNFORM) UNFORM = Y
      IF(Z.LT.UNFORM) UNFORM = Z
      GC TO 9288
8988  UNFORM = X
      IF(Y.GT.UNFORM) UNFORM = Y
      IF(Z.GT.UNFORM) UNFORM = Z
      GO TO 9288
8999  UNFORM = (X*Y)+Z
      GC TO 9288
9000  UNFORM = (X*Y)-Z
      GO TO 9288
9011  IF(Y.EQ.0.0) GO TO 8766
      UNFORM = (X/Y)+Z
      GO TO 9288
9022  IF(Y.EQ.0.0) GO TO 8766
      UNFORM = (X/Y)-Z
      GO TO 9288
9033  UNFORM = (X+Y)*Z
      GO TO 9288
9044  IF(Z.EQ.0.0) GO TO 8766
      UNFORM = (X+Y)/Z
      GO TO 9288
9055  UNFORM = (X-Y)*Z
      GO TO 9288
9066  IF(Z.EQ.0.0) GO TO 8766
      UNFORM = (X-Y)/Z
      GC TO 9288
9077  UNFORM = X + (Y*Z)
      GC TO 9288
9088  UNFORM = X - (Y*Z)
```

```
      GO TO 9288
 9099 IF(Z.EQ.0.0) GO TO 8766
      UNFORM = X + (Y/Z)
      GO TO 9288
 9100 IF(Z.EQ.0.0) GO TO 8766
      UNFORM = X - (Y/Z)
      GO TO 9288
 9111 UNFORM = -X -Y + Z
      GO TO 9288
 9122 UNFORM = -X +Y + Z
      GO TO 9288
 9133 IF(Y.EQ.0.0.OR.Z.EQ.C.0) GO TO 8766
      UNFORM = X/Y/Z
      GO TO 9288
C
C     INTEGERIZE THE TABLE LOOK-UP INDICES, THEN DO THE LOOK-UP
C
 9144 IX = X + 0.5
      IY = Y + 0.5
      IZ = Z + 0.5
      M = M - 37
      IF(M.LE.7) GC TO 9147
      UNFORM = 0.0
      GO TO 9150
 9147 IF(IX.LT.LXCK(M).OR.IY.LT.LYCK(M).OR.IZ.LT.LZCK(M)) GO TO 8766
      IF(IX.GT.MXCK(M).OR.IY.GT.MYCK(M).OR.IZ.GT.MZCK(M)) GO TO 8766
 9150 GO TO (9155,9166,9177,9188,9199,9200,9211,9222,9233,9244,9255,
     19266,9277),M
 9155 UNFORM = TAB1(IX,IY,IZ)
      GO TO 9288
 9166 UNFORM = TAB2(IX,IY,IZ)
      GO TO 9288
 9177 UNFORM = TAB3(IX,IY,IZ)
      GC TO 9288
 9188 UNFORM = TAB4(IX,IY,IZ)
      GO TO 9288
 9199 UNFORM = TAB5(IX,IY,IZ)
      GC TO 9288
 9200 UNFORM = TAB6(IX,IY,IZ)
      GO TO 9288
 9211 UNFORM = TAB7(IX,IY,IZ)
      GO TO 9288
 9222 CALL SUB1(X,Y,Z,UNFORM,KNT(1))
      GO TO 9288
 9233 CALL SUB2(X,Y,Z,UNFORM,KNT(2))
      GO TO 9288
 9244 CALL SUB3(X,Y,Z,UNFORM,KNT(3))
      GC TO 9288
 9255 CALL SUB4(X,Y,Z,UNFORM,KNT(4))
      GC TO 9288
 9266 CALL SUB5(X,Y,Z,UNFORM,KNT(5))
      GO TO 9288
 9277 CALL SUB6(X,Y,Z,UNFORM,KNT(6))
C
C     SHOULD THE RESULTS CF THIS TRANSFORMATION BE SUMED INTO THE TIME,
C        COST OR PERFORMANCE VALUE BEING GENERATED FOR THIS ARC.
C
 9288 IF(MM.LE.50) GO TO 9299
      MMM = UNFORM
      UNFORM = MMM
 9299 IF(ASTORE(N-7).GT.0.0) TIMESM = TIMESM + UNFORM
      L = L + 1
```

```
      ISAVE = 0
      WORK(L) = UNFORM
 9300 CONTINUE
C
C     STORE THE VARIABLE CONTRIBUTION
C
      IF(K.EQ.1) UTIMEA(I) = UTIMEA(I) + TIMESM
      IF(K.EQ.2) UCOSTA(I) = UCOSTA(I) + TIMESM
      IF(K.EQ.3) UPERFA(I) = UPERFA(I) + TIMESM
 9311 CONTINUE
 9322 CONTINUE
      IF(NODEI(I).EQ.0) GO TO 9400
C
C     STORE THE CUMULATIVE VALUES AND DETERMINE THIS ARC'S STATUS
C
 9333 N = NODEI(I)
      UNFORM = TIMEN(N)
      IF(LOGI(N).LT.5) GO TO 9344
      M = NSTORE(NPOINT(N)+NARCI(N)+NARCO(N)-1)
      IF(M.EQ.I) GO TO 9344
      IF(TIMEA(I).NE.SETUP) UNFORM = TIMEA(I)
 9344 TIMEA(I) = UTIMEA(I) + UNFORM
      IF(INFC.AND.INFP) GO TO 9377
      IF(IDSC - 2)9355,9366,9366
 9355 IF(IDSP - 2)9377,9366,9366
 9366 PROB1 = FINT**((UTIMEA(I)/2.0 + UNFORM - BTIME)/FACTOR)
      IF(IDSC.GE.2) UCOSTA(I) = UCOSTA(I)*PROB1
      IF(IDSP.GE.2) UPERFA(I) = UPERFA(I)*PROB1
 9377 IF(DISC.AND.DISP) GO TO 9399
      IF(IDSC.EQ.1) GO TO 9388
      IF(IDSC.EQ.3) GO TO 9388
      IF(IDSP.EQ.1) GO TO 9388
      IF(IDSP.NE.3) GO TO 9399
 9388 PROB1 = DINT**((UTIMEA(I)/2.0 + UNFORM - BTIME)/FACTOR)
      IF(IDSC.EQ.1.OR.IDSC.EQ.3) UCOSTA(I) = UCOSTA(I)/PROB1
      IF(IDSP.EQ.1.OR.IDSP.EQ.3) UPERFA(I) = UPERFA(I)/PROB1
 9399 COSTA(I) = UCOSTA(I) + COSTN(N)
      PERFA(I) = UPERFA(I) + PERFN(N)
      IF(J.LE.0) GO TO 9400
      IF(ASTORE(J).EQ.1.0) GO TO 9400
      CALL RANDOM
      IF(UNFORM.LE.ASTORE(J)) GO TO 9400
      ISTATE(I) = 2
      RETURN
 9400 ISTATE(I) = 3
      RETURN
      END
      SUBROUTINE SVECT
      IMPLICIT INTEGER*4(I-N)
     0COMMON EMARK(11),SETUP,FINT,DINT,FACTOR,STIME,SCOST,SPERF,CTIME,
     1CCOST,CPERF,BTIME,BCOST,BPERF,TIMESM,UNFORM,ITRIP1(22),ITRIP2(22),
     2KNT(6),INPT,IOUT,IPNH,IWF1,IWF2,IWF3,IWF4,IBLK,ITRACE,ICUT,ISEED,
     3IRROR,ISAVE,NODESM,MAXTAG,ICOUNT,JCOUNT,NTRIP,ICORR,LCOMP,NWARN,
     4NRK,NRKS,SLAK,MODE,INFC,INFP,DISC,DISP,CRIT,KGEN,MED,SINGLE,DOUBLE
     0COMMON/ARCS/ASTORE( 2800),UTIMEA( 350),TIMEA( 350),UCOSTA( 350),
     1COSTA( 350),UPERFA( 350),PERFA( 350),WORK( 350),ISTATE( 350),
     2NODEI( 350),NODEO( 350),ICRITA( 350),KEEPC( 350),KEEPP( 350),
     3IARC1( 350),IARC2( 350),IPOINT( 350),JPOINT( 350),ISLAK( 350),
     4KARC,LARC,MARC,NARC,ITALC,ITALP,ISTAR
     0COMMON/NODES/TIMEN( 200),COSTN( 200),PERFN( 200),NSTORE( 5400),
     1NODE1( 200),NODE2( 200),LOGI( 200),LOGO( 200),NSTATE( 200),
     2NARCI( 200),NARCO( 200),ISTAT( 200),INSTAT( 200),ICRITN( 200),
```

```
     3NPOINT( 200),NSLAK( 200),JUMP( 200),KNODE,LNODE,MNODE,NNODE,MTAG,
     4NTAG
      LOGICAL SLAK,MODE,INFC,INFP,DISC,DISP,CRIT,KGEN,MED,SINGLE,DOUBLE
C
C     THE FUNCTION THIS SUBROUTINE PERFORMS IS THAT OF STORING ARC
C         ADDRESSES USED IN CALCULATING NODE COSTS AND PERFORMANCES
C
      ISTAR = J
      ISUM = ITALC + ITALP + 2
      IF(ISUM.LE.2) RETURN
      I = ISUM + NTAG
      IF(I.LE.MTAG) GO TO 9455
C
C     THIS SEGMENT REORGANIZES THE AREA OF NSTORE DEDICATED TO STORING
C         ARC ADDRESSES USED IN CALCULATING NODE COSTS AND PERFORMANCES
C
      ISTAR = KNODE + 1
      IEND = NTAG
      NTAG = KNODE
      KEY = 2
      DO 9433 I=ISTAR,IEND
      DO 9411 J=1,NARC
      IF(ISTATE(J).LE.1) GO TO 9411
      IF(I.NE.JPOINT(J)) GO TO 9411
      JPOINT(J) = NTAG + 1
      KEY = 1
 9411 CONTINUE
      IF(NSTORE(I).EQ.-100) GO TO 9433
      GO TO (9422,9433),KEY
 9422 NTAG = NTAG + 1
      NSTORE(NTAG) = NSTORE(I)
 9433 IF(NSTORE(I).EQ.-992) KEY = 2
      I = ISUM + NTAG
      IF(I.LE.MTAG) GO TO 9455
      CALL ERROR (9444,IBLK,IBLK,IBLK)
      RETURN
C
C     LOAD THE VECTOR OF ARC ADDRESSES
C
 9455 ISTAR = NTAG + 1
      IF(ITALC.LE.0) GO TO 9477
      DO 9466 I=1,ITALC
      NTAG = NTAG + 1
 9466 NSTORE(NTAG) = KEEPC(I)
 9477 NTAG = NTAG + 1
      NSTORE(NTAG) = -991
      IF(ITALP.LE.0) GO TO 9499
      DO 9488 I=1,ITALP
      NTAG = NTAG + 1
 9488 NSTORE(NTAG) = KEEPP(I)
 9499 NTAG = NTAG + 1
      NSTORE(NTAG) = -992
      RETURN
      END
      SUBROUTINE INITAL(L)
      IMPLICIT INTEGER*4(I-N)
      0COMMON/ARCS/ASTORE( 2800),UTIMEA( 350),TIMEA( 350),UCOSTA( 350),
     1COSTA( 350),UPERFA( 350),PERFA( 350),WORK( 350),ISTATE( 350),
     2NODEI( 350),NODEO( 350),ICRITA( 350),KEEPC( 350),KEEPP( 350),
     3IARC1( 350),IARC2( 350),IPOINT( 350),JPOINT( 350),ISLAK( 350),
     4KARC,LARC,MARC,NARC,ITALC,ITALP,ISTAR
      0COMMON/NODES/TIMEN( 200),COSTN( 200),PERFN( 200),NSTORE( 5400),
```

```
     1NODE1( 200),NODE2( 200),LOGI( 200),LOGO( 200),NSTATE( 200),
     2NARCI( 200),NARCO( 200),ISTAT( 200),INSTAT( 200),ICRITN( 200),
     3NPOINT( 200),NSLAK( 200),JUMP( 200),KNODE,LNODE,MNODE,NNODE,MTAG,
     4NTAG
C
C     INITAL INITIALIZES AN ARC AND ITS OUTPUT NODE, IF THIS ARC WAS
C        PROCESSED AND IS NOW BEING CONSIDERED FOR REPROCESSING, ELIMI-
C        NATE IT FROM THE COST-PERFORMANCE CALCULATIONS VECTOR
C
      M = NODEO(L)
      IF(ISTATE(L).NE.-1) NSTATE(M) = 2
      IF(ISTATE(L).EQ.1) GO TO 1122
      M = KNODE + 1
      DO 1111 I=M,NTAG
 1111 IF(NSTORE(I).EQ.L) NSTORE(I) = -100
 1122 JPOINT(L) = 0
      ISTATE(L) = 0
      RETURN
      END
      SUBROUTINE DVECT (KEY,IFIRST,L,XSAVE,TSAVE)
      IMPLICIT INTEGER*4(I-N)
      COMMON EMARK(11),SETUP,FINT,DINT,FACTOR,STIME,SCOST,SPERF,CTIME,
     1CCOST,CPERF,BTIME,BCOST,BPERF,TIMESM,UNFORM,ITRIP1(22),ITRIP2(22),
     2KNT(6),INPT,IOUT,IPNH,IWF1,IWF2,IWF3,IWF4,IBLK,ITRACE,ICUT,ISEED,
     3IRROR,ISAVE,NODESM,MAXTAG,ICCUNT,JCOUNT,NTRIP,ICORR,LCOMP,NWARN,
     4NRK,NRKS,SLAK,MODE,INFC,INFP,DISC,DISP,CRIT,KGEN,MED,SINGLE,DOUBLE
      COMMON/ARCS/ASTORE( 2800),UTIMEA( 350),TIMEA( 350),UCOSTA( 350),
     1COSTA( 350),UPERFA( 350),PERFA( 350),WORK( 350),ISTATE( 350),
     2NODEI( 350),NODEO( 350),ICRITA( 350),KEEPC( 350),KEEPP( 350),
     3IARC1( 350),IARC2( 350),IPOINT( 350),JPOINT( 350),ISLAK( 350),
     4KARC,LARC,MARC,NARC,ITALC,ITALP,ISTAR
      COMMON/NODES/TIMEN( 200),COSTN( 200),PERFN( 200),NSTORE( 5400),
     1NODE1( 200),NODE2( 200),LOGI( 200),LOGO( 200),NSTATE( 200),
     2NARCI( 200),NARCO( 200),ISTAT( 200),INSTAT( 200),ICRITN( 200),
     3NPOINT( 200),NSLAK( 200),JUMP( 200),KNODE,LNODE,MNODE,NNODE,MTAG,
     4NTAG
      LOGICAL SLAK,MODE,INFC,INFP,DISC,DISP,CRIT,KGEN,MED,SINGLE,DOUBLE
      INTEGER*2KEY
C
C     THIS SUBROUTINE PRUNES ARCS AND DEVELOPS COST (FIRST HALF) AND
C        PERFORMANCE (SECOND HALF) VECTORS (SEE SUBROUTINE HELPDV)
C
      GO TO (1133,1211),KEY
 1133 IF(ICUT.GT.1) GO TO 1144
      IF(TIMEA(L)-UTIMEA(L).LT.TSAVE) GO TO 1144
      ISTATE(L) = -1
      GO TO 1155
 1144 IF(UCOSTA(L).NE.0.0) CALL HELPDV (1,L,TSAVE,XSAVE)
 1155 I = JPOINT(L)
      XSAVE = XSAVE + BCOST
      IF(I.LE.0) RETURN
 1166 K = NSTORE(I)
      IF(K.EQ.-100) GO TO 1200
      IF(K.EQ.-991) RETURN
      IF(ITALC.EQ.0.OR.IFIRST.EQ.1) GO TO 1188
      DO 1177 J=1,ITALC
      IF(K.EQ.KEEPC(J)) GO TO 1200
 1177 CONTINUE
 1188 IF(ICUT.GT.1) GO TO 1199
      IF(TIMEA(K)-UTIMEA(K).LT.TSAVE) GO TO 1199
      ISAVE = 1
      CALL INITAL(K)
```

```
      GO TO 1200
 1199 CALL HELPDV (1,K,TSAVE,XSAVE)
 1200 I = I + 1
      GO TO 1166
 1211 IF(ICUT.GT.1) GO TO 1222
      IF(TIMEA(L)-UTIMEA(L).LT.TSAVE) GO TO 1222
      ISTATE(L) = -1
      GO TO 1233
 1222 IF(UPERFA(L).NE.0.0) CALL HELPDV (2,L,TSAVE,XSAVE)
 1233 I = JPOINT(L)
      XSAVE = XSAVE + BPERF
      IF(I.LE.0) RETURN
 1244 IF(NSTORE(I).EQ.-991) GO TO 1299
      I = I + 1
      GO TO 1244
 1255 K = NSTORE(I)
      IF(K.EQ.-100) GO TO 1299
      IF(K.EQ.-992) RETURN
      IF(ITALP.EQ.0.OR.IFIRST.EQ.1) GO TO 1277
      DO 1266 J=1,ITALP
      IF(K.EQ.KEEPP(J)) GO TO 1299
 1266 CONTINUE
 1277 IF(ICUT.GT.1) GO TO 1288
      IF(TIMEA(K)-UTIMEA(K).LT.TSAVE) GO TO 1288
      ISAVE = 1
      CALL INITAL(K)
      GO TO 1299
 1288 CALL HELPDV (2,K,TSAVE,XSAVE)
 1299 I = I + 1
      GO TO 1255
      END
      SUBROUTINE HELPDV (KEY,L,TSAVE,XSAVE)
      IMPLICIT INTEGER*4(I-N)
     0COMMON EMARK(11),SETUP,FINT,DINT,FACTOR,STIME,SCOST,SPERF,CTIME,
     1CCOST,CPERF,BTIME,BCOST,BPERF,TIMESM,UNFORM,ITRIP1(22),ITRIP2(22),
     2KNT(6),INPT,IOUT,IPNF,IWF1,IWF2,IWF3,IWF4,IBLK,ITRACE,ICUT,ISEED,
     3IRROR,ISAVE,NODESM,MAXTAG,ICOUNT,JCOUNT,NTRIP,ICORR,LCOMP,NWARN,
     4NRK,NRKS,SLAK,MODE,INFC,INFP,DISC,DISP,CRIT,KGEN,MED,SINGLE,DOUBLE
     0COMMON/ARCS/ASTORE( 2800),UTIMEA( 350),TIMEA( 350),UCOSTA( 350),
     1COSTA( 350),UPERFA( 350),PERFA( 350),WORK( 350),ISTATE( 350),
     2NODEI( 350),NODEO( 350),ICRITA( 350),KEEPC( 350),KEEPP( 350),
     3IARC1( 350),IARC2( 350),IPOINT( 350),JPOINT( 350),ISLAK( 350),
     4KARC,LARC,MARC,NARC,ITALC,ITALP,ISTAR
     0COMMON/NODES/TIMEN( 200),COSTN( 200),PERFN( 200),NSTORE( 5400),
     1NODE1( 200),NODE2( 200),LOGI( 200),LOGO( 200),NSTATE( 200),
     2NARCI( 200),NARCO( 200),ISTAT( 200),INSTAT( 200),ICRITN( 200),
     3NPOINT( 200),NSLAK( 200),JUMP( 200),KNODE,LNODE,MNODE,NNODE,MTAG,
     4NTAG
      LOGICAL SLAK,MODE,INFC,INFP,DISC,DISP,CRIT,KGEN,MED,SINGLE,DOUBLE
C
C     THIS SUBROUTINE ADJUST ARC UNIT COST AND PERFORMANCES FOR PARTIAL
C         VALUES FOR THOSE ARCS WHOSE COMPLETION TIME EXCEEDS THE CUTOFF
C         TIME.  IT ALSO CREATES VECTORS OF COST AND PERFORMANCE ELEMENTS
C         OF ARCS REQUIRED TO BE COMPLETED BEFORE PROCESSING THE ARC
C         UNDER CONSIDERATION (COST FIRST HALF AND PERFORMANCE SECOND
C         HALF)
C
      GO TO (1300,1322),KEY
 1300 IF(ICUT.EQ.0.OR.ICUT.EQ.2.OR.TIMEA(L).LE.TSAVE) GO TO 1311
      COSTA(L) = COSTA(L) - UCOSTA(L)
      UCOSTA(L) = UCOSTA(L)*(1.0 - ((TIMEA(L) - TSAVE)/UTIMEA(L)))
      TIMEA(L) = TSAVE
```

```
      COSTA(L) = CCSTA(L) + UCOSTA(L)
 1311 ITALC = ITALC + 1
      KEEPC(ITALC) = L
      XSAVE = XSAVE + UCCSTA(L)
      RETURN
 1322 IF(ICUT.EQ.0.OR.ICUT.EQ.2.OR.TIMEA(L).LE.TSAVE) GO TO 1333
      PERFA(L) = PERFA(L) - UPERFA(L)
      UPERFA(L) = UPERFA(L)*(1.0 - ((TIMEA(L) - TSAVE)/UTIMEA(L)))
      TIMEA(L) = TSAVE
      PERFA(L) = PERFA(L) + UPERFA(L)
 1333 ITALP = ITALP + 1
      KEEPP(ITALP) = L
      XSAVE = XSAVE + UPERFA(L)
      RETURN
      END
      SUBROUTINE OCOST (TCRIT)
      IMPLICIT INTEGER*4(I-N)
      COMMON EMARK(11),SETUP,FINT,DINT,FACTOR,STIME,SCOST,SPERF,CTIME,
     1CCOST,CPERF,BTIME,BCCST,BPERF,TIMESM,UNFORM,ITRIP1(22),ITRIP2(22),
     2KNT(6),INPT,IOUT,IPNH,IWF1,IWF2,IWF3,IWF4,IBLK,ITRACE,ICUT,ISEED,
     3IRROR,ISAVE,NODESM,MAXTAG,ICOUNT,JCOUNT,NTRIP,ICORR,LCOMP,NWARN,
     4NRK,NRKS,SLAK,MODE,INFC,INFP,DISC,DISP,CRIT,KGEN,MED,SINGLE,DOUBLE
      COMMON/ARCS/ASTORE( 2800),UTIMEA( 350),TIMEA( 350),UCOSTA( 350),
     1COSTA( 350),UPERFA( 350),PERFA( 350),WORK( 350),ISTATE( 350),
     2NODEI( 350),NODEO( 350),ICRITA( 350),KEEPC( 350),KEEPP( 350),
     3IARC1( 350),IARC2( 350),IPOINT( 350),JPOINT( 350),ISLAK( 350),
     4KARC,LARC,MARC,NARC,ITALC,ITALP,ISTAR
      COMMON/NODES/TIMEN( 200),COSTN( 200),PERFN( 200),NSTORE( 5400),
     1NODE1( 200),NODE2( 200),LOGI( 200),LOGO( 200),NSTATE( 200),
     2NARCI( 200),NARCO( 200),ISTAT( 200),INSTAT( 200),ICRITN( 200),
     3NPOINT( 200),NSLAK( 200),JUMP( 200),KNODE,LNODE,MNODE,NNODE,MTAG,
     4NTAG
      LOGICAL SLAK,MODE,INFC,INFP,DISC,DISP,CRIT,KGEN,MED,SINGLE,DOUBLE
C
C
C     THIS SUBROUTINE CALCULTES WHAT IS REFERRED TO AS OVERALL COST.
C        THIS IS CALCULATED FOR NODES AT THE TIME THEY ARE COMPLETED AND
C        FOR THE TIME AN ITERATION IS COMPLETED.  ITERATION COMPLETION
C        TIME IS NOT NECESSARILY THE TIME AT WHICH THE WINNING TERMINAL
C        NODE WAS COMPLETED.  TCRIT IS THE NODE OR ITERATION TIME.
C
      UNFORM = BCOST
      TIMESM = BPERF
      DO 1377 I=1,NARC
      IF(NODEI(I).EQ.0) GO TO 1377
      IF(ISTATE(I).LE.0) GO TO 1377
      J = NODEI(I)
      IF(LOGI(J).LE.4) START = TIMEN(J)
      IF(LOGI(J).GT.4) START = TIMEA(I) - UTIMEA(I)
      IF(START - TCRIT)1355,1344,1377
 1344 IF(UTIMEA(I).GT.0.0) GO TO 1377
 1355 RATIO = 1.0
      IF(TIMEA(I).LE.TCRIT.OR.ICUT.EQ.0.OR.ICUT.EQ.2) GO TO 1366
      RATIO = (TCRIT - START)/UTIMEA(I)
 1366 XCOST = UCOSTA(I)*RATIO
      XPERF = UPERFA(I)*RATIO
      UNFORM = UNFORM + XCOST
      TIMESM = TIMESM + XPERF
 1377 CONTINUE
      RETURN
      END
      SUBROUTINE RANDOM
      IMPLICIT INTEGER*4(I-N)
```

```
      OCOMMON EMARK(11),SETUP,FINT,DINT,FACTOR,STIME,SCOST,SPERF,CTIME,
     1CCOST,CPERF,BTIME,BCOST,BPERF,TIMESM,UNFORM,ITRIP1(22),ITRIP2(22),
     2KNT(6),INPT,IOUT,IPNH,IWF1,IWF2,IWF3,IWF4,IBLK,ITRACE,ICUT,ISEED,
     3IRROR,ISAVE,NODESM,MAXTAG,ICOUNT,JCOUNT,NTRIP,ICORR,LCOMP,NWARN,
     4NRK,NRKS,SLAK,MODE,INFC,INFP,DISC,DISP,CRIT,KGEN,MED,SINGLE,DOUBLE
      LOGICAL SLAK,MODE,INFC,INFP,DISC,DISP,CRIT,KGEN,MED,SINGLE,DOUBLE
C
C
C     THIS SUBROUTINE GENERATES UNIFORM DEVIATES ON AN IBM 360 MACHINE
C        IN THE EVENT THIS PROGRAM IS PUT ON SOMETHING OTHER THAN 360
C        GEAR, ANOTHER UNIFORM GENERATOR MAY BE REQUIRED.  ISEED CARRIES
C        THE SEED UTILIZED TO MAINTAIN AND CREATE NEW UNIFORM DEVIATES.
C        UNFORM CARRIES THE UNIFORM DEVIATE.  (THIS IS IBM'S RANDU.)
C
      IF(KGEN) GO TO 1399
      ISEED = ISEED*65539
      IF(ISEED.GE.0) GO TO 1388
      ISEED = ISEED + 2147483647 + 1
 1388 UNFORM = ISEED
      UNFORM = UNFORM*.4656613E-9
      RETURN
C
C     PLACE A SECOND 0 TO 1 UNIFORM RANDOM NUMBER GENERATOR HERE
C
 1399 UNFORM = URAN(1)
      RETURN
      END
      SUBROUTINE NORM
      IMPLICIT INTEGER*4(I-N)
      OCOMMON EMARK(11),SETUP,FINT,DINT,FACTOR,STIME,SCOST,SPERF,CTIME,
     1CCOST,CPERF,BTIME,BCOST,BPERF,TIMESM,UNFORM,ITRIP1(22),ITRIP2(22),
     2KNT(6),INPT,IOUT,IPNH,IWF1,IWF2,IWF3,IWF4,IBLK,ITRACE,ICUT,ISEED,
     3IRROR,ISAVE,NODESM,MAXTAG,ICCUNT,JCOUNT,NTRIP,ICORR,LCOMP,NWARN,
     4NRK,NRKS,SLAK,MODE,INFC,INFP,DISC,DISP,CRIT,KGEN,MED,SINGLE,DOUBLE
      LOGICAL SLAK,MODE,INFC,INFP,DISC,DISP,CRIT,KGEN,MED,SINGLE,DOUBLE
      CALL RANDOM
      TIMESM = UNFORM
      CALL RANDOM
      UNFORM = ((-2.0*ALOG(TIMESM))**0.5)*(COS(6.283*UNFORM))
      RETURN
      END
      SUBROUTINE GAM
      IMPLICIT INTEGER*4(I-N)
      OCOMMON EMARK(11),SETUP,FINT,DINT,FACTOR,STIME,SCOST,SPERF,CTIME,
     1CCOST,CPERF,BTIME,BCOST,BPERF,TIMESM,UNFORM,ITRIP1(22),ITRIP2(22),
     2KNT(6),INPT,IOUT,IPNH,IWF1,IWF2,IWF3,IWF4,IBLK,ITRACE,ICUT,ISEED,
     3IRROR,ISAVE,NODESM,MAXTAG,ICCUNT,JCOUNT,NTRIP,ICORR,LCOMP,NWARN,
     4NRK,NRKS,SLAK,MODE,INFC,INFP,DISC,DISP,CRIT,KGEN,MED,SINGLE,DOUBLE
      LOGICAL SLAK,MODE,INFC,INFP,DISC,DISP,CRIT,KGEN,MED,SINGLE,DOUBLE
      TIMESM = 0.0
      DO 1411 I=1,ISAVE
 1400 CALL RANDOM
      IF(UNFORM.LT..1D-77) GO TO 1400
      UNFORM = ALOG(UNFORM)
 1411 TIMESM = TIMESM - UNFORM
      RETURN
      END
      SUBROUTINE OUTFLO
      IMPLICIT INTEGER*4(I-N)
      OCOMMON EMARK(11),SETUP,FINT,DINT,FACTOR,STIME,SCOST,SPERF,CTIME,
     1CCOST,CPERF,BTIME,BCOST,BPERF,TIMESM,UNFORM,ITRIP1(22),ITRIP2(22),
     2KNT(6),INPT,IOUT,IPNH,IWF1,IWF2,IWF3,IWF4,IBLK,ITRACE,ICUT,ISEED,
     3IRROR,ISAVE,NODESM,MAXTAG,ICCUNT,JCOUNT,NTRIP,ICORR,LCOMP,NWARN,
```

```
4NRK,NRKS,SLAK,MODE,INFC,INFP,DISC,DISP,CRIT,KGEN,MED,SINGLE,DOUBLE
  COMMON/TRUCK/PDF(24),CDF(24),SIDE(25),AVE,STD,TOS
 OCOMMON/TRIALS/STORET( 1000,4),TERM(10,8),KPOINT(10),NODET( 1000),
 1MTERM,NTERM,MITER,ITER
 OCOMMON/ARCS/ASTORE( 2800),UTIMEA( 350),TIMEA( 350),UCOSTA( 350),
 1COSTA( 350),UPERFA( 350),PERFA( 350),WORK( 350),ISTATE( 350),
 2NODEI( 350),NODEO( 350),ICRITA( 350),KEEPC( 350),KEEPP( 350),
 3IARC1( 350),IARC2( 350),IPOINT( 350),JPOINT( 350),ISLAK( 350),
 4KARC,LARC,MARC,NARC,ITALC,ITALP,ISTAR
 OCOMMON/NODES/TIMEN( 200),COSTN( 200),PERFN( 200),NSTORE( 5400),
 1NODE1( 200),NODE2( 200),LOGI( 200),LOGO( 200),NSTATE( 200),
 2NARCI( 200),NARCO( 200),ISTAT( 200),INSTAT( 200),ICRITN( 200),
 3NPOINT( 200),NSLAK( 200),JUMP( 200),KNODE,LNODE,MNODE,NNODE,MTAG,
 4NTAG
 CCOMMON/INTERN/XMIN(20,4),XMAX(20,4),HMIN(20,4),HMAX(20,4),
 1HAVE(20,4),ICBS(20),MHIST,NHIST
 OCOMMON/SLACK/RMIN(20),RMAX(20),SMIN(20),SMAX(20),SAVE(20),
 1JOBS(20),MSLACK,NSLACK
 OCOMMON/CPGAP/T1(10),T2(10),CONFL(10),CONFS(10),CAVE(10),CSMIN(10),
 1CSMAX(10),CHMIN(10),CHMAX(10),PAVE(10),PSMIN(10),PSMAX(10),
 2PHMIN(10),PHMAX(10),KCOBS(10),KPOBS(10),MCPGAP,NCPGAP,ICPGAP
  COMMON/MEDAN/Z( 1000),XMED,KCUNT
 CDIMENSION HSTD(20,4), HTOS(20,4), HIST(20,4,25), ABSA(10),
 1ITCP(4), ITAB(17,4)
  LOGICAL SLAK,MODE,INFC,INFP,DISC,DISP,CRIT,KGEN,MED,SINGLE,DOUBLE
  DATA IASTR,ITCP/1H*,4HTIME,4HCOST,4HCOST,4HPERF/
  DATA ABSA,IC1,IC2/.1,.2,.3,.4,.5,.6,.7,.8,.9,1.0,4HCOMP,4HSITE/
C
C
C     LOAD THE TABLE FOR SELECTIVELY PRINTING TIME AND/OR PATH COST
C         AND/OR OVERALL COST AND/OR PERFORMANCE FOR REQUESTED INTERNAL
C         AND TERMINAL NODES
C
      DO 1422 I=1,17
      DO 1422 J=1,4
 1422 ITAB(I,J) = 1
      ITAB(1,2) = 0
      ITAB(1,3) = 0
      ITAB(1,4) = 0
      ITAB(2,1) = 0
      ITAB(2,3) = 0
      ITAB(2,4) = 0
      ITAB(3,1) = 0
      ITAB(3,2) = 0
      ITAB(3,4) = 0
      ITAB(4,1) = 0
      ITAB(4,2) = 0
      ITAB(4,3) = 0
      ITAB(5,3) = 0
      ITAB(5,4) = 0
      ITAB(6,2) = 0
      ITAB(6,4) = 0
      ITAB(7,2) = 0
      ITAB(7,3) = 0
      ITAB(8,1) = 0
      ITAB(8,4) = 0
      ITAB(9,1) = 0
      ITAB(9,3) = 0
      ITAB(10,1) = 0
      ITAB(10,2) = 0
      ITAB(11,4) = 0
      ITAB(12,3) = 0
      ITAB(13,2) = 0
```

```
      ITAB(14,1) = 0
      ITAB(15,1) = 0
      ITAB(15,2) = 0
      ITAB(15,3) = 0
      ITAB(15,4) = 0
C
C     THIS SUBROUTINE DIRECTS THE LISTING OF THE SIMULATION RESULTS
C        FIRST TASK, IF THIS RUN ISN'T A DEBUG RUN, FINISH THE SLACKS
C
      IF(ITRACE.EQ.3) GO TO 2622
      ICUT = 0
      CRIT = .FALSE.
      IF(SLAK) GO TO 1622
      DO 1444 I=1,NSLACK
      IF(JOBS(I).LE.0) GO TO 1444
      DO 1433 J=1,25
 1433 HIST(I,1,J) = C.0
      HSTD(I,1) = C.0
      HTOS(I,1) = C.0
      IF(SMIN(I).EQ.+SETUP) SMIN(I) = RMIN(I)
      IF(SMAX(I).EQ.-SETUP) SMAX(I) = RMAX(I)
      SAVE(I) = SAVE(I)/FLOAT(JOBS(I))
      IF(RMIN(I).EQ.RMAX(I))SAVE(I) = RMIN(I)
 1444 CONTINUE
C
C     COMPUTE THE SQUARE FROM THE MEAN AND FILL THE HISTOGRAMS
C
      IF(ITRACE.EQ.C.OR.NRKS.EQ.0) GO TO 1500
      ITALC = 0
      REWIND IWF4
 1455 READ (IWF4) M, X
      IF(RMIN(M).EQ.RMAX(M)) GO TO 1488
      L = 1
      IF(X.LT.SMIN(M)) GO TO 1477
      L = 24
      IF(X.GT.SMAX(M)) GO TO 1477
      UNFORM = (SMAX(M) - SMIN(M))/22.0
      TIMESM = SMIN(M) + UNFORM
      DO 1466 ISAVE=2,23
      L = ISAVE
      IF(X.LT.TIMESM) GO TO 1477
 1466 TIMESM = TIMESM + UNFORM
 1477 HIST(M,1,L) = HIST(M,1,L) + 1.0
      UNFORM = SAVE(M) - X
      UNFORM = UNFORM*UNFORM
      HSTD(M,1) = HSTD(M,1) + UNFORM
      HTOS(M,1) = HTOS(M,1) + UNFORM*UNFORM
 1488 ITALC = ITALC + 1
      IF(ITALC.LT.NRKS) GO TO 1455
C
C     COMPUTE THE STANDARD DEVIATION AND KURTOSIS
C
      DO 1499 I=1,NSLACK
      IF(JOBS(I).LE.1) GO TO 1499
      X = JOBS(I) - 1
      UNFORM = HSTD(I,1)/X
      HSTD(I,1) = SQRT(UNFORM)
      IF(UNFORM.NE.C.0) HTOS(I,1) = HTOS(I,1)/(UNFORM*UNFORM*X)
      IF(RMIN(I).EQ.RMAX(I)) HSTD(I,1) = 0.0
 1499 CONTINUE
C
C     PRINT THE SLACK INFO
```

```
C
 1500 DO 1611 I=1,NSLACK
      IF(JOBS(I).EC.C) GC TC 1611
      SIDE(1) = RMIN(I)
      SIDE(2) = SMIN(I)
      SIDE(24)= SMAX(I)
      SIDE(25)= RMAX(I)
      IF(ITRACE.EQ.0) GO TC 1533
      DO 1511 J=1,24
      PDF(J) = HIST(I,1,J)/FLCAT(JCBS(I))
      CDF(J) = PDF(J)
 1511 IF(J.GT.1) CDF(J) = CDF(J) + CDF(J-1)
      IF(PDF(24).EC.C.0) CDF(24) = C.C
      UNFORM = (SMAX(I) - SMIN(I))/22.0
      DO 1522 J=2,22
 1522 SIDE(J+1) = SIDE(J) + UNFORM
      TOS = HTOS(I,1)
      STD = HSTD(I,1)
 1533 AVE = SAVE(I)
      IF(SIDE(1).EC.SIDE(25).AND.AVE.EC.C.C) GO TC 1611
      IRROR = JOBS(I)
C
C     IF THE MEDIAN HAS BEEN REQUESTED, LOAD MEDIAN STORAGE
C
      IF(MED) GO TC 1566
      IF(NRKS.EQ.C) GO TO 1566
      ITALC = C
      REWIND IWF4
      KOUNT = C
 1544 READ (IWF4) M, X
      IF(M.NE.I) GC TO 1555
      KOUNT = KOUNT + 1
      Z(KOUNT) = X
 1555 ITALC = ITALC + 1
      IF(ITALC.LT.NRKS) GO TO 1544
      CALL MEDIAN
C
C     FIND WHICH ARC CR NCDE
C
 1566 DO 1577 J=1,NARC
      IF(ISLAK(J).NE.I) GC TO 1577
      M = IARC1(J)
      N = IARC2(J)
      JTITLE = 4
      GO TO 1600
 1577 CCNTINUE
      DO 1588 J=1,NNODE
      K = J
      IF(NSLAK(J).EQ.I) GO TO 1599
 1588 CONTINUE
 1599 M = NODE1(K)
      N = NODE2(K)
      JTITLE = 3
 1600 CALL HISTO (M,N,IBLK,IBLK,XMED,5,JTITLE,0)
 1611 CONTINUE
C
C     FINISH THE CCST-PERFCRMANCE TIME INTERVALS HISTOGRAMS
C
 1622 IF(ICPGAP.EQ.C) GC TC 2066
      ITALC = NCPGAP + 1
      ITALP = NCPGAP + NCPGAP
      IF(ICPGAP - 2)1633,2C44,1633
```

```
 1633 ISTAR = 0
      ITITLE = 6
 1644 DO 1666 I=1,NCPGAP
      DO 1655 J=1,24
 1655 HIST(I,1,J) = 0.0
      HSTD(I,1) = 0.0
      HTOS(I,1) = 0.0
      IF(CHMIN(I).EQ.+SETUP) CHMIN(I) = CSMIN(I)
      IF(CHMAX(I).EQ.-SETUP) CHMAX(I) = CSMAX(I)
      IF(KCOBS(I).EQ.0) GC TO 1666
      CAVE(I) = CAVE(I)/FLOAT(KCOBS(I))
 1666 IF(CSMIN(I).EQ.CSMAX(I)) CAVE(I) = CSMIN(I)
C
C     COMPUTE THE SQUARE FROM THE MEAN AND FILL THE HISTOGRAMS
C
      IF(ITRACE.EQ.0) GC TC 1744
      REWIND IWF2
      DO 1700 I=1,ITER
      READ (IWF2) (WORK(J), J=1,NCPGAP), (WORK(J), J=ITALC,ITALP)
      DO 1699 M=1,NCPGAP
      X = WORK(M+ISTAR)
      IF(CSMIN(M).EQ.CSMAX(M).OR.X.LE.0.0) GO TO 1699
      L = 1
      IF(X.LT.CHMIN(M)) GO TO 1688
      L = 24
      IF(X.GT.CHMAX(M)) GC TO 1688
      UNFORM = (CHMAX(M) - CHMIN(M))/22.0
      TIMESM = CHMIN(M) + UNFORM
      DC 1677 ISAVE=2,23
      L = ISAVE
      IF(X.LT.TIMESM) GO TC 1688
 1677 TIMESM = TIMESM + UNFCRM
 1688 HIST(M,1,L) = HIST(M,1,L) + 1.0
      UNFORM = CAVE(M) - X
      UNFORM = UNFORM*UNFORM
      HSTD(M,1) = HSTD(M,1) + UNFORM
      HTOS(M,1) = HTOS(M,1) + UNFORM*UNFORM
 1699 CONTINUE
 1700 CONTINUE
C
C     COMPUTE THE STANDARD DEVIATION AND KURTOSIS
C
      DO 1733 I=1,NCPGAP
      X = KCOBS(I) - 1
      IF(X.GT.0.0) GO TO 1722
      HTOS(I,1) = 0.0
 1711 HSTD(I,1) = C.C
      GO TO 1733
 1722 UNFORM = HSTD(I,1)/X
      HSTD(I,1) = SQRT(UNFCRM)
      IF(UNFORM.NE.C.0) HTCS(I,1) = HTCS(I,1)/(UNFORM*UNFORM*X)
      IF(CSMIN(I).EQ.CSMAX(I)) GO TO 1711
 1733 CONTINUE
C
C     PRINT THE COST POINT DATA
C
 1744 DC 1811 ISAVE=1,NCPGAP
      IRROR = KCOBS(ISAVE)
      IF(IRROR.EQ.0) GO TC 1811
      SIDE(1) = CSMIN(ISAVE)
      SIDE(2) = CHMIN(ISAVE)
      SIDE(24)= CHMAX(ISAVE)
```

```
      SIDE(25)= CSMAX(ISAVE)
      IF(SIDE(1).EQ.SIDE(25).OR.ITRACE.EQ.0) GO TO 1788
      DO 1755 J=1,24
      PDF(J) = HIST(ISAVE,1,J)/FLOAT(IRROR)
      CDF(J) = PDF(J)
      IF(J.GT.1) CDF(J) = CDF(J) + CDF(J-1)
 1755 HIST(ISAVE,2,J) = CDF(J)
      UNFORM = (CHMAX(ISAVE) - CHMIN(ISAVE))/22.0
      DO 1766 J=2,22
 1766 SIDE(J+1) = SIDE(J) + UNFORM
      DO 1777 J=1,25
 1777 HIST(ISAVE,3,J) = SIDE(J)
      TOS = HTOS(ISAVE,1)
      STD = HSTD(ISAVE,1)
 1788 AVE = CAVE(ISAVE)
      IF(SIDE(1).EQ.SIDE(25).AND.AVE.EQ.0.0) GO TO 1811
C
C     IF THE MEDIAN HAS BEEN REQUESTED, LOAD MEDIAN STORAGE WITH ALL THE
C        OBSERVATIONS
C
      IF(MED) GO TO 1800
      KOUNT = 0
      REWIND IWF2
      DO 1799 L=1,ITER
      READ (IWF2) (WORK(J), J=1,NCPGAP), (WORK(J), J=ITALC,ITALP)
      X = WORK(ISAVE+ISTAR)
      IF(X.LE.0.0) GO TO 1799
      KOUNT = KOUNT + 1
      Z(KOUNT) = X
 1799 CONTINUE
      CALL MEDIAN
 1800 CALL HISTO (IBLK,IBLK,IBLK,IBLK,XMED,ITITLE,5,0)
 1811 CONTINUE
      IF(ITITLE.EQ.7) GO TO 2066
C
C     DETERMINE THE INDIVIDUAL PERIOD EXPENDITURES NEEDED TO MEET THE
C        OVERALL RISK LEVEL DESIRED.  IDIR INDICATES THE DIRECTION ALL
C        THE INDIVIDUAL TIME PERIODS ARE BEING ADJUSTED TO YIELD THE
C        CONFIDENCE OF THE ENTIRE TIME PERIOD AT THE TRACE COST
C        IDIR = +1 - MOVING UP
C        IDIR = -1 - MOVING DOWN
C        CONFL - CARRIES EACH PERIODS CONFIDENCE LEVEL
C        CONFS - CARRIES UNIT CONFIDENCE ADJUSTMENTS (STEPS UP OR DOWN)
C
      N = 0
      IDIR = 0
      LESS1 = NCPGAP - 1
 1822 N = N + 1
      IF(N.LT.2000) GO TO 1844
      IRROR = 0
      CALL ERROR (1833,IBLK,IBLK,IBLK)
      GO TO 2033
 1844 UNFORM = 0.0
      DO 1933 I=1,NCPGAP
      IF(N.GT.1) GO TO 1866
      IF(CONFL(I) - 0.0)2033,2033,1855
 1855 IF(CONFL(I) - 1.0)1899,1888,2033
 1866 IF(CONFL(I).GT.0.0) GO TO 1877
      X = HIST(I,3,1)
      GO TO 1933
 1877 IF(CONFL(I).LT.1.0) GO TO 1899
 1888 X = HIST(I,3,25)
```

```
      GO TO 1933
C
C        FIND THE COST CELL WHERE THE CONFIDENCE VALUE CURRENTLY LIES
C
 1899 DO 1900 J=1,24
      IPT = J
      IF(HIST(I,2,J).EQ.1.0) GO TO 1911
      IF(HIST(I,2,J).GE.CCNFL(I)) GO TO 1911
 1900 CONTINUE
C
C        INTERPOLATE WITHIN THIS COST CELL AND THUS DETERMINE THE COST
C           WHICH CORRESPONDS TO THE GIVEN CONFIDENCE LEVEL
C
 1911 BOT = 0.0
      TOP = HIST(I,2,IPT)
      IF(IPT.EQ.1) GO TO 1922
      BOT = HIST(I,2,IPT-1)
 1922 X = (CCNFL(I) - BOT)/(TOP - BOT)
      X = HIST(I,3,IPT) + X*(HIST(I,3,IPT+1) - HIST(I,3,IPT))
      HIST(I,4,1) = X
 1933 IF(I.LT.NCPGAP) UNFORM = UNFORM + X
C
C        DOES THE COMBINED COST OF ALL THE PARTS EXCEED OR FALL SHORT OF
C           THE COMBINED TOTAL COST
C
      IF(UNFORM - X)1944,1988,1955
 1944 IF(IDIR.EQ.-1) GO TO 1988
      IDIR =+1
      GO TO 1966
 1955 IF(IDIR.EQ.+1) GO TO 1988
      IDIR =-1
 1966 DO 1977 I=1,LESS1
      CONFL(I) = CONFL(I) + CCNES(I)*FLOAT(IDIR)
      IF(CONFL(I).LT.0.0) CONFL(I) = 0.0
 1977 IF(CONFL(I).GT.1.0) CCNFL(I) = 1.0
      GO TO 1822
C
C        LIST THE REFORMULATED CONFIDENCES
C
 1988 WRITE (IOUT,1999) N
 1999 FORMAT (52H1COST CONFIDENCE BALANCE AMONG SELECTED TIME PERIODS//
     119H NO OF ITERATIONS =, I6// 71H CFD NO   TIME INTERVAL COVERED  C
     2ONFIDENCES   COST INTERPOLATED FOR THE/ 1H , 11X, 5HSTART, 10X,
     34HSTOP, 5X, 8HCOMPUTED, 7X, 20HCONFIDENCES COMPUTED)
      DO 2011 I=1,NCPGAP
      IF(I.EQ.NCPGAP) WRITE (6,2000) UNFORM
 2000 FORMAT (1H , 70(1H-)/ 1H , 25X, 18HSUM OF ABOVE COSTS, F27.8)
 2011 WRITE (IOUT,2022) I, T1(I), T2(I), CUNFL(I), HIST(I,4,1)
 2022 FORMAT (1H0, I4, 2H. , F10.2, 2H -, F12.2, F13.2, F27.8)
C
C        LOAD PERFORMANCE INTO COST IF PERFORMANCE IS WANTED
C
 2033 IF(ICPGAP - 2)2066,2044,2044
 2044 ISTAR = NCPGAP
      ITITLE = 7
      DO 2055 I=1,NCPGAP
      CSMIN(I) = PSMIN(I)
      CSMAX(I) = PSMAX(I)
      CHMIN(I) = PHMIN(I)
      CHMAX(I) = PHMAX(I)
      CAVE(I)  = PAVE(I)
 2055 KCOBS(I) = KPOBS(I)
```

```
      GO TO 1644
C
C     FINISH THE INTERNAL NODE STATISTICS
C
 2066 IF(MODE) GO TO 2277
      DO 2077 I=1,NHIST
      DO 2077 J=1,4
      HSTD(I,J) = 0.0
      HTOS(I,J) = 0.0
      DO 2077 K=1,24
 2077 HIST(I,J,K) = 0.0
C
C     CHECK FOR USER SCALE AND COMPUTE THE AVERAGE
C
      DO 2099 K=1,NHIST
      IF(IOBS(K).LE.0) GO TO 2099
      DO 2088 L=1,4
      IF(HMIN(K,L).EQ.SETUP) HMIN(K,L) = XMIN(K,L)
      IF(HMAX(K,L).EQ.-SETUP)HMAX(K,L) = XMAX(K,L)
      HAVE(K,L) = HAVE(K,L)/FLOAT(IOBS(K))
 2088 IF(XMIN(K,L).EQ.XMAX(K,L)) HAVE(K,L) = XMIN(K,L)
 2099 CONTINUE
C
C     COMPUTE THE SQUARE FROM THE MEAN AND FILL THE HISTOGRAMS
C
      IF(ITRACE.EQ.0.OR.NRK.EQ.0) GO TO 2166
      ITALC = 0
      REWIND IWF3
 2100 READ (IWF3) M, (WORK(K), K=1,4)
      DO 2133 K=1,4
      IF(XMIN(M,K).EQ.XMAX(M,K)) GO TO 2133
      L = 1
      IF(WORK(K).LT.HMIN(M,K)) GO TO 2122
      L = 24
      IF(WORK(K).GT.HMAX(M,K)) GO TO 2122
      UNFORM = (HMAX(M,K) - HMIN(M,K))/22.0
      TIMESM = HMIN(M,K) + UNFORM
      DO 2111 ISAVE=2,23
      L = ISAVE
      IF(WORK(K).LT.TIMESM) GO TO 2122
 2111 TIMESM = TIMESM + UNFORM
 2122 HIST(M,K,L) = HIST(M,K,L) + 1.0
      UNFORM = HAVE(M,K) - WORK(K)
      UNFORM = UNFORM*UNFORM
      HSTD(M,K) = HSTD(M,K) + UNFORM
      HTOS(M,K) = HTOS(M,K) + UNFORM*UNFORM
 2133 CONTINUE
      ITALC = ITALC + 1
      IF(ITALC.LT.NRK) GO TO 2100
C
C     COMPUTE THE STANDARD DEVIATION AND KURTOSIS
C
      DO 2155 K=1,NHIST
      IF(IOBS(K).LE.1) GO TO 2155
      TIMESM = IOBS(K) - 1
      DO 2144 L=1,4
      UNFORM = HSTD(K,L)/TIMESM
      HSTD(K,L) = SQRT(UNFORM)
      IF(UNFORM.NE.0.0) HTOS(K,L) = HTOS(K,L)/(UNFORM*UNFORM*TIMESM)
 2144 IF(XMIN(K,L).EQ.XMAX(K,L)) HSTD(K,L) = 0.0
 2155 CONTINUE
C
```

```
C       LIST THE INTERNAL NODE STATISTICS
C
 2166 DO 2266 I=1,NHIST
      IF(IOBS(I).EQ.0) GO TO 2266
      K = 0
      M = IBLK
      N = IBLK
      DO 2199 J=1,NNODE
      IF(ISTAT(J).NE.I) GO TO 2199
      K = K + 1
      IF(K - 2)2177,2188,2188
 2177 L = J
      JTITLE = 3
      GO TO 2199
 2188 M = NODE1(J)
      N = NODE2(J)
      JTITLE = 2
 2199 CONTINUE
      IRROR = IOBS(I)
      DO 2255 K=1,4
      SIDE(1) = XMIN(I,K)
      SIDE(2) = HMIN(I,K)
      SIDE(24)= HMAX(I,K)
      SIDE(25)= XMAX(I,K)
      DO 2200 J=1,24
      PDF(J) = HIST(I,K,J)/FLOAT(IOBS(I))
      CDF(J) = PDF(J)
 2200 IF(J.GT.1) CDF(J) = CDF(J) + CDF(J-1)
      UNFORM = (HMAX(I,K) - HMIN(I,K))/22.0
      DO 2211 J=2,22
 2211 SIDE(J+1) = SIDE(J) + UNFORM
      TCS = HTCS(I,K)
      STD = HSTD(I,K)
      AVE = HAVE(I,K)
      IF(SIDE(1).EQ.SIDE(25).AND.AVE.EQ.0.0) GO TO 2255
C
C       IF THE MEDIAN HAS BEEN REQUESTED, LOAD MEDIAN STORAGE
C
      IF(MED) GO TO 2244
      IF(NRK.EQ.0) GO TO 2244
      KOUNT = 0
      ITALC = 0
      REWIND IWF3
 2222 READ (IWF3) M, (WORK(J), J=1,4)
      IF(M.NE.I) GO TO 2233
      KOUNT = KOUNT + 1
      Z(KOUNT) = WORK(K)
 2233 ITALC = ITALC + 1
      IF(ITALC.LT.NRK) GO TO 2222
      CALL MEDIAN
 2244 ITALC = INSTAT(L)
      IF(ITAB(ITALC,K).EQ.0) GO TO 2255
      J = 0
      IF(ITALC.EQ.17) J = ITCP(K)
      CALL HISTO (NODE1(L),NODE2(L),M,N,XMED,K,JTITLE,J)
 2255 CONTINUE
 2266 CONTINUE
C
C       LIST EACH TERMINAL NODE
C
 2277 NRK = 0
      ICUT = 1
```

```
       ITAL = 0
       ISTAR = 0
       MED  = .FALSE.
       MODE = .FALSE.
       REWIND IWF2
       DO 2311 I=1,NNODE
       IF(LOGI(I).GE.5) GO TO 2311
       IF(LOGO(I).GE.11)GO TO 2288
       IF(LOGO(I).NE.1) GO TO 2311
 2288 IF(ISTAT(I).LT.0) IOUT = 2
       NODESM = I
       DO 2300 J=1,4
       CALL MS(J)
       IF(IRROR.EQ.0) GO TO 2311
       K = NRK + J
       TIMEA(K) = SIDE(1)
       COSTA(K) = AVE
       PERFA(K) = SIDE(25)
       IF(SIDE(1).EQ.SIDE(25).AND.AVE.EQ.0.0) GO TO 2300
       IF(ISTAT(I).NE.-2) GO TO 2299
       ITAL = ITAL + 1
       WRITE (IWF2) NODE1(I), NODE2(I), ITCP(J), SIDE, PDF
 2299 CALL MEDIAN
       ITALC = INSTAT(I)
       IF(ITAB(ITALC,J).EQ.0) GO TO 2300
       M = 0
       IF(ITALC.EQ.17) M = ITCP(J)
       CALL HISTO (NODE1(I),NODE2(I),IBLK,IBLK,XMED,J,3,M)
 2300 CONTINUE
       IF(IRROR.GT.1) NRK = NRK + 4
       ISTAR = ISTAR + 1
 2311 CONTINUE
C
C      LIST THE COMPOSITE NODE
C
       IF(ISTAR.LE.1) GO TO 2400
       MODE = .TRUE.
       DO 2333 J=1,4
       CALL MS(J)
       K = NRK + J
       TIMEA(K) = SIDE(1)
       COSTA(K) = AVE
       PERFA(K) = SIDE(25)
       IF(ITAL.EQ.0) GO TO 2322
       IF(SIDE(1).EQ.SIDE(25).AND.AVE.EQ.0.0) GO TO 2333
       ITAL = ITAL + 1
       WRITE (IWF2) IC1, IC2, ITCP(J), SIDE, PDF
 2322 CALL MEDIAN
       IF(ITAB(LCOMP,J).EQ.0) GO TO 2333
       M = 0
       IF(LCOMP.EQ.17) M = ITCP(J)
       CALL HISTO (IC1,IC2,IBLK,IBLK,XMED,J,1,M)
 2333 CONTINUE
C
C      LIST THE TERMINAL NODE INDEX
C
 2400 IF(ISTAR.LE.1) GO TO 2500
       WRITE (IOUT,2411) ITER
 2411 FORMAT (47H1OPTIMUM TERMINAL NODE INDEX - NO. ITERATIONS =, I5)
       WRITE (IOUT,2422) ABSA
 2422 FORMAT (1H , 18X, 10F5.1/ 1H , 16X, 1HI, 10(5H----+), 1HI)
       DO 2488 J=1,NNODE
```

```
      IF(LOGI(J).GE.5) GO TO 2488
      IF(LOGO(J).GE.11)GO TO 2433
      IF(LOGO(J).NE.1) GO TO 2488
 2433 ICOUNT = 0
      DO 2444 K=1,ITER
 2444 IF(NODET(K).EQ.J) ICOUNT = ICOUNT + 1
      IF(ICOUNT.EQ.0) GO TO 2488
      UNFORM = FLOAT(ICOUNT)/FLOAT(ITER)
      K = UNFORM*50.0 + 0.5
      DO 2455 L=1,50
      ISTATE(L) = IBLK
 2455 IF(L.LE.K) ISTATE(L) = IASTR
      WRITE (IOUT,2466) NODE1(J), NODE2(J), UNFORM, (ISTATE(L), L=1,50)
 2466 FORMAT (1H , 2A4, F7.4, 2H I, 50A1, 1HI)
      IF(J.NE.NNODE) WRITE (IOUT,2477)
 2477 FORMAT (1H , 16X, 1HI, 10(4X,1H+), 1HI)
 2488 CONTINUE
      WRITE (IOUT,2499) ABSA
 2499 FORMAT (1H , 16X, 1HI, 10(5H----+), 1HI/ 1H , 18X, 10F5.1)
 2500 IF(JCOUNT.EQ.0) GO TO 2744
C
C     LIST TERMINAL NODES EXCLUDED IN THE CRITICAL-OPTIMUM PATH ANALYSIS
C
      GO TO (2555,2511),IOUT
 2511 WRITE (IOUT,2522)
 25220FORMAT (74H0THESE TERMINAL NODES ARE EXCLUDED FROM THE CRITICAL-OP
     1TIMUM PATH ANALYSIS///)
      DO 2533 I=1,NNODE
 2533 IF(ISTAT(I).EQ.-1) WRITE (IOUT,2544) NODE1(I), NODE2(I)
 2544 FORMAT (1H , 10X, 2A4)
C
C     LIST CRITICAL-OPTIMUM PATH INDEXES FOR THE NODES
C
 2555 WRITE (IOUT,2566) JCOUNT
 2566 FORMAT (48H1NODES CRITICAL-OPTIMUM PATH INDEX - NO. PATHS =, I5)
      WRITE (IOUT,2422) ABSA
      DO 2588 I=1,NNODE
      IF(ICRITN(I).EQ.0) GO TO 2588
      UNFORM = FLOAT(ICRITN(I))/FLOAT(JCOUNT)
      K = UNFORM*50.0 + 0.5
      DO 2577 L=1,50
      ISTATE(L) = IBLK
 2577 IF(L.LE.K) ISTATE(L) = IASTR
      WRITE (IOUT,2466) NODE1(I), NODE2(I), UNFORM, (ISTATE(L), L=1,50)
      IF(I.NE.NNODE) WRITE (IOUT,2477)
 2588 CONTINUE
      WRITE (IOUT,2499) ABSA
C
C     LIST CRITICAL-OPTIMUM PATH INDEXES FOR THE ARCS
C
      WRITE (IOUT,2599) JCOUNT
 2599 FORMAT (47H1ARCS CRITICAL-OPTIMUM PATH INDEX - NO. PATHS =, I5)
      WRITE (IOUT,2422) ABSA
      DO 2611 I=1,NARC
      IF(ICRITA(I).EQ.0) GO TO 2611
      UNFORM = FLOAT(ICRITA(I))/FLOAT(JCOUNT)
      K = UNFORM*50.0 + 0.5
      DO 2600 L=1,50
      ISTATE(L) = IBLK
 2600 IF(L.LE.K) ISTATE(L) = IASTR
      WRITE (IOUT,2466) IARC1(I), IARC2(I), UNFORM, (ISTATE(L), L=1,50)
      IF(I.NE.NARC) WRITE (IOUT,2477)
```

```
 2611 CONTINUE
      WRITE (IOUT,2499) ABSA
      GO TO 2744
C
C     LIST STORAGE UTILIZATION AND ENDING SEED
C
 2622 I = (MITER - ITER)*5
      J = (MARC  - NARC)*16
      K =   LARC  - KARC
      L = (MNODE - NNODE)*13
      M =   LNODE - KNODE
      N =   MAXTAG- M
      IF(N.LT.0) N = 0
      ISTAR = MTAG  - N
      ITALC = (MHIST - NHIST)*141
      ITALP = (MTERM - NTERM)*9
      ISAVE = (MSLACK - NSLACK)*6
      ICUT = (MCPGAP - NCPGAP)*7
      ICOUNT = I + J + K + L + M + ISTAR + ITALC + ITALP + ISAVE + ICUT
      OJCOUNT = MITER*5 + MARC*16 + LARC + MNODE*13 + LNODE + MTAG +
     1MHIST*141 + MTERM*9 + MSLACK*6 + MCPGAP*7
      UNFORM = 100.0*(1.0 - FLOAT(ICOUNT)/FLOAT(JCOUNT))
      WRITE (IOUT,2633) MITER, ITER, I
 2633 FORMAT (1H1, 10X, 32HCORE DATA STORAGE USAGE ANALYSIS/ 58H0A1. CUR
     1RENT NUMBER OF ITERATIONS ALLOWED (VALUE OF MITER), 17(1H-), I6/
     235H A2. NUMBER OF ITERATIONS REQUESTED, 40(1H-), I6/ 49H A3. ITERA
     3TION STORAGE SPACE NOT USED ((A1-A2)*5), 26(1H-), I12)
      WRITE (IOUT,2644) MARC, NARC, J
 2644 FORMAT (51HCB1. CURRENT NUMBER OF ARCS ALLOWED (VALUE OF MARC),
     124(1H-), I6/ 27H B2. NUMBER OF ARCS ENTERED, 48(1H-), I6/ 44H B3.
     2ARC STORAGE SPACE NOT USED ((B1-B2)*16), 31(1H-), I12)
      WRITE (IOUT,2655) LARC, KARC, K
 2655 FORMAT (60HOC1. VARIABLE ARC STORAGE (ASTORE) AVAILABLE (VALUE OF
     1LARC), 15(1H-), I6/ 30H C2. VARIABLE ARC STORAGE USED, 45(1H-),I6/
     248H C3. VARIABLE ARC STORAGE SPACE NOT USED (C1-C2), 27(1H-), I12)
      WRITE  (IOUT,2666) MNODE, NNODE, L
 2666 FORMAT (53HCD1. CURRENT NUMBER OF NODES ALLOWED (VALUE OF MNODE),
     1122(1H-), I6/ 28H D2. NUMBER OF NODES ENTERED, 47(1H-), I6/ 45H D3.
     2 NODE STORAGE SPACE NOT USED ((D1-D2)*13), 30(1H-), I12)
      WRITE (IOUT,2677) LNODE, KNODE, M
 2677 FORMAT (73HOE1. VARIABLE NODE STORAGE (FRONT PART NSTORE) AVAILABL
     1E (VALUE OF LNODE), 2(1H-), I6/ 31H E2. VARIABLE NODE STORAGE USED
     2, 44(1H-), I6/ 49H E3. VARIABLE NODE STORAGE SPACE NOT USED (E1-E2
     3), 26(1H-), I12)
      WRITE (IOUT,2688) MTAG, N, ISTAR
 2688 FORMAT (74HOF1. COST-PERF VECTOR STORAGE (BACK PART NSTORE) AVAILA
     1BLE (VALUE OF MTAG), 1(1H-), I6/ 34H F2. COST-PERF VECTOR STORAGE
     2USED, 41(1H-), I6/ 52H F3. COST-PERF VECTOR STORAGE SPACE NOT USED
     3 (F1-F2), 23(1H-), I12)
      WRITE (IOUT,2699) MHIST, NHIST, ITALC
 2699 FORMAT (72HOG1. CURRENT NUMBER OF INTERNAL NODE STATISTICS ALLOWED
     1 (VALUE OF MHIST),  3(1H-), I6/ 49H G2. NUMBER OF INTERNAL NODE ST
     2ATISTICS REQUESTED, 26(1H-), I6/ 66H G3. INTERNAL NODE STATISTICS
     3STORAGE SPACE NOT USED ((G1-G2)*141),  9(1H-), I12)
      WRITE (IOUT,2700) MTERM, NTERM, ITALP
 2700 FORMAT (72HOH1. CURRENT NUMBER OF TERMINAL NODE HISTOGRAMS ALLOWED
     1 (VALUE OF MTERM),  3(1H-), I6/ 47H H2. NUMBER OF TERMINAL NODE HI
     2STOGRAMS ENTERED, 28(1H-), I6/ 63H H3. TERMINAL NODE HISTOGRAM STO
     3RAGE SPACE NOT USED ((H1-H2)*9), 12(1H-), I12)
      WRITE (IOUT,2711) MSLACK, NSLACK, ISAVE
 2711 FORMAT (65HOI1. CURRENT NUMBER OF SLACK HISTOGRAMS ALLOWED (VALUE
     1OF MSLACK), 1 (1H-), I6/ 41H I2. NUMBER OF SLACK HISTOGRAMS REQUES
```

```
     2TED, 34(1H-), I6/ 55H I3. SLACK HISTOGRAM STORAGE SPACE NOT USED (
     3(I1-I2)*6), 20(1H-), I12)
      WRITE (ICUT,2722) MCPGAP, NCPGAP, ICUT
27220FORMAT (75H0J1. CURRENT NUMBER OF COST-PERF TIME INTERVALS AVAILAB
     1LE (VALUE OF MCPGAP), I6/ 49H J2. NUMBER OF COST-PERF TIME INTERVA
     2LS REQUESTED, 26(1H-), I6/ 50H J3. COST-PERF TIME INTERVALS NOT US
     3ED ((J1-J2)*6), 25(1H-), I12)
      WRITE (IOUT,2733) JCOUNT, ICOUNT, UNFORM
27330FORMAT (33H0     TOTAL CORE STORAGE AVAILABLE, 42(1H-), I6/ 32H0
     1 TOTAL CORE STORAGE NOT USED, 43(1H-), I12/ 37H0     OVERALL CORE S
     2TORAGE UTILIZATION, 38(1H-), F12.2, 1H%)
 2744 WRITE (IOUT,2755) ISEED
27550FORMAT (/30H0     LAST RANDOM NUMBER SEED =, I13)
      RETURN
      END
      SUBROUTINE MEDIAN
      IMPLICIT INTEGER*4(I-N)
     0COMMON EMARK(11),SETUP,FINT,DINT,FACTOR,STIME,SCOST,SPERF,CTIME,
     1CCOST,CPERF,BTIME,BCOST,BPERF,TIMESM,UNFORM,ITRIP1(22),ITRIP2(22),
     2KNT(6),INPT,IOUT,IPNH,IWF1,IWF2,IWF3,IWF4,IBLK,ITRACE,ICUT,ISEED,
     3IRROR,ISAVE,NODESM,MAXTAG,ICOUNT,JCOUNT,NTRIP,ICORR,LCOMP,NWARN,
     4NRK,NRKS,SLAK,MODE,INFC,INFP,DISC,DISP,CRIT,KGEN,MED,SINGLE,DOUBLE
      COMMON/MEDAN/Z( 1000),XMED,KOUNT
      LOGICAL SLAK,MODE,INFC,INFP,DISC,DISP,CRIT,KGEN,MED,SINGLE,DOUBLE
C
C     THIS SUBROUTINE COMPUTES THE MEDIAN OF THE OBS. STORED IN Z
C
      XMED = 0.0
      IF(KOUNT - 1)2777,2766,2788
 2766 XMED = Z(1)
 2777 RETURN
C
C     ORDER THE ELEMENTS STORED IN VECTOR Z
C
 2788 MIDPT = FLOAT(KOUNT)/2.0 + 0.6
      MIDPT1 = MIDPT + 1
      DO 2800 I=1,MIDPT1
      Y = SETUP
      DO 2799 J=I,KOUNT
      IF(Z(J).GE.Y) GO TO 2799
      K = J
      Y = Z(J)
 2799 CONTINUE
      Z(K) = Z(I)
 2800 Z(I) = Y
C
C     COMPUTE THE ACTUAL MEDIAN
C
      XMED = Z(MIDPT)
      I = MIDPT + MIDPT
      IF(I.GT.KOUNT) GO TO 2811
      XMED = (Z(MIDPT+1) + XMED)/2.0
 2811 RETURN
      END
      SUBROUTINE MS(J)
      IMPLICIT INTEGER*4(I-N)
     0COMMON EMARK(11),SETUP,FINT,DINT,FACTOR,STIME,SCOST,SPERF,CTIME,
     1CCOST,CPERF,BTIME,BCOST,BPERF,TIMESM,UNFORM,ITRIP1(22),ITRIP2(22),
     2KNT(6),INPT,IOUT,IPNH,IWF1,IWF2,IWF3,IWF4,IBLK,ITRACE,ICUT,ISEED,
     3IRROR,ISAVE,NODESM,MAXTAG,ICOUNT,JCOUNT,NTRIP,ICORR,LCOMP,NWARN,
     4NRK,NRKS,SLAK,MODE,INFC,INFP,DISC,DISP,CRIT,KGEN,MED,SINGLE,DOUBLE
      COMMON/TRUCK/PDF(24),CDF(24),SIDE(25),AVE,STD,TOS
```

```
      0COMMON/TRIALS/STORET( 1000,4),TERM(10,8),KPOINT(10),NODET( 1000),
     1MTERM,NTERM,MITER,ITER
      COMMON/MEDAN/Z( 1000),XMED,KOUNT
      LOGICAL SLAK,MODE,INFC,INFP,DISC,DISP,CRIT,KGEN,MED,SINGLE,DOUBLE
C
C     THIS SUBROUTINE CALCULATES MEANS, STANDARD DEVIATIONS AND FILLS
C        HISTOGRAMS FOR TERMINAL NODES
C
      K = 0
      IF(MODE) GO TO 2833
      DO 2822 I=1,NTERM
 2822 IF(NODESM.EQ.KPOINT(I)) K = I
      GO TO 2844
 2833 IF(KPOINT(1).LT.0) K = 1
 2844 SIDE(1) = SETUP
      SIDE(25)=-SETUP
      AVE = 0.0
      STD = 0.0
      TCS = 0.0
      IRROR = 0
      UNFORM = 0.0
      DO 2866 I=1,ITER
      IF(MODE) GO TO 2855
      IF(NODESM.NE.NODET(I)) GO TO 2866
 2855 IRROR = IRROR + 1
      Z(IRROR) = STORET(I,J)
      UNFORM = UNFORM + STORET(I,J)
      IF(STORET(I,J).LT.SIDE(1)) SIDE(1) = STORET(I,J)
      IF(STORET(I,J).GT.SIDE(25))SIDE(25)= STORET(I,J)
 2866 CONTINUE
      KOUNT = IRROR
      IF(IRROR.EQ.1) AVE = UNFORM
      IF(IRROR.LE.1) RETURN
      AVE = UNFORM/FLOAT(IRROR)
      IF(ITRACE.EQ.0) RETURN
      IF(K.EQ.0) GO TO 2877
      M = J*2
      L = M-1
      IF(TERM(K,L).EQ.SETUP.AND.TERM(K,M).EQ.-SETUP) GO TO 2877
      SIDE(2) = TERM(K,L)
      SIDE(24)= TERM(K,M)
 1444 GO TO 2888
 2877 SIDE(2) = SIDE(1)
      SIDE(24)= SIDE(25)
 2888 UNFORM = (SIDE(24) - SIDE(2))/22.0
      DO 2899 I=2,22
 2899 SIDE(I+1) = SIDE(I) + UNFORM
      DO 2900 I=1,24
 2900 PDF(I) = 0.0
      DO 2944 I=1,ITER
      IF(MODE) GO TO 2911
      IF(NODESM.NE.NODET(I)) GO TO 2944
 2911 L = 1
      IF(STORET(I,J).LT.SIDE(2)) GO TO 2933
      L = 24
      IF(STORET(I,J).GT.SIDE(24))GO TO 2933
      DO 2922 ISAVE=3,24
      L = ISAVE - 1
      IF(STORET(I,J).LT.SIDE(ISAVE)) GO TO 2933
 2922 CONTINUE
 2933 PDF(L) = PDF(L) + 1.0
      UNFORM = AVE - STORET(I,J)
```

```fortran
      UNFORM = UNFORM*UNFORM
      STD = STD + UNFORM
      TOS = TOS + UNFORM*UNFORM
 2944 CONTINUE
      IF(STD.EQ.0) RETURN
      UNFORM = STD/FLOAT(IRROR-1)
      STD = SQRT(UNFORM)
      TOS = TOS/(UNFORM*UNFORM*FLOAT(IRROR-1))
      DO 2955 I=1,24
      PDF(I) = PDF(I)/FLOAT(IRROR)
      CDF(I) = PDF(I)
 2955 IF(I.GT.1) CDF(I) = CDF(I) + CDF(I-1)
      RETURN
      END
      SUBROUTINE HISTO (M1,M2,N1,N2,XMED,ITITLE,JTITLE,ITYPE)
      IMPLICIT INTEGER*4(I-N)
     0COMMON EMARK(11),SETUP,FINT,DINT,FACTOR,STIME,SCOST,SPERF,CTIME,
     1CCOST,CPERF,BTIME,BCOST,BPERF,TIMESM,UNFORM,ITRIP1(22),ITRIP2(22),
     2KNT(6),INPT,IOUT,IPNF,IWF1,IWF2,IWF3,IWF4,IBLK,ITRACE,ICUT,ISEED,
     3IRROR,ISAVE,NODESM,MAXTAG,ICOUNT,JCOUNT,NTRIP,ICORR,LCOMP,NWARN,
     4NRK,NRKS,SLAK,MODE,INFC,INFP,DISC,DISP,CRIT,KGEN,MED,SINGLE,DOUBLE
      COMMON/TRUCK/PDF(24),CDF(24),SIDE(25),AVE,STD,TOS
     0COMMON/CPGAP/T1(10),T2(10),CCNFL(10),CONFS(10),CAVE(10),CSMIN(10),
     1CSMAX(10),CHMIN(10),CHMAX(10),PAVE(10),PSMIN(10),PSMAX(10),
     2PHMIN(10),PHMAX(10),KCOBS(10),KPOBS(10),MCPGAP,NCPGAP,ICPGAP
      DIMENSION ABSA(15), LINEP(25), LINEC(50)
      LOGICAL SLAK,MODE,INFC,INFP,DISC,DISP,CRIT,KGEN,MED,SINGLE,DOUBLE
      DATA ABSA,IA/.1,.2,.3,.4,.5,.6,.7,.8,.9,1.0,.05,.1,.15,.2,.25,1H*/
C
C     THIS SUBROUTINE CONSTRUCTS HISTOGRAMS WITH ACCOMPANYING STATISTICS
C         OR A ONE LINER CONSISTING OF THE MIN, AVE AND MAX.  FIRST TASK
C         CONSISTS OF LISTING THE FIRST HALF OF THE TITLE
C
      IF(SIDE(1).EQ.SIDE(25)) GO TO 2966
      IF(ITRACE.NE.0) GO TO 2988
 2966 WRITE (IOUT,2977)
 2977 FORMAT (//1H )
      GO TO 3000
 2988 WRITE (IOUT,2999)
 2999 FORMAT (1H1)
 3000 GO TO (3011,3033,3055,3077,3099,3111,3133),ITITLE
 3011 WRITE (IOUT,3022)
 3022 FORMAT (1H+, 4X, 13HNETWORK TIME )
      GO TO 3233
 3033 WRITE (IOUT,3044)
 3044 FORMAT (1H+, 7X, 10HPATH COST )
      GO TO 3199
 3055 WRITE (IOUT,3066)
 3066 FORMAT (1H+, 4X, 13HOVERALL COST )
      GO TO 3199
 3077 WRITE (IOUT,3088)
 3088 FORMAT (18H+PATH PERFORMANCE )
      GO TO 3155
 3099 WRITE (IOUT,3100)
 3100 FORMAT (1H+, 6X, 11HSLACK TIME )
      GO TO 3233
 3111 WRITE (IOUT,3122) T1(ISAVE), T2(ISAVE)
 31220FORMAT (1H+, 5X, 38HPOSITIVE COST INCURRED BETWEEN PERIODS,
     1F10.2, 2H -, F12.2)
      GO TO 3199
 3133 WRITE (IOUT,3144) T1(ISAVE), T2(ISAVE)
 31440FORMAT (1H+, 5X, 43HPOSITIVE PERFORMANCE GAINED BETWEEN PERIODS,
```

```
      1F10.2, 2H -, F12.2)
C
C        DOES PERFORMANCE NEED TC BE INFLATED OR DISCOUNTED
C
 3155 IF(INFP) GO TO 3177
      WRITE (IOUT,3166)
 3166 FORMAT (21H PERFORMANCE INFLATED)
 3177 IF(DISP) GO TO 3233
      WRITE (IOUT,3188)
 3188 FORMAT (1H+, 25X, 22HPERFORMANCE DISCOUNTED)
      GO TO 3233
C
C        DOES COST NEED TO BE INFLATED OR DISCOUNTED
C
 3199 IF(INFC) GO TO 3211
      WRITE (IOUT,3200)
 3200 FORMAT (14H COST INFLATED)
 3211 IF(DISC) GO TO 3233
      WRITE (IOUT,3222)
 3222 FORMAT (1H+, 15X, 15HCOST DISCOUNTED)
C
C        IS THERE A SECOND HALF OF THE TITLE NEEDED
C
 3233 GO TO (3244,3266,3288,3300,3322),JTITLE
 3244 WRITE (IOUT,3255)
 3255 FORMAT (1H+, 17X, 31HFOR THE COMPOSITE TERMINAL NODE)
      GO TO 3322
 3266 WRITE (IOUT,3277) M1, M2, N1, N2
 3277 FORMAT (1H+, 17X, 26HFOR THE GAP BETWEEN NODES , 2A4, 5H AND ,2A4)
      GO TO 3322
 3288 WRITE (IOUT,3299) M1, M2
 3299 FORMAT (1H+, 17X, 9HFOR NODE , 2A4)
      GO TO 3322
 3300 WRITE (IOUT,3311) M1, M2
 3311 FORMAT (1H+, 17X, 8HFOR ARC , 2A4)
C
C        A ONE LINER
C
 3322 IF(SIDE(1).EQ.SIDE(25)) GO TO 3333
      IF(ITRACE.NE.0) GO TO 3355
 3333 WRITE (IOUT,3344) IRRCR, AVE, SIDE(1), SIDE(25)
 3344 FORMAT (9H NO OBS =, I17, 7H  AVE =, G20.6/ 6H MIN =, G20.6,
     1 17H   MAX =, G20.6)
      IF(SIDE(1).NE.SIDE(25)) GO TO 3567
      IF(ITYPE.EQ.0) GO TO 3572
      N = 1
      WRITE (IPNH,3350) M1, M2, ITYPE, N, AVE
 3350 FORMAT (2A4, 1HD, A4, I2, 3X, 3H1.0, 4X, F10.3)
      RETURN
C
C        A HISTOGRAM
C
 3355 IF(ITRACE.EQ.1) GO TO 3377
      WRITE (IOUT,3366) ABSA, SIDE(1), SIDE(1)
 3366 FORMAT (1H0, 12X, 3HCFD, 1X, 10(F4.1,1X), 15X, 4HRFD , 5F5.2/ 1H ,
     1 1G12.4, 2H I, 10(5H----I), 4H MIN, G13.4, 2H I, 5(5H----I), 4H MIN)
      GO TO 3399
 3377 WRITE (IOUT,3388) (ABSA(L), L=1,10), SIDE(1)
 3388 FORMAT (1H0, 12X, 3HCFD, 1X, 10(F4.1,1X)/ 1H , G12.4, 2H I,
     1 10(5H----I), 4H MIN, G13.4)
 3399 UNFORM = 0.0
      DO 3455 I=1,24
```

```
        IF(PDF(I).GT.UNFORM) UNFORM = PDF(I)
     ,  DO 3400 J=1,25
        LINEP(J) = IBLK
        X = FLOAT(J)/100.0
 3400 IF(X.LE.PCF(I)) LINEP(J) = IA
        DO 3411 J=2,100,2
        L = J/2
        LINEC(L) = IBLK
        X = FLOAT(J)/100.0 - 0.01
 3411 IF(X.LE.CDF(I)) LINEC(L) = IA
        IF(ITRACE.EQ.1) GO TC 3433
        WRITE (IOUT,3422) LINEC,CDF(I),LINEP,PDF(I),SIDE(I+1),SIDE(I+1)
34220FORMAT (1H , 12X, 2H I, 50A1, F5.3, 12X, 2H I, 25A1, F5.3/ 1H ,
     1G12.4, 2H I, 55X, G12.4, 2H I)
        GC TO 3455
 3433 WRITE (IOUT,3444) LINEC, CDF(I), SIDE(I+1)
 3444 FORMAT (1H , 12X, 2H I, 50A1, F5.3/ 1H , G12.4, 2H I)
 3455 CONTINUE
        IF(ITRACE.EQ.1) GO TC 3477
        WRITE (IOUT,3466)
 3466 FORMAT (1H+, 14X, 10(5H----I), 4H MAX, 15X, 5(5H----I), 4H MAX)
        GC TO 3499
 3477 WRITE (IOUT,3488)
 3488 FORMAT (1H+, 14X, 10(5H----I), 4H MAX)
 3499 X = 0.0
        IF(AVE.NE.0.0) X = STD/AVE
        WRITE (IOUT,3500) IRROR, STD, X, AVE, TOS
35000FORMAT (1H0, 13X, 6HNO OBS, 12(1H-), I8, 2X, 10HSTD ERROR-, G18.4/
     11H , 13X, 18HCOEF CF VARIATION-, F8.2, 2X, 10HMEAN------, G18.4/
     21H , 13X, 18HKURTOSIS (BETA 2)-, F8.2)
        IF(MED) GO TC 3522
        WRITE (IOUT,3511) XMED
 3511 FORMAT (1H+, 41X, 10HMEDIAN----, G18.4)
C
C       FIND THE MODE, COMPUTE AND LIST THE PEARSONIAN SKEW
C
 3522 M = 99
        N =-99
        DO 3533 I=1,24
        IF(PDF(I).NE.UNFORM) GO TO 3533
        IF(I.LT.M) M = I
        IF(I.GT.N) N = I
 3533 CONTINUE
        IF(M.EQ.N) GO TO 3555
        WRITE (IOUT,3544)
 3544 FORMAT (1H , 13X, 23HMULTIMODAL DISTRIBUTION)
        GO TO 3567
 3555 F1 = 0.0
        IF(M.GT.1) F1 = PCF(M-1)
        F2 = 0.0
        IF(N.LT.24)F2 = PDF(N+1)
       0XMODE = SIDE(M) + ((PCF(M) - F1)/(2.0*PDF(M) - F2 - F1))*
     1(SIDE(10) - SIDE(9))
        PS = 0.0
        IF(STD.GT.0.0) PS = ABS((AVE - XMODE)/STD)
        WRITE (IOUT,3566) PS, XMODE
 3566 FORMAT (1H ,13X,18HPEARSONIAN SKEW---,F8.2,2X,10HMODE------,G18.4)
C
C       LIST THE DISTRIBUTION CARDS IF WANTED
C
 3567 IF(ITYPE.EQ.0) GO TO 3572
        DO 3569 L=3,23,3
```

```
      K = L - 2
      N = L/3
 3569 WRITE (IPNH,3570) M1, M2, ITYPE, N, (SIDE(M), PDF(M), M=K,L)
 3570 FORMAT (2A4, 1HH, A4, I2, 3(F10.3, F10.8))
      N = 9
      WRITE (IPNH,3570) M1, M2, ITYPE, N, SIDE(25)
 3572 RETURN
      END
      SUBROUTINE CORR
      IMPLICIT INTEGER*4(I-N)
     0COMMON EMARK(11),SETUP,FINT,DINT,FACTOR,STIME,SCOST,SPERF,CTIME,
     1CCOST,CPERF,BTIME,BCCST,BPERF,TIMESM,UNFORM,ITRIP1(22),ITRIP2(22),
     2KNT(6),INPT,IOUT,IPNH,IWF1,IWF2,IWF3,IWF4,IBLK,ITRACE,ICUT,ISEED,
     3IRROR,ISAVE,NODESM,MAXTAG,ICCUNT,JCOUNT,NTRIP,ICORR,LCOMP,NWARN,
     4NRK,NRKS,SLAK,MODE,INFC,INFP,DISC,DISP,CRIT,KGEN,MED,SINGLE,DOUBLE
     0COMMON/TRIALS/STORET( 1000,4),TERM(10,8),KPOINT(10),NODET( 1000),
     1MTERM,NTCRM,MITER,ITER
     0COMMON/ARCS/ASTORE( 2800),UTIMEA( 350),TIMEA( 350),UCOSTA( 350),
     1COSTA( 350),UPERFA( 350),PERFA( 350),WORK( 350),ISTATE( 350),
     2NODEI( 350),NODEO( 350),ICRITA( 350),KEEPC( 350),KEEPP( 350),
     3IARC1( 350),IARC2( 350),IPOINT( 350),JPOINT( 350),ISLAK( 350),
     4KARC,LARC,MARC,NARC,ITALC,ITALP,ISTAR
     0COMMON/NCDES/TIMEN( 200),COSTN( 200),PERFN( 2C0),NSTORE( 5400),
     1NODE1( 2C0),NODE2( 2C0),LOGI( 200),LOGO( 200),NSTATE( 200),
     2NARCI( 2C0),NARCO( 2C0),ISTAT( 200),INSTAT( 2C0),ICRITN( 200),
     3NPOINT( 200),NSLAK( 200),JUMP( 200),KNODE,LNODE,MNODE,NNODE,MTAG,
     4NTAG
      DIMENSION R(2,2), B(2), XM(2), NUMB(11), ISTORE(55,110)
      LOGICAL SLAK,MODE,INFC,INFP,CISC,DISP,CRIT,KGEN,MED,SINGLE,DOUBLE
      DOUBLE PRECISION R, B, XM
      DATA NUMB/1H1,1H2,1H3,1H4,1H5,1H6,1H7,1H8,1H9,1H ,1H*/
C
C
C     THIS SUBROUTINE COMPUTES THE CORRELATION AND PLOTS 2 TERMINAL NODE
C        VARIABLES AT A TIME.  FIRST, INITIALIZE FOR INDEXING THE NODES.
C
      NRK = -4
      ICUT = 1
      ISAVE = 0
 3577 ISAVE = ISAVE + 1
      IF(ISAVE.LE.NNODE) GC TO 3588
      ICUT = 2
      GO TO 3600
 3588 IF(LOGI(ISAVE).GE.5) GO TO 3577
      IF(LOGC(ISAVE).GE.11)GC TO 3599
      IF(LOGO(ISAVE).NE.1) GO TO 3577
 3599 IF(ISTAT(ISAVE).LT.0)GO TO 3577
 3600 NRK = NRK + 4
      DO 3888 L=11,ICCRR
      M = ITRIP1(L)
      IF(M.EQ.0) GC TO 3888
      N = ITRIP2(L)
      ITALC = M + NRK
      ITALP = N + NRK
C
C     CALCULATE SUMS CF SQUARES AND CRCSS PRODUCTS
C
      XM(1) = COSTA(ITALC)
      XM(2) = CCSTA(ITALP)
      DO 3611 I=1,2
      DC 3611 J=1,2
 3611 R(I,J) = 0.0
      ISTAR = 0
```

```
      DO 3644 K=1,ITER
      IF(ICUT.EQ.2) GO TO 3622
      IF(ISAVE.NE.NODET(K)) GO TO 3644
 3622 B(1) = STORET(K,M)
      B(2) = STORET(K,N)
      DO 3633 I=1,2
      DO 3633 J=1,2
 3633 R(I,J) = R(I,J) + (B(I) - XM(I))*(B(J) - XM(J))
      ISTAR = ISTAR + 1
 3644 CONTINUE
      IF(ISTAR.GT.1) GO TO 3655
      NRK = NRK - 4
      GO TO 3899
C
C     COMPUTE THE VALUE OF THE VARIANCE-COVARIANCE MATRIXAND PRINT THE
C        TITLE BLOCK AT THE TOP OF THE PAGE
C
 3655 DO 3666 I=1,2
 3666 B(I) = DSQRT(R(I,I))
      IF(B(1).EQ.0.0.OR.B(2).EQ.0.0) GO TO 3888
      R(1,2) = R(1,2)/(B(1)*B(2))
      IF(M.EQ.1) WRITE (IOUT,3677)
 3677 FORMAT (1H1, 55X, 11HTIME (Y) VS)
      IF(M.EQ.2) WRITE (IOUT,3688)
 3688 FORMAT (1H1, 50X, 16HPATH COST (Y) VS)
      IF(M.EQ.3) WRITE (IOUT,3699)
 3699 FORMAT (1H1, 47X, 19HOVERALL COST (Y) VS)
      IF(M.EQ.4) WRITE (IOUT,3700)
 3700 FORMAT (1H1, 48X, 18HPERFORMANCE (Y) VS)
      IF(N.EQ.1) WRITE (IOUT,3711)
 3711 FORMAT (1H+, 67X, 8HTIME (X))
      IF(N.EQ.2) WRITE (IOUT,3722)
 3722 FORMAT (1H+, 67X, 13HPATH COST (X))
      IF(N.EQ.3) WRITE (IOUT,3733)
 3733 FORMAT (1H+, 67X, 16HOVERALL COST (X))
      IF(N.EQ.4) WRITE (IOUT,3744)
 3744 FORMAT (1H+, 67X, 15HPERFORMANCE (X))
      GO TO (3755,3777),ICUT
 3755 WRITE (IOUT,3766) NODE1(ISAVE), NODE2(ISAVE)
 3766 FORMAT (1H+, 85X, 16HTERMINAL NODE = , 2A4)
      GO TO 3799
 3777 WRITE (IOUT,3788)
 3788 FORMAT (1H+, 85X, 23HCOMPOSITE TERMINAL NODE)
 3799 WRITE (IOUT,3800) R(1,2), PERFA(ITALC)
 3800 FORMAT (1H , 10X, 25HCORRELATION COEFFICIENT =, F5.2, 2X, 79H* MEA
     1NS PLOTTING 10 OR MORE POINTS PER UNIT CELL - (110 X 55 OR      CE
     2LL PLOT)/ 1H , 10X, F20.3, 14H (MAX Y VALUE), 77(1HX))
C
C     LOAD THE DATA ARRAYS, FIRST INITIALIZE
C
      DO 3811 I=1,55
      DO 3811 J=1,110
 3811 ISTORE(I,J) = 0
      XM(1) = PERFA(ITALC) - TIMEA(ITALC)
      XM(2) = PERFA(ITALP) - TIMEA(ITALP)
      DO 3833 I=1,ITER
      IF(ICUT.EQ.2) GO TO 3822
      IF(ISAVE.NE.NODET(I)) GO TO 3833
 3822 J=56-(56.0-(((STORET(I,N) - TIMEA(ITALP))/XM(2))* 54.0 + 1.0))+0.5
      K = (111.0-(((STORET(I,M) - TIMEA(ITALC))/XM(1))*109.0 + 1.0))+0.5
      ISTORE(J,K) = ISTORE(J,K) + 1
 3833 CONTINUE
```

```
C
C          CONVERT STORAGE TO PRINT CHARACTERS AND PLOT
C
           DO 3855 I=1,55
           DO 3844 J=1,110
           K = ISTORE(I,J)
           IF(K.GE.10) K = 11
           IF(K.LE.0) K = 10
 3844 ISTORE(I,J) = NUMB(K)
 3855 WRITE (IOUT,3866) (ISTORE(I,J), J=1,110)
 3866 FORMAT (1H , 9X, 1HY, 110A1, 1HY)
           WRITE (IOUT,3877) TIMEA(ITALC), TIMEA(ITALP), PERFA(ITALP)
 3877 0FORMAT (1H , 10X,   F20.3, 14H (MIN Y VALUE), 77(1HX)/ 1H , 10X,
      1F20.3, 14H (MIN X VALUE), 44X, 14H (MAX X VALUE), F20.3)
 3888 CONTINUE
 3899 GO TO (3577,3900),IOUT
 3900 RETURN
           END
           SUBROUTINE SUB1 (X,Y,Z,R,KOUNT)
           IMPLICIT INTEGER*4(I-N)
C
C          SUBROUTINE SUB1 THRU SUB6 ARE SETUP FOR THE USER TO CREATE HIS OWN
C             TRANSFORMATIONS.  THE RESULTNAT VALUE CREATED MUST BE PLACED IN
C             R.  KOUNT IS A COUNTER WHICH ENABLES THE USER TO MAKE MULTIPLE
C             DATA TRANSFERRING CALLS TO THIS SUBROUTINE BEFORE COMPUTING A
C             VALUE FOR R.  THIS EXAMPLE FINDS THE MIDDLE VALUE OF X, Y, & Z.
C
           AMIN = X
           IF(Y.LT.AMIN) AMIN = Y
           IF(Z.LT.AMIN) AMIN = Z
           AMAX = X
           IF(Y.GT.AMAX) AMAX = Y
           IF(Z.GT.AMAX) AMAX = Z
           IF(AMIN.NE.X.AND.AMAX.NE.X) R = X
           IF(AMIN.NE.Y.AND.AMAX.NE.Y) R = Y
           IF(AMIN.NE.Z.AND.AMAX.NE.Z) R = Z
           RETURN
           END
           SUBROUTINE SUB2 (X,Y,Z,R,KOUNT)
           IMPLICIT INTEGER*4(I-N)
C
C          SUBROUTINE SUB1 THRU SUB6 ARE SETUP FOR THE USER TO CREATE HIS OWN
C             TRANSFORMATIONS.  THE RESULTNAT VALUE CREATED MUST BE PLACED IN
C             R.  KOUNT IS A COUNTER WHICH ENABLES THE USER TO MAKE MULTIPLE
C             DATA TRANSFERRING CALLS TO THIS SUBROUTINE BEFORE COMPUTING A
C             VALUE FOR R.  THIS EXAMPLE EVALUATES THE EXPRESSION BELOW.
C
           R = (X**3 + Y**4)/Z**0.3
           RETURN
           END
           SUBROUTINE SUB3 (X,Y,Z,R,KOUNT)
           IMPLICIT INTEGER*4(I-N)
C
C          SUBROUTINE SUB1 THRU SUB6 ARE SETUP FOR THE USER TO CREATE HIS OWN
C             TRANSFORMATIONS.  THE RESULTNAT VALUE CREATED MUST BE PLACED IN
C             R.  KOUNT IS A COUNTER WHICH ENABLES THE USER TO MAKE MULTIPLE
C             DATA TRANSFERRING CALLS TO THIS SUBROUTINE BEFORE COMPUTING A
C             VALUE FOR R.  THIS EXAMPLE SUBROUTINE FINDS THE ABSOLUTE VALUE
C             OF THE PRODUCT OF X, Y, , AND Z.
C
           R = ABS(X*Y*Z)
           RETURN
```

```
      END
      SUBROUTINE SUB4 (X,Y,Z,R,KOUNT)
      IMPLICIT INTEGER*4(I-N)
C
C     SUBROUTINE SUB1 THRU SUB6 ARE SETUP FOR THE USER TO CREATE HIS OWN
C        TRANSFORMATIONS.  THE RESULTNAT VALUE CREATED MUST BE PLACED IN
C        R.  KOUNT IS A COUNTER WHICH ENABLES THE USER TO MAKE MULTIPLE
C        DATA TRANSFERRING CALLS TO THIS SUBROUTINE BEFORE COMPUTING A
C        VALUE FOR R.  THIS EXAMPLE SUBROUTINE FINDS THE ABSOLUTE VALUE
C        OF THE SUM OF X, Y, AND Z.
C
      R = ABS(X+Y+Z)
      RETURN
      END
      SUBROUTINE SUB5 (X,Y,Z,R,KOUNT)
      IMPLICIT INTEGER*4(I-N)
C
C     SUBROUTINE SUB1 THRU SUB6 ARE SETUP FOR THE USER TO CREATE HIS OWN
C        TRANSFORMATIONS.  THE RESULTNAT VALUE CREATED MUST BE PLACED IN
C        R.  KOUNT IS A COUNTER WHICH ENABLES THE USER TO MAKE MULTIPLE
C        DATA TRANSFERRING CALLS TO THIS SUBROUTINE BEFORE COMPUTING A
C        VALUE FOR R.  THIS EXAMPLE SUBROUTINE FINDS THE SUM OF THE
C        ABSOLUTE CALUES OF X, Y, AND Z.
C
      R = ABS(X) + ABS(Y) + ABS(Z)
      RETURN
      END
      SUBROUTINE SUB6 (X,Y,Z,R,KOUNT)
      IMPLICIT INTEGER*4(I-N)
      COMMON/SIMS/FX(251),SPACE,SUM,NUMBFX
C
C     SUBROUTINE SUB1 THRU SUB6 ARE SETUP FOR THE USER TO CREATE HIS OWN
C        TRANSFORMATIONS.  THE RESULTNAT VALUE CREATED MUST BE PLACED IN
C        R.  KOUNT IS A COUNTER WHICH ENABLES THE USER TO MAKE MULTIPLE
C        DATA TRANSFERRING CALLS TO THIS SUBROUTINE BEFORE COMPUTING A
C        VALUE FOR R.  THIS EXAMPLE SUBROUTINE WITH SUBROUTINE
C        SIMSON, COMPUTE THE INTEGRAL OF E**T DT OVER THE INTEGRAL OF
C        X TO Y.
C
      MAXFX = 251
      SPACE = 0.01
      NUMBFX = (Y-X)/SPACE + 1.99999
      I = (NUMBFX/2)*2
      IF(I.EQ.NUMBFX) NUMBFX = NUMBFX + 1
      IF(NUMBFX.GT.MAXFX) RETURN
      SPACE = (Y-X)/FLOAT(NUMBFX-1)
      DO 3911 I=1,NUMBFX
      FX(I) = EXP(X)
 3911 X = X + SPACE
      CALL SIMSON
      R = SUM
      RETURN
      END
      SUBROUTINE SIMSON
      IMPLICIT INTEGER*4(I-N)
      COMMON/SIMS/FX(251),SPACE,SUM,NUMBFX
C
C     NUMBFX= NUMBER OF POINTS TO BE INTEGRATED (MUST BE AN ODD NUMBER)
C     SPACE = SPACING BETWEEN THE POINTS OF INTEGRATION ON THE X-AXIS
C     SUM   = FINAL VALUE OF THE INTEGRATION
C     FX    = ARRAY OF POINTS TO BE INTEGRATED ON THE F(X)-AXIS
C
```

```
      N = NUMBFX - 3
      SUM = 0.0
      DO 3922 I=2,N,2
 3922 SUM = SUM + 4.0*FX(I) + 2.0*FX(I+1)
      SUM = (SUM + FX(1) + 4.0*FX(NUMBFX-1) + FX(NUMBFX))*SPACE/3.0
      RETURN
      END
```

REFERENCES

Archibald, R. D. 1976. *Managing High Technology Programs and Projects.* New York: Wiley.

Beer, S. 1968. *Management Science.* Garden City, N.Y.: Doubleday.

Brown, E. L. 1975. "An Application of Simulation Networking Techniques in Operational Test Design and Evaluation." Unpublished masters thesis. Georgia Institute of Technology, Atlanta.

A Course of Instruction in Risk Analysis. 1972. ALM-3315-H. United States Army Logistics Management Center, Fort Lee, Va.

Crowstone, W. B., and G. L. Thompson. 1967. "Decision CPM: A Method for Simultaneous Planning, Scheduling, and Control of Projects." *Operations Research* 15:407-26.

Dalkey, N. 1969. "An Experimental Study of Group Opinion, The Delphi Method." *Futures* (5):408-26.

Digman, L. A. 1967. "PERT/LOB: Life-Cycle Technique." *Journal of Industrial Engineering* 28 (2):154-58.

________. 1980. "A Decision Analysis of the Airline Coupon Strategy." *Interfaces* 10 (2):97-101

________, and G. I. Green. 1981. "A Framework for Evaluating Network Planning and Control Techniques." *Research Management* 24 (1):10-17.

DOD/NASA Guide to PERT/COST Systems Design. 1962. Washington, D.C.: Office of the Secretary of Defense and National Aeronautics and Space Administration (June).

Eisner, H. 1962. "A Generalized Network Approach to the Planning and Scheduling of a Research Program." *Operations Research* 10:115–25.

Elmaghraby, S. E. 1964. "An Algebra for the Analysis of Generalized Activity Network." *Management Science* 10:494–514.

Hespos, R. F., and P. A. Strassman. 1965. "Stochastic Decision Trees for the Analysis of Investment Decisions." *Management Science,* series B, 11 (10):244–59.

"Improvement in Weapon Systems Acquisition." 1969. Memorandum from the Deputy Secretary of Defense to the Secretaries of the Army, Navy, and Air Force, Washington, D.C. (July 31).

Juran, J. M. 1964. *Managerial Breakthrough: A New Concept of the Manager's Job.* New York: McGraw-Hill.

Lee, S. M. 1981. *Management by Multiple Objectives.* Princeton, N.J.: Petrocelli Books.

______, and L. J. Moore. 1975. *Introduction to Decision Science.* New York: Petrocelli/Charter.

______, ______, and B. W. Taylor. 1978. "Analysis of Multi-Objective Project Crashing Models." *AIIE Transactions* 10 (2):163–69.

______, ______, and ______. 1981. *Management Science.* Dubuque, Iowa: Wm. C. Brown.

______, O. Park, and S. Economides. 1978. "Optimal Resource Planning for Multiple Projects." *Decision Sciences* 9 (1):49–67.

Lindblom, C. E. 1959. "The Science of Muddling Through." *Public Administration Review* 20 (Spring):79–88.

Magee, J. F. 1964a. "Decision Trees for Decision Making." *Harvard Business Review* 42 (4):126.

______. 1964b. "How to Use Decision Trees in Capital Investment." *Harvard Business Review* 42 (5):79.

Malcolm, D. G., et al. 1959. "Application of a Technique for Research and Development Program Evaluation." *Operations Research* 7:646–69.

Maslow, A. 1965. *Eupsychian Management: A Journal.* Homewood, Ill.: Irwin-Dorsey.

Mathematica. 1970. *Mathnet,* preliminary ed. Princeton, N.J.

McLuhan, Marshall. 1964. *Understanding Media.* New York: McGraw-Hill.

Mintzberg, Henry. 1975. "The Manager's Job: Folklore and Fact." *Harvard Business Review* 53 (4):49.

Moder, J. J., and C. R. Philips. 1964. *Project Management with CPM and PERT.* New York: Reinhold.

Moeller, G. L. 1971. *Statnet.* Rock Island, Ill.: U.S. Army Management Engineering Training Agency.

______. 1972. "VERT." Technical papers of the American Institute of Industrial Engineers, 23rd conference and convention, Anaheim, Calif.

______, and L. A. Digman. 1978. "VERT: A Technique to Assess Risks." Vol. 2, proceedings of the American Institute for Decision Sciences, 10th annual conference, St. Louis, Mo. (October).

______, and ______. 1981*a*. "Operations Planning with VERT." *Operations Research* 29 (4).

______, and ______. 1981*b*. "A Stochastic Networking Approach to Strategic Planning." *Long Range Planning* (forthcoming).

Moore, Laurence J., and Edward R. Clayton. 1976. *GERT Modeling and Simulation: Fundamentals and Applications.* New York: Petrocelli/Charter.

Northrop, G. M., et al. 1970. "Use of Quantified Expert Judgment in Cost Effectiveness Studies." Paper presented at the 38th national meeting of the Operations Research Society of America, Detroit, Mich.

Percy, Stephen R. 1973. "Advanced Solvnet: A Network Analyzer Program." Report no. Papas-14. Prepared by Plans Office, Picatinny Arsenal, Dover, N.J. (January).

Pritsker, A. A. B. 1977. *Modeling and Analysis Using Q-GERT Networks.* New York: Halsted Press.

______, and W. W. Happ. 1966. "GERT: Graphical Evaluation and Review Technique—Part I." *Journal of Industrial Engineering* 18 (5): 267–74.

______, and C. D. Pegden. 1979. *Introduction to Simulation and SLAM.* New York: Halsted Press

______, and G. G. Whitehouse. 1966. "GERT: Graphical Evaluation and Review Technique—Part II." *Journal of Industrial Engineering* 18 (6):293–301.

Project Management System (PMS) 360: Application Description Manual, 2nd ed., no. GH200210. 1968. White Plains, N.Y.: IBM Corporation.

Quinn, J. B. 1977. "Strategic Goals: Process and Politics." *Sloan Management Review* 19 (1):21.

Raiffa, H. 1968. *Decision Analysis: Introductory Lectures on Choices under Uncertainty.* Reading, Mass.: Addison-Wesley.

Rappaport, A. 1979. "Strategic Analysis for More Profitable Acquisitions." *Harvard Business Review* 57 (4):99–110.

Salter, M. S., and W. A. Weinhold. 1981. "Choosing Compatible Acquisitions." *Harvard Business Review* 59 (1):117–27.

Schoderbek, P., and L. A. Digman. 1967*a*. "Third Generation, PERT/LOB." *Harvard Business Review* 45 (5):100–10.

______, and ______. 1967*b*. "Third Generation in PERT Systems." In R. W. Millman and M. P. Hottenstein, eds., *Promising Research Directions: Academy of Management Proceedings.* Washington, D.C.

Thomas, Maj. T. N. 1977. "VERT—A Risk Analysis Technique for Program Managers." Unpublished report. Defense Systems Management College, Ft. Belvoir, Va.

Uyterhoeven, H. T., R. W. Ackerman, and J. W. Rosenblum. 1977. *Strategy and Organization,* rev. ed. Homewood, Ill.: Irwin.

Weston, J.F., and E.F. Brigham. 1969. *Essentials of Managerial Finance,* 3rd ed. New York: Holt, Rinehart & Winston.

Wiest, Jerome D., and Ferdinand K. Levy. 1977. *A Management Guide to PERT/CPM,* 2nd ed. Englewood Cliffs, N.J.: Prentice-Hall.

Work Scheduling Techniques. 1968. U.S. Department of Army pamphlet no. 1-54 (March).

INDEX